Optimal Networked Control Systems with MATLAB®

AUTOMATION AND CONTROL ENGINEERING

A Series of Reference Books and Textbooks

Series Editors

FRANK L. LEWIS, Ph.D.,
Fellow IEEE, Fellow IFAC
Professor
The Univeristy of Texas Research Institute
The University of Texas at Arlington

SHUZHI SAM GE, Ph.D.,
Fellow IEEE
Professor
Interactive Digital Media Institute
The National University of Singapore

PUBLISHED TITLES

Optimal Networked Control Systems with MATLAB®, *Jagannathan Sarangapani; Hao Xu*

Cooperative Control of Multi-agent Systems: A Consensus Region Approach, *Zhongkui Li; Zhisheng Duan*

Nonlinear Control of Dynamic Networks, *Tengfei Liu; Zhong-Ping Jiang; David J. Hill*

Modeling and Control for Micro/Nano Devices and Systems, *Ning Xi; Mingjun Zhang; Guangyong Li*

Linear Control System Analysis and Design with MATLAB®, Sixth Edition, *Constantine H. Houpis; Stuart N. Sheldon*

Real-Time Rendering: Computer Graphics with Control Engineering, *Gabriyel Wong; Jianliang Wang*

Anti-Disturbance Control for Systems with Multiple Disturbances, *Lei Guo; Songyin Cao*

Tensor Product Model Transformation in Polytopic Model-Based Control, *Péter Baranyi; Yeung Yam; Péter Várlaki*

Fundamentals in Modeling and Control of Mobile Manipulators, *Zhijun Li; Shuzhi Sam Ge*

Optimal and Robust Scheduling for Networked Control Systems, *Stefano Longo; Tingli Su; Guido Herrmann; Phil Barber*

Advances in Missile Guidance, Control, and Estimation, *S.N. Balakrishna; Antonios Tsourdos; B.A. White*

End to End Adaptive Congestion Control in TCP/IP Networks, *Christos N. Houmkozlis; George A Rovithakis*

Robot Manipulator Control: Theory and Practice, *Frank L. Lewis; Darren M Dawson; Chaouki T. Abdallah*

Quantitative Process Control Theory, *Weidong Zhang*

Classical Feedback Control: With MATLAB® and Simulink®, Second Edition, *Boris Lurie; Paul Enright*

Intelligent Diagnosis and Prognosis of Industrial Networked Systems, *Chee Khiang Pang; Frank L. Lewis; Tong Heng Lee; Zhao Yang Dong*

Synchronization and Control of Multiagent Systems, *Dong Sun*

Subspace Learning of Neural Networks, *Jian Cheng; Zhang Yi; Jiliu Zhou*

Reliable Control and Filtering of Linear Systems with Adaptive Mechanisms, *Guang-Hong Yang; Dan Ye*

Reinforcement Learning and Dynamic Programming Using Function Approximators, *Lucian Busoniu; Robert Babuska; Bart De Schutter; Damien Ernst*

Modeling and Control of Vibration in Mechanical Systems, *Chunling Du; Lihua Xie*

Analysis and Synthesis of Fuzzy Control Systems: A Model-Based Approach, *Gang Feng*

Lyapunov-Based Control of Robotic Systems, *Aman Behal; Warren Dixon; Darren M. Dawson; Bin Xian*

System Modeling and Control with Resource-Oriented Petri Nets, *MengChu Zhou; Naiqi Wu*

Sliding Mode Control in Electro-Mechanical Systems, Second Edition, *Vadim Utkin; Juergen Guldner; Jingxin Shi*

Autonomous Mobile Robots: Sensing, Control, Decision Making and Applications, *Shuzhi Sam Ge; Frank L. Lewis*

Linear Control Theory: Structure, Robustness, and Optimization, *Shankar P. Bhattacharyya; Aniruddha Datta; Lee H.Keel*

Optimal Control: Weakly Coupled Systems and Applications, *Zoran Gajic*

Deterministic Learning Theory for Identification, Recognition, and Control, *Cong Wang; David J. Hill*

Intelligent Systems: Modeling, Optimization, and Control, *Yung C. Shin; Myo-Taeg Lim; Dobrila Skataric; Wu-Chung Su; Vojislav Kecman*

Optimal Networked Control Systems with MATLAB®

Jagannathan Sarangapani
Missouri University of Science and Technology
Rolla, Missouri, USA

Hao Xu
Texas A&M University - Corpus Christi
Corpus Christi, Texas, USA

CRC Press
Taylor & Francis Group
Boca Raton London New York

CRC Press is an imprint of the
Taylor & Francis Group, an **informa** business

CRC Press
Taylor & Francis Group
6000 Broken Sound Parkway NW, Suite 300
Boca Raton, FL 33487-2742

First issued in paperback 2020

ISBN-13: 978-1-4822-3525-8 (hbk)
ISBN-13: 978-0-367-77867-5 (pbk)

Library of Congress Cataloging-in-Publication Data

Sarangapani, Jagannathan, 1965- author.
Optimal networked control systems with MATLAB / Jagannathan Sarangapani and Hao Xu.
pages cm. -- (Automation and control engineering)
Includes bibliographical references and index.
ISBN 978-1-4822-3525-8 (alk. paper)
1. Feedback control systems--Computer-aided design. 2. Mathematical optimization. 3. MATLAB. I. Xu, Hao (Engineer), author. II. Title.

TJ216.S2835 2015
629.8'6553--dc23 2015025712

Visit the Taylor & Francis Web site at
http://www.taylorandfrancis.com

and the CRC Press Web site at
http://www.crcpress.com

Jagannathan Sarangapani dedicates the book to Sarangapani (father), Janaki (mother), and Sandhya (wife).

Hao Xu dedicates the book to Heqing Xu (father), Ning Ge (mother), Ran Cai (wife), and Isaac Xu (son).

Contents

Preface

In the past decade, significant advances in theoretical and applied research have occurred at the intersection of computation, communication, and control. Modern feedback control systems have been responsible for major successes in the fields of aerospace engineering, automotive technology, defense, and industrial systems. The function of a feedback controller is to alter the behavior of the system in order to meet a desired level of performance. Most recently, a communication network is inserted within the feedback loop to form a networked control system (NCS). This novel NCS concept is considered to be a third-generation control system. In NCS, a communication packet carries the reference input, plant output, and control input, which are exchanged by using a communication network among control system components such as a sensor, controller, and actuators.

Compared to the traditional control systems, an NCS reduces system wiring with ease of system diagnosis and maintenance while increasing system agility. Because of these advantages, the NCS has been implemented in manufacturing and power industries and continues to appear in other industries despite increased system complexity. The complexity of such man-made systems has placed severe constraints on existing feedback design techniques. More stringent performance requirements in both speed and accuracy in the face of system uncertainties and unknown environments such as a communication network within the control loop have challenged the limits of modern control. Operating a complex system in different regimes with an imperfect communication network consisting of random delays, packet losses, and quantization errors inside its feedback control loop requires that the controller be intelligent with adaptive and learning capabilities in the presence of unknown disturbances, unmodeled dynamics, and unstructured uncertainties.

Intelligent control systems, which are modeled after biological systems and human cognitive capabilities, possess learning and adaptation, and render optimal performance. As a result, these so-called intelligent controllers provide hope of improved performance for today's complex systems. In this book, we explore optimal controller design using the Q-function for linear systems, and artificial neural networks for nonlinear systems since neural networks capture the parallel processing, adaptive, and learning capabilities of biological nervous systems. The application of neural networks in closed-loop networked feedback control systems has only recently been rigorously studied.

Controllers designed in discrete time have the important advantages that they can be directly implemented in digital form on modern-day embedded hardware. Unfortunately, discrete-time design is far more complex than continuous-time design when Lyapunov stability analysis is used since the first difference in Lyapunov function is quadratic in the states, not linear as in the case of continuous time.

This book presents for the first time the learning controller design in discrete time for NCSs. Several powerful modern control techniques in discrete time are used in the book for the design of intelligent controllers for such NCSs. Thorough development, rigorous stability proofs, and simulation examples are presented in each case.

Chapter 1 provides the background on NCSs along with networked imperfections while Chapter 2 provides background information on dynamical systems, stability theory, and stochastic discrete-time optimal adaptive controllers for linear and nonlinear systems. Chapter 3 lays the foundation of traditional Q-learning-based optimal adaptive controllers used in the book for finite and infinite horizons. In Chapter 4, we introduce quantization effects for linear and nonlinear NCSs and use learning to design stochastic controllers for a class of linear and nonlinear systems enclosed by a communication network with magnitude constraints on the control input.

Chapter 5 confronts the additional complexity introduced by uncertainty in the disturbance control input and presents a two-player zero-sum game-theoretic formulation for linear system in input–output form enclosed by a communication network. Mainly an output feedback controller is derived. Chapter 6 presents the stochastic optimal control of nonlinear NCSs by using neurodynamic programming. In Chapter 7, we discuss stochastic optimal design for nonlinear two-player zero-sum game under communication constraints. Chapter 8 covers distributed joint optimal network scheduling and control design for wireless NCSs. So far, the controller designs presented in the previous chapters use periodic sampling intervals. In order to minimize transmission of state and control signals within the feedback loop via the communication network, Chapter 9 treats an event-sampled distributed NCS. Finally, Chapter 10 describes the effect of network protocols on the NCS controller design. The appendices at the end of each chapter include analytical proofs for the controllers and MATLAB® code needed to build intelligent controllers for the above class of nonlinear systems and for real-time control applications. Additional material is available from the CRC website http://www.crcpress.co/product/isbn/9781482225258.

This book has been written for graduate students in a college curriculum, for practicing engineers in industry, and for university researchers. Detailed derivations, stability analysis, and computer simulations show how to understand neural network controllers as well as how to build them.

Jagannathan Sarangapani acknowledges and is grateful to his teacher F.L. Lewis who gave him inspiration and passion and taught him persistence and attention to detail. The first author wants to convey special thanks to all his students, in particular Hao Xu, Qiming Zhao, Avimanyu Sahoo, and Behzad Talaei who compelled him to take the work seriously and become completely involved.

Hao Xu acknowledges and is grateful to his mentor Jagannathan Sarangapani who inspired and guided him on this research. Hao Xu wants to convey special thanks to his wife and parents and rest of his family for their love and support.

This research work is supported in part by the National Science Foundation under grant ECCS#1128281 and the Intelligent Systems Center at the Missouri University of Science and Technology.

Jagannathan Sarangapani
Rolla, Missouri

Hao Xu
Corpus Christi, Texas

MATLAB® is a registered trademark of The MathWorks, Inc. For product information, please contact:

The MathWorks, Inc.
3 Apple Hill Drive
Natick, MA 01760-2098 USA
Tel: 508 647 7000
Fax: 508-647-7001
E-mail: info@mathworks.com
Web: www.mathworks.com

Authors

Jagannathan Sarangapani (referred to as S. Jagannathan) is at the Missouri University of Science and Technology (formerly the University of Missouri-Rolla) where he is a Rutledge-Emerson endowed chair professor of electrical and computer engineering and site director for the NSF (National Science Foundation) Industry/University Cooperative Research Center on Intelligent Maintenance Systems. He has coauthored 134 peer-reviewed journal articles most of them in various *IEEE Transactions* and 238 refereed IEEE conference articles, several book chapters, 3 books with this being the fourth, coedited an additional one, and holds 20 US patents. He supervised to graduation around 20 doctoral and 29 MS students with his funding an excess of $16 million from various US federal and industrial members. His research interests include neural network control and adaptive dynamic programming, secure networked control systems, RFID (radio-frequency identification) sensors, prognostics, and autonomous systems/robotics. He was the coeditor for the IET book series on control from 2010 through 2013 and has served on many editorial boards. He received many awards and has been on organizing committees of several IEEE conferences. Currently he is the IEEE CSS Tech Committee chair on intelligent control. He is a fellow of the Institute of Measurement and Control, UK and the Institution of Engineering and Technology (IET), UK.

Hao Xu earned his master's in electrical engineering from Southeast University in 2009, and a PhD from the Missouri University of Science and Technology (formerly, the University of Missouri-Rolla), Rolla in 2012. Currently, he is at Texas A&M University – Corpus Christi where he is an assistant professor at the College of Science and Engineering and director of the Unmanned Systems Research Laboratory. His research interests included autonomous unmanned aircraft systems, multi-agent systems, wireless passive sensor networks, localization, detection, networked control systems, cyber-physical systems, distributed network protocol designs, optimal control and adaptive control.

1 Introduction to Networked Control Systems

In the past decade, significant advances in theoretical and applied researches have occurred in computation, communication, and control areas (Liou and Ray 1991). Control systems have made great strides from analog control (first generation) to digital (second generation) (Kumar 2009) with the appearance of digital computers in the 1940s. Similarly, wireless communication is preferred over wired communication as it allows mobility. In recent control applications, reinforcement learning, which is used for computational intelligence, has been introduced in complex control system design. Most recently, a communication network is combined with a modern control system to form a networked feedback control system due to the presence of a real-time communication network. This networked control system (NCS) (Halevi and Ray 1988; Branicky et al. 2000; Antsaklis and Baillieul 2004; Hespanha et al. 2007) concept is considered a third-generation control system (Kumar 2009). In NCS, a communication packet carries the reference input, plant output, and control input, which are exchanged by using a communication network among control system components such as sensor, controller, and actuators as shown in Figure 1.1.

Compared with traditional control systems, an NCS reduces system wiring with ease of system diagnosis and maintenance while increasing system agility. Because of these advantages, NCSs have been implemented in manufacturing. Multiple devices sense data from controlled plants by using embedded sensors and then packetize the data and transmit the sensed data to remote controllers through the wireless network. When the respective controllers receive information from the controlled plant, suitable control inputs can be designed based on that information and transmitted back to the respective devices through the network.

Similarly, an NCS is implemented on the smart grid, which is considered as the next-generation power system. The sensors can report consumer demand to the smart grid processor, which can decide how to deal with these demands, whether to request/increase power generation or to utilize stored energy. Compared with traditional power systems, a network control-based smart grid can manage power resources more efficiently.

NCSs have been finding application in a broad range of areas such as mobile sensor networks, remote surgery, haptics collaboration over the Internet, automated highway systems, and unmanned aerial vehicles (UAV) (see, for instance, Hespanha et al. 2007) and in network-enabled or cyber manufacturing (Song et al. 2006). However, the use of a shared communication network—in contrast to using several

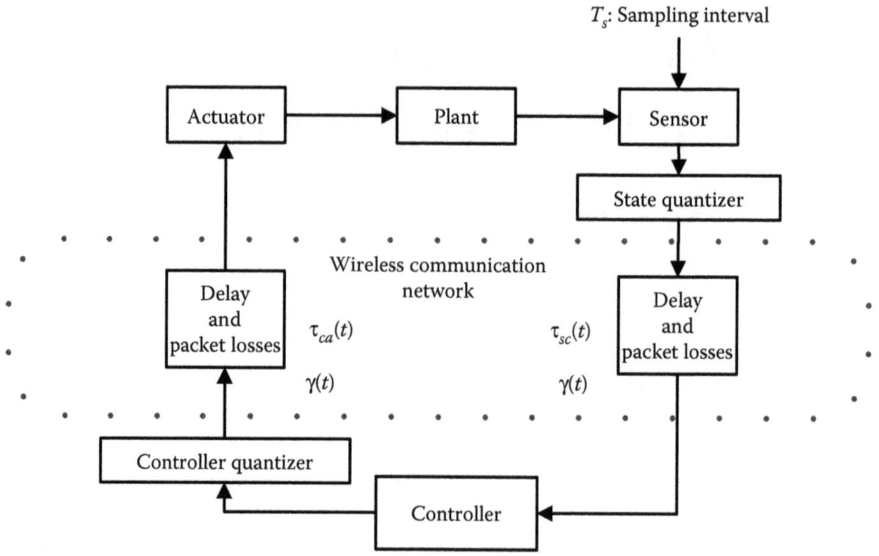

FIGURE 1.1 Networked control systems.

independent connections—introduces new challenges, and control over networks is identified as one of the key future directions for control.

However, every coin has two sides. Because of unreliable communication networks, NCSs have many challenging issues to be solved before reaping their benefits. The first issue is the network-induced delay that occurs while exchanging data among devices connected to shared communication media. This delay, either constant or random, can degrade the performance of the control system and even destabilize the system when the delay is not explicitly considered in the design process (Zhang et al. 2001). The second issue is packet losses due to unreliable network transmission, which can cause a loss in data exchange resulting in instability (Wu and Chen 2007). Because of limited network capacity, sensed plant data and designed control inputs need to be quantized prior to transmission, which may lead to quantization errors for both measured states and control inputs (Hespanha et al. 2007; Heemels et al. 2010). Since these quantization errors can cause the instability of an NCS, this is considered as the third issue.

Next, an overview of current methodologies for NCS design is presented, and their shortcomings are discussed. Subsequently, the organization and the contributions of this book are introduced.

1.1 OVERVIEW OF NETWORKED CONTROL TECHNIQUES

As introduced in the above section, although the NCS can offer several advantages, it also brings many challenging issues (e.g., network-induced delays and packet losses) due to the presence of a communication network and its associated network protocol utilized for packet transmission. For the NCS shown in Figure 1.1, researchers analyzed the stability of such an NCS starting in the 1990s. Branicky et al. (2000), Hespanha et al. (2007), and Walsh et al. (1999) evaluated the stability and performance of an NCS with constant network-induced delay and derived a stability region of linear NCS. Selecting a conventional stable controller with constant gains, related maximum allowable transfer interval (MATI), and maximum allowable delay (MAD) can be calculated based on the stability region of NCS (Walsh et al. 1999; Zhang et al. 2001). In addition, the effect of packet losses on the NCS has been analyzed in Schenato et al. (2007).

Similar to network-induced delay, Schenato et al. (2007) derived a stability region for packet losses based on stochastic control. Hu and Zhu (2003) and Zhang et al. (2001) conducted the stability analysis of the NCS in the presence of packet losses and delays and proposed the region of stability. Experimental studies by Zhang et al. (2001) and Walsh et al. (1999) have illustrated that a conventional controller can still maintain an NCS stable in the mean when network-induced delays and packet losses fall within the region of stability.

On the other hand, optimal control design (Lian et al. 2003) is also pursued in NCS research. Nilsson et al. (1998) introduced the optimal design problem and derived an optimal controller for the NCS with a short network-induced delay (i.e., delay less than one sampling interval). Nilsson et al. (1998) represented the NCS dynamics with augment states. Then, the optimal controller has been derived using standard Riccati equation-based optimal control theory (Stengel 1986).

Recently, Hu and Zhu (2003) extended linear NCS optimal controller design with network-induced delays of over several sampling intervals (i.e., delay is more than one sampling interval). Compared with previous NCS schemes, Schenato et al. (2007) considered specific network protocols such as a transmission control protocol (TCP) and user datagram protocol (UDP), and derived optimal control design for NCS under TCP and UDP. However, all these methodologies required full knowledge of system dynamics and network imperfections (i.e., network-induced delay and packet losses), which are not known beforehand in practical NCS. Therefore, methods developed by Wu and Chen (2007) and Cloosterman et al. (2009) may not be suitable to yield the best performance during implementation. Also, the literature on NCS focuses only on linear dynamic systems. However, practical systems are inherently nonlinear. Therefore, the control design for such nonlinear NCS is important and necessary.

In addition, network protocol design is critical for NCS development (Walsh et al. 1999; Hespanha et al. 2007). At present, limited effort has been in place to understand the effect of protocols and most of them merely evaluate the behavior of existing network protocols by separating the controller and network protocol design. However, since controller and network protocol designs are related to each other closely, they cannot be separated. Thus, in this book, novel controller and network protocol designs are introduced jointly to address the drawbacks described above. Additionally, stability guarantees are provided by comparing the proposed schemes with that of the existing NCS.

1.2 CHALLENGES IN NETWORKED CONTROL SYSTEMS

Although an NCS introduces enormous advantages such as reducing installing cost, upgrading system flexibility, and so on, it also causes several serious challenges that have to be addressed carefully. The challenges in an NCS result due to the following issues:

- Network imperfections such as network-induced delay and packet dropouts
- Quantization
- Uncertainties in the NCS dynamics
- Network protocol behavior

Next, we will explain these challenges. By understanding and mitigating the above challenges, we can find suitable solutions successfully.

1.2.1 Network Imperfections

Network is an important component in NCS. According to Hespanha et al. (2007) and due to the inevitable network infrastructure limitation, several imperfections need to be seriously considered. In this book, we will mainly focus on the above-mentioned challenges. The details are given as follows.

1.2.1.1 Network-Induced Delay

It is well known that transmission delay is a big challenge in communication networks. According to the wireless communication theory (Kumar 2014), transmission delay depends on the packet size and wireless channel data rate, which depends on wireless channel quality (i.e., signal-to-interference-plus-noise ratio, SINR).

Recalling Figure 1.1, since the system plant and controller are connected through a communication network, there exist two types of network-induced delays between the plant and the controller, that is, sensor-to-controller delay and controller-to-actuator delay. These delays will degrade system performance significantly if they have not been considered carefully. Figure 1.2 presents the delay caused by the network when inserted within the feedback loop of the control system.

1.2.1.2 Packet Dropouts

While the network quality becomes worse, information between plant and controller cannot be exchanged successfully. Generally, we consider this scenario as

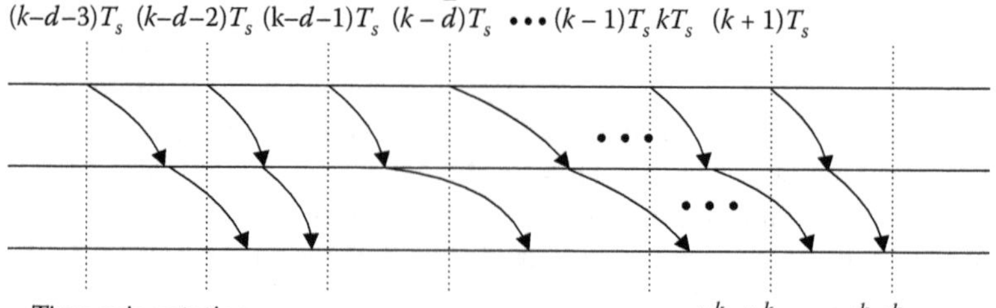

Time axis notation:
1. Top line: Sensor packet sent
2. Middle line: Controller received packet and computed control action transmitted
3. Bottom line: Actuator received control action

FIGURE 1.2 Timing diagram of signals in NCS.

network-induced packet dropout. For NCS, this type of effect is fatal since the NCS could become an open-loop system while control inputs and sensed system states information are lost.

Since packet dropout happens based on real-time network quality, it is not controllable. To reduce this deficiency, researchers introduced network protocols such as transmission control protocol/Internet protocol (TCP/IP). In TCP/IP, an acknowledgment scheme has been included. When the packet is lost, an acknowledgment is not received, which will prompt the transmitter to send the message again until the packet has been successfully received.

Although TCP/IP can relieve the effects of network-induced packet dropout, network traffic will be increased significantly due to the acknowledgment information exchange, which may result in network congestion, ultimately causing instability to the NCS. In addition, retransmitting a feedback signal or control command may not be useful in a closed-loop control system from the stability point of view. Therefore, selecting a proper network protocol to reduce both packet dropouts and network traffic is still an open issue for NCS. The derivation of NCS dynamics after incorporating the effect of network imperfections resulting from network protocol is given in Section 1.2.3.

1.2.2 Quantization

In traditional feedback control systems, it is quite common to assume that the measured signals are transmitted to the controller and the control inputs are delivered back to the plant with arbitrarily high precision via dedicated connections. However, in practice, the interface between the plant and the controller is often connected via analog-to-digital (A/D) and digital-to-analog (D/A) devices, which quantize the signals. In an NCS, the signals are quantized due to the presence of a communication network within the control loop as shown in Figure 1.3. As a result, the quantized networked control system (QNCS) has attracted a great deal of attention from control researchers since the quantization process always existed in computer-based control systems but now with a communication network within the feedback loop. The quantization can cause instability to the control system unless it is carefully dealt.

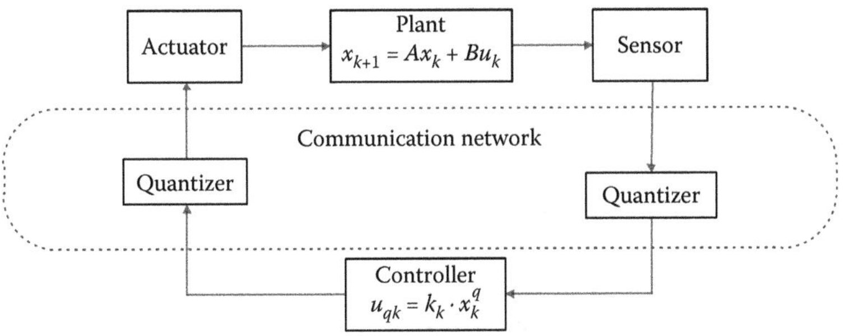

FIGURE 1.3 Block diagram of the QNCS with a lossless network.

In the past literature, the study on the effect of quantization in feedback control systems is normally categorized based on whether the quantizer is static or dynamic. In static quantizers, the quantization region does not change with time and it was first analyzed for unstable linear systems by means of quantized state feedback. Later, it was pointed out that logarithmic quantizers are preferred.

In the case of dynamic quantizers, the quantization region can be adjusted over time based on the idea of scaling quantization levels. A "zoom" parameter is utilized (see Chapter 4) to extend the idea of changing the sensitivity of the quantizer to both linear and nonlinear systems. For these systems, however, the stabilization of the closed-loop system in the presence of quantization was the major issue addressed when system dynamics are known whereas quantization effects in the presence of uncertain system dynamics and optimal control designs for such systems have not yet been satisfactorily addressed.

1.2.3 Network Protocol Effects

In the past NCS literature, researchers assumed that network imperfections were deterministic and/or random. However, network imperfections (e.g., delay, packet dropouts) vary with the network protocol under consideration. Therefore, the evaluation of network protocol that is utilized in the communication is crucial for the NCS.

One major benefit of understanding network protocol behavior is to estimate protocol effects and in some cases predict their behavior for real-time control (Xu and Jagannathan 2012). This could relieve network imperfection impacts significantly. For TCP or UDP, an observer is normally required in the controller design, which can complicate the controller design and stability analysis of the NCS. Owing to the presence of a communication network as shown in Figure 1.1 within the feedback loop, the following network-induced delays and packet losses are observed: (1) $\tau_{sc}(t)$: sensor-to-controller delay, (2) $\tau_{ca}(t)$: controller-to-actuator delay, (3) $\gamma(t)$: indicator of packet lost at controller, and (4) $\upsilon(t)$: indicator of packet lost at actuator.

After incorporating the network-induced delays and packet losses, the original linear time-invariant plant $\dot{x}(t) = Ax(t) + Bu(t)$, $y(t) = Cx(t)$ can be expressed as

$$\begin{aligned} \dot{x}(t) &= Ax(t) + Bu^a(t) \\ u^a(t) &= \upsilon(t)u^c(t - \tau_{ca}(t)) \\ y(t) &= \gamma(t)Cx(t - \tau_{sc}(t)) \end{aligned} \tag{1.1}$$

where

$$\gamma(t) = \begin{cases} \mathbf{I}^{n\times n} & \text{if controller received } x(t) \text{ from the plant at time } t \\ \mathbf{0}^{n\times n} & \text{otherwise} \end{cases}$$

$$\upsilon(t) = \begin{cases} \mathbf{I}^{m\times m} & \text{if actuator received control command from controller at } t \\ \mathbf{0}^{m\times m} & \text{otherwise} \end{cases}$$

with $x(t) \in \mathbb{R}^{n\times n}, u^c(t), u^a(t) \in \mathbb{R}^{m\times m}$, and $y(t) \in \mathbb{R}^{n\times n}$ representing the system state, control inputs computed at the controller and received at the actuator, and output of the plant, respectively, and $A \in \mathbb{R}^{n\times n}$ and $B \in \mathbb{R}^{n\times m}$ denoting system matrices. Assume that the sum of network-induced delays is considered to be bounded, that is, $\tau_{sc}(t) + \tau_{ca}(t) < bT_s$ where b denotes the delay bound while T_s is the sampling interval.

Since the actuator is event-driven, the control input received by the actuator $u^a(t)$ to the plant is a piecewise constant. According to NCS under TCP or UDP protocols, at most b number of current and previous control inputs can be received at the actuator at the same time, and only the latest control input is allowed to be applied on the plant during any sampling interval (i.e., $[kT_s, (k+1)T_s)$, $\forall k$), and other previous control inputs are ignored. It is important to note that since the controller and actuator are event-driven, the plant can implement control inputs at the time instant $kT_s + t_i^k, i = 0,1,2,\ldots,b$ and $t_i^k < t_{i-1}^k$, where $t_i^k = \tau_i^k - iT_s$ as shown in Figure 1.3.

Since the controller is event-driven, the integration of Equation 1.1 over a sampling interval $[kT_s, (k+1)T_s)$ yields

$$\begin{aligned} x_{k+1} &= A_s x_k + B_k^1 u_{k-1}^a + B_k^2 u_{k-2}^a + \cdots + B_k^b u_{k-b}^a + B_k^0 u_k^a \\ u_{k-i}^a &= \upsilon_{k-i} u_{k-i}^c \quad \forall i = 0,1,\ldots,b, \quad \forall k = 0,1,\ldots \\ y_k &= \gamma_k C x_k \quad \forall k = 0,1,2,\ldots \end{aligned} \tag{1.2}$$

where

$$x_k = x(kT_s),\ A_s = e^{AT_s},\ B_k^0 = \int_{\tau_0^k}^{T_s} e^{A(T_s - s)} ds B \delta(T_s - \tau_0^k)$$

$$B_k^i = \int_{\tau_i^k - iT_s}^{\tau_{i-1}^k - (i-1)T_s} e^{A(T_s - s)} ds B \delta(T_s + \tau_{i-1}^k - \tau_i^k) \cdot \delta(\tau_i^k - iT_s)$$

$\forall i = 1,2,\ldots,b,\quad \delta(x) = \begin{cases} 1, & x \geq 0 \\ 0, & x < 0 \end{cases}$, u_k^a is the control input received at the actuator and at time kT_s while u_k^c is the control input computed at the controller and at time kT_s, and γ_k, υ_k are the packet loss indicators at the controller and actuator, respectively, which are also independent and identically distributed Bernoulli random variables with $P(\gamma_k = 1) = \bar{\gamma}$ and $P(\upsilon_k = 1) = \bar{\upsilon}$, where $P(.)$ is the probability of (.).

For simplifying the NCS representation (1.2), a new augment state variable consisting of current state and previous control inputs (i.e., $z_k = [x_k^T\ u_{k-1}^{cT}\ u_{k-2}^{cT} \ldots u_{k-b}^{cT}]^T$

$\in \mathbb{R}^{l=n+bm}$) is introduced. Equation 1.2 becomes a stochastic time-varying system, which can be rewritten as

$$z_{k+1} = A_{zk}z_k + B_{zk}u_k^c,\ y_k = \Gamma_k z_k \quad \forall k = 0,1,2,\ldots \tag{1.3}$$

where time-varying system matrices are given by

$$A_{zk} = \begin{bmatrix} A_s & \upsilon_{k-1}B_k^1 & \cdots & \cdots & \upsilon_{k-b+1}B_k^{b-1} & \upsilon_{k-b}B_k^b \\ 0 & 0 & \cdots & \cdots & 0 & 0 \\ 0 & I_m & \cdots & \cdots & 0 & 0 \\ \vdots & \vdots & \ddots & & \vdots & \vdots \\ \vdots & \vdots & & \ddots & \vdots & \vdots \\ 0 & 0 & \cdots & \cdots & I_m & 0 \end{bmatrix}, \quad B_{zk} = \begin{bmatrix} \upsilon_k B_k^0 \\ I_m \\ 0 \\ 0 \\ \vdots \\ 0 \end{bmatrix}$$

$$\Gamma_k = [C\gamma_k \quad 0 \quad 0 \quad \cdots \quad 0]$$

Here, the system matrices are uncertain due to the presence of network imperfections caused by the communication protocol TCP or UDP with output vector alone being measurable.

The block diagram representation of NCS under TCP or UDP is shown in Figures 1.4 and 1.5, respectively. Compared with UDP, TCP uses acknowledgments to indicate the reception of a packet (i.e., υ_k−1). Therefore, the following network information set Ψ_k, ζ_k can be defined for NCS under TCP or UDP, respectively, as (Schenato et al. 2007)

$$\psi_k = \{\mathbf{y}_k, \gamma_k, \tau_{k-1}, \upsilon_{k-1}\}$$
$$\zeta_k = \{\mathbf{y}_k, \gamma_k, \tau_{k-1}\} \tag{1.4}$$

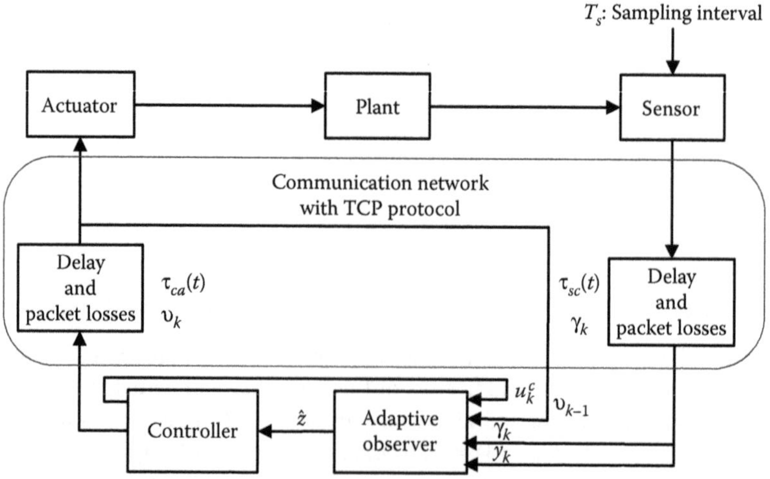

FIGURE 1.4 Block diagram representation of NCS under TCP.

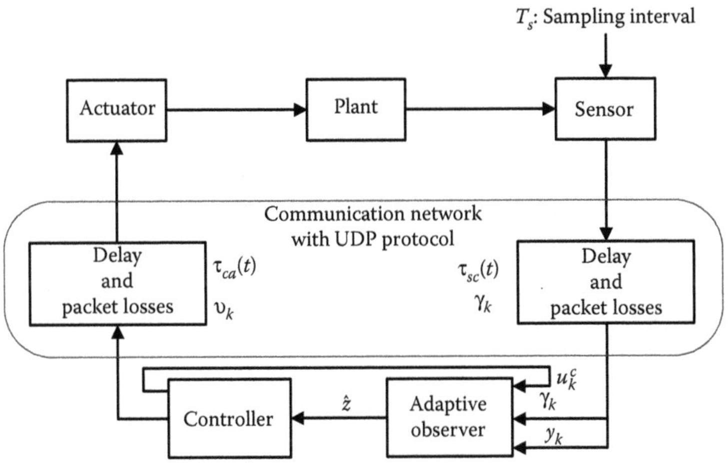

FIGURE 1.5 Block diagram representation of NCS under UDP.

where $\mathbf{y}_k = \{y_k, y_{k-1}, \ldots, y_1\}$, $\gamma_k = \{\gamma_k, \gamma_{k-1}, \ldots, \gamma_1\}$, $\tau_k = \{\tau_k, \tau_{k-1}, \ldots, \tau_1\}$, and $\upsilon_k = \{\upsilon_k, \upsilon_{k-1}, \ldots, \upsilon_1\}$ represent current and previous outputs, packet loss indicators at the controller and the actuator, and network-induced delays, respectively.

Equation 1.3 is a linear representation of NCS after incorporating network imperfections resulting from protocol behavior. A similar representation can be obtained when the original system is nonlinear or when a quantizer is included in the feedback loop. Based on the representation given in Equation 1.3, one has to design satisfactory linear stochastic controller schemes for NCS.

The modeling of a dynamic system with a communication network inside its feedback loop is quite different from the dynamic system acted upon a Gaussian random disturbance and measurement noise. Both these scenarios result in stochastic systems. In the former case, the ultimate system is stochastic time-varying, whereas in the latter case, it becomes stochastic due to the disturbance input and measurement noise. Even for the proposed stochastic time-varying system, one can still include zero mean random disturbance and measurement noise. This addition further makes the overall closed-loop system involved. In other words, additional terms appear in the stochastic Riccati equation. This book covers a suite of certainty-equivalent stochastic linear and nonlinear optimal control schemes in the absence of random disturbance inputs and measurement noise.

In this book, based on the representation of NCS under unreliable communication networks (TCP or UDP) (1.3), the stochastic optimal control of NCS under TCP or UDP, respectively, is derived by minimizing the related value function

$$
\begin{aligned}
V^*(z_k) &= \min_{u_m} E\left[\sum_{m=k}^{\infty} (z_m^T O z_m + u_m^T R u_m) \mid \psi_k\right] \quad \text{NCS under TCP} \\
V^*(z_k) &= \min_{u_m} E\left[\sum_{m=k}^{\infty} (z_m^T O z_m + u_m^T R u_m) \mid \zeta_k\right] \quad \text{NCS under UDP}
\end{aligned}
\tag{1.5}
$$

where $V^*(z_k)$, is the optimal value or cost function, O and R are symmetric positive semidefinite and symmetric positive definite constant matrices, respectively, with $E[(\cdot)\,|\,\psi_k]$, $E[(\cdot)\,|\,\zeta_k]$ are the expectation operator (i.e., mean value) of $\sum_{m=k}^{\infty}(z_m^T O z_m + u_m^T R u_m)$ based on the TCP information set Ψ_k or UDP information set ζ_k defined in Equation 1.4.

The NCS can also be expressed by using available measured data, that is, current and historical input and output sequences. Consider the NCS dynamics given by Equation 1.3 as $z_{k+1} = A_{zk} z_k + B_{zk} u_k^c$, $y_k = \Gamma_k z_k$, where (A_{zk}, B_{zk}) is controllable and (A_{zk}, Γ_k) is observable. It is shown that when the original time-invariant system is controllable and observable, under mild assumptions on network imperfections, the NCS system is also controllable and observable.

According to the observability property of (A_{zk}, Γ_k), the full system states z_k can be reconstructed by using observations of NCS output y_k over a long time horizon. For current time k, NCS dynamics on the time horizon $[k-N, k]$ can be written as

$$z_k = \prod_{i=k-N}^{k-1} A_{zi} z_{k-N} + \left[B_{zk-1} \; A_{zk-1} B_{zk-2} \cdots \left(\prod_{i=k-N+1}^{k-1} A_{zi} \right) B_{zk-N} \right] \begin{bmatrix} u_{k-1}^c \\ u_{k-2}^c \\ \vdots \\ u_{k-N}^c \end{bmatrix}$$

$$\begin{bmatrix} y_{k-1} \\ y_{k-2} \\ \vdots \\ y_{k-N} \end{bmatrix} = \begin{bmatrix} \Gamma_{k-1} \prod_{i=k-N}^{k-2} A_{zi} \\ \Gamma_{k-2} \prod_{i=k-N}^{k-3} A_{zi} \\ \vdots \\ \Gamma_{k-N} \end{bmatrix} z_{k-N}$$

$$+ \begin{bmatrix} 0 & \Gamma_{k-1} B_{zk-2} & \Gamma_{k-1} A_{zk-2} B_{zk-3} & \cdots & \Gamma_{k-1} \left(\prod_{i=k-N+1}^{k-2} A_{zi} \right) B_{zk-N} \\ 0 & 0 & \Gamma_{k-2} B_{zk-2} & \cdots & \Gamma_{k-2} \left(\prod_{i=k-N+1}^{k-3} A_{zi} \right) B_{zk-N} \\ \vdots & \vdots & \vdots & \ddots & \vdots \\ 0 & 0 & 0 & \cdots & \Gamma_{k-N} B_{zk-N} \\ 0 & 0 & 0 & \cdots & 0 \end{bmatrix} \begin{bmatrix} u_{k-1}^c \\ u_{k-2}^c \\ \vdots \\ u_{k-N}^c \end{bmatrix} \quad (1.6)$$

Define controllability and observability matrices of NCS as

$$F_{Nk}^{o} = \left[B_{zk-1} A_{zk-1} B_{zk-2} \ldots \left(\prod_{i=k-N+1}^{k-1} A_{zi} \right) B_{zk-N} \right]$$
$$H_{Nk}^{o} = \left[\left(\Gamma_{k-1} \prod_{i=k-N}^{k-2} A_{zi} \right)^{T} \left(\Gamma_{k-2} \prod_{i=k-N}^{k-3} A_{zi} \right)^{T} \ldots \Gamma_{k-N}^{T} \right]^{T} \tag{1.7}$$

Meanwhile, Toeplitz matrix of Markov parameter and the available measured data (i.e., input and output sequences) over time horizon $[k-1, k-N]$ can be defined as

$$G_{kN}^{o} = \begin{bmatrix} 0 & \Gamma_{k-1} B_{zk-2} & \Gamma_{k-1} A_{zk-2} B_{zk-3} & \cdots & \Gamma_{k-1} \left(\prod_{i=k-N+1}^{k-2} A_{zi} \right) B_{zk-N} \\ 0 & 0 & \Gamma_{k-2} B_{zk-2} & \cdots & \Gamma_{k-2} \left(\prod_{i=k-N+1}^{k-3} A_{zi} \right) B_{zk-N} \\ \vdots & \vdots & \vdots & \ddots & \vdots \\ 0 & 0 & 0 & \cdots & \Gamma_{k-N} B_{zk-N} \\ 0 & 0 & 0 & \cdots & 0 \end{bmatrix}$$

$$\mathbf{y}_{k-1}^{o} = \begin{bmatrix} y_{k-1} \\ y_{k-2} \\ \vdots \\ y_{k-N} \end{bmatrix}, \text{ and } \mathbf{u}_{k-1} = \begin{bmatrix} u_{k-1}^{c} \\ u_{k-2}^{c} \\ \vdots \\ u_{k-N}^{c} \end{bmatrix} \tag{1.8}$$

Using Equations 1.7 and 1.8, Equation 1.6 can be represented as

$$z_k = \prod_{i=k-N}^{k-1} A_{zi} z_{k-N} + F_{Nk}^{o} \mathbf{u}_{k-1} \tag{1.9}$$

$$\mathbf{y}_{k-1}^{o} = H_{Nk}^{o} z_{k-N} + G_{Nk}^{o} \mathbf{u}_{k-1} \tag{1.10}$$

Since (A_{zk}, Γ_k) is observable, there exists an observability index l such that H_{Nk}^{o} is full rank when $N \geq l$. Therefore, let $N \geq l$, then left inverse of H_{Nk}^{o} is given as

$$(H_{Nk}^{o})^{+} = (H_{Nk}^{oT} H_{Nk}^{o})^{-1} H_{Nk}^{oT} \tag{1.11}$$

Multiplying the left inverse of H_{Nk}^o on both sides of Equation 1.9, $z_{k-N}=(H_{Nk}^o)^+ y_{k-1}^o-(H_{Nk}^o)^+G_{Nk}^o u_{k-1}$. Substituting z_{k-N} into Equation 1.9, NCS states z_k can be expressed as

$$z_k=\prod_{i=k-N}^{k-1}A_{zi}[(H_{Nk}^o)^+y_{k-1}^o-(H_{Nk}^o)^+G_{Nk}^o\mathbf{u}_{k-1}]+F_{Nk}^o\mathbf{u}_{k-1} \tag{1.12}$$

$$=\left(\prod_{i=k-N}^{k-1}A_{zi}\right)(H_{Nk}^o)^+\mathbf{y}_{k-1}^o+\left[F_{Nk}^o-\left(\prod_{i=k-N}^{k-1}A_{zi}\right)(H_{Nk}^o)^+G_{Nk}^o\right]\mathbf{u}_{k-1}$$

$$=D_y^o\mathbf{y}_{k-1}^o+D_u^o\mathbf{u}_{k-1}=\begin{bmatrix}D_y^o & D_u^o\end{bmatrix}\begin{bmatrix}\mathbf{y}_{k-1}^o\\ \mathbf{u}_{k-1}\end{bmatrix}$$

where

$$D_y^o=\left(\prod_{i=k-N}^{k-1}A_{zi}\right)(H_{Nk}^o)^+ \quad \text{and} \quad D_u^o=F_{Nk}^o-\left(\prod_{i=k-N}^{k-1}A_{zi}\right)\times(H_{Nk}^o)^+G_{Nk}^o$$

Also, since (A_{zk},Γ_k) is observable and $N\geq l$, D_y^o is full column rank and the left inverse of D_y^o can be expressed as

$$(D_y^o)^+=(D_y^{oT}D_y^o)^{-1}D_y^{oT} \tag{1.13}$$

On the other hand, it is important to note that $\|D_y^o\|\leq D_M$ with D_M known since D_y^o is composed of NCS dynamics A_{zk}, B_{zk}, Γ_k, which are considered to be bounded as $\|A_{zk}\|\leq A_M$, $\|B_{zk}\|\leq B_M$, $\|\Gamma_k\|\leq\Gamma_M$ with the bounds A_M, B_M, Γ_M considered known. No matter which representation, either Equation 1.3 or Equations 1.9 and 1.10, is used, the dynamics will become stochastic and uncertain.

Many existing control schemes such as traditional control, nonlinear control, as well as optimal control require system dynamics. However, in practice, system dynamics are not known accurately due to the uncertainties such as unmodeled dynamics. Without system dynamics, traditional control schemes cannot be designed.

In NCS, owing to the presence of network imperfections, the closed-loop linear and nonlinear systems will become time-varying stochastic systems with uncertain dynamics (see Xu and Jagannathan, 2012, Chapters 3 and 5). However, a majority of the reported techniques in the literature assume information about network imperfections while only a probability density function of the network imperfections is normally known. Therefore, extending the control techniques to the stochastic domain (Stengel 1986) with the system dynamics being uncertain is a major challenge.

1.3 CURRENT RESEARCH

During the past several decades, great strides have been made such that novel communication networks and control designs provide benefits such as robustness, optimality, and flexibility. Here, we want to summarize these recent trends in the area of network and controller design. The three critical trends in networking aspects include energy efficiency, spectrum management, and game theory. These are discussed below.

1.3.1 Energy Efficiency

Energy consumption has been considered as one of the critical challenges in advanced communication network design. Especially for wireless sensor networks (WSN), an energy-efficient protocol is necessary for saving energy and prolonging network lifetime. Wei et al. (2002) proposed a novel medium-access control (MAC) for WSN, which reduces energy consumption significantly while achieving scalability and avoiding collision from interfering nodes. Recently, researchers have demonstrated that clustering is an efficient technique to increase network lifetime and scalability (Lim and Lee 1997; Zhang et al. 2008). Kumar (2014) proposed and evaluated the single-hop and multihop energy-efficient clustering protocol for WSN.

1.3.2 Spectrum Management

Owing to limited spectrum resource availability and with increasing users on wireless networks, it is necessary to manage spectrum resources efficiently. Recently, cognitive radio (CR) technology has been envisaged to solve spectrum management challenges. Salem et al. (2014) developed an opportunistic spectrum access protocol for CR ad hoc networks that can improve spectrum usage significantly.

1.3.3 Game Theory

Owing to the distributed and autonomous nature of existing wireless networks, users might behave selfishly to maximize their own benefits, thus affecting the entire network performance. Therefore, a proper distributed network protocol is needed to manage network resources among different users. Meanwhile, game theory has been considered as a promising technique to address multiplayer resource allocation problems. Engaging game theory with network protocol design is a future direction for wireless network research. Recently, Guan et al. (2011) introduced a novel routing algorithm to solve the obstacle problem in WSN based on a game-theory model. Nie and Comaniciu (2006) proposed a game-theoretic framework to analyze the behavior of CRs for distributed adaptive channel allocation.

On the other hand, modern control theory has been developing quickly during the past few decades to address challenges in such systems with constrained resources. Of all the methods, optimal control and event-triggered control are two important and relevant options for discussion. These are briefly introduced.

1.3.4 Optimal Control

Optimality (Lewis and Syrmos 1995) is more preferred than stability alone for a dynamic system. However, according to Werbos (1991), optimal control of nonlinear systems is a very difficult and even impossible feat when system dynamics are unknown. Inspired from computational intelligence, in this book, adaptive/approximate dynamic programming (ADP) (Stengel 1986; Werbos 1991) has been introduced to overcome this deficiency. For solving the optimal control problem, one has to find the solution to either the Bellman or Hamilton–Jacobi–Bellman (HJB) equations. Unfortunately, no closed-form solution can be determined for the Bellman or HJB equations and one has to resort to iterative methods.

Recently, Wang et al. (2011), Zhang et al. (2011), and Heydari and Balakrishnan (2011) proposed policy and/or value iteration-based ADP schemes to attain near-optimal control without requiring system dynamics. However, these approaches require a significant number of iterations within a sampling interval, which is difficult to realize in practice. To overcome this deficiency, Dierks and Jagannathan (2012) introduced a time-based ADP algorithm to solve the optimal control problem by using system historical information instead of iteration-based information. However, these traditional methods mentioned here are periodically time-driven and they all sample the state vector at each sampling instant. This also implies that the state vector and its associated control input vector have to be quantized and transmitted over the communication network. As mentioned before in this chapter, the sampling rate is normally selected at a high level keeping stability in mind. Unfortunately, with high sampling rates that are periodic, significant network resources are required. In addition, with increased network transmission and limited network bandwidth, congestion may be observed causing a drop in performance, ultimately leading to packet dropouts. To mitigate this issue, in the recent literature, event-sampled control has been proposed.

1.3.5 Event-Sampled Control

Event-sampled or event-triggered control has emerged as an alternate method to reduce network communication and controller execution. In an earlier work for a continuous-time system, Tabuada (2007) ensured the system is input-to-state stable (ISS) with respect to an event-triggering condition that decides when to sample the state and input vector. Recently, a variety of event-triggered control techniques (e.g., zero-order-hold scheme (Donkers and Heemels 2012) and model-based scheme (Garcia and Antsaklis 2013)) have been developed for both linear and nonlinear systems demonstrating the effectiveness of event-triggered control over periodically time-driven control schemes.

Owing to the explosive growth in these areas, an interesting idea surfaced in the community, which is to combine network protocol design and control theory (Liu and Goldsmith 2004). This novel class of systems is referred to as distributed networked control systems (DNCS). The basic structure of a DNCS is described in Figure 1.6, where numerous systems communicate to their corresponding controllers through a shared communication network such as an IEEE 802.1x (Xia et al. 2011). It is considered as a DNCS since multiple subsystems are geographically distributed

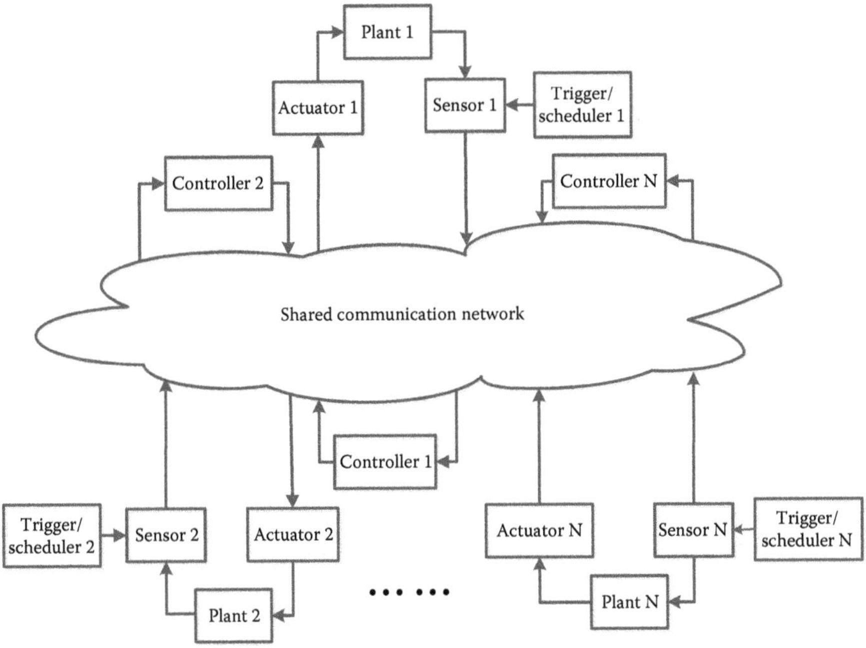

FIGURE 1.6 Distributed NCS.

without any interconnection terms among them in the physical layer. While a shared communication network among multiple systems can enhance the usage of the network, it may also cause a bottleneck unless scheduling of data within the network from these multiple subsystems is done carefully. It is clear that the shared communication network will affect the performance of the control system and the NCS under consideration cannot maintain stability since the information from the physical system cannot be transmitted to their respective controllers.

REFERENCES

Antsaklis, P. and Baillieul, J. 2004. Special issue on networked control systems. *IEEE Transactions on Automatic Control*, 49, 1421–1423.

Branicky, M.S., Phillips, S.M., and Zhang, W. 2000. Stability of networked control systems: Explicit analysis of delay. *Proceedings of the American Control Conference*, Chicago, IL, USA, vol. 4, pp. 2352–2357.

Cloosterman, M.B.G., van de Wouw, N., Heemels, W.P.M.H., and Nijmeijer, H. 2009. Stability of networked control systems with uncertain time-varying delays. *IEEE Transactions on Automatic Control*, 54, 1575–1580.

Dierks, T. and Jagannathan, S. 2012. Online optimal control of affine nonlinear discrete-time systems with unknown internal dynamics by using time-based policy update. *IEEE Transactions on Neural Networks and Learning Systems*, 23, 1118–1129.

Donkers, M. and Heemels, W. 2012. Output-based event-triggered control with guaranteed L_∞-gain and improved and decentralised event-triggering. *IEEE Transactions on Automatic Control*, 57, 1362–1376.

Garcia, E. and Antsaklis, P.J. 2013. Model-based event-triggered control for systems with quantization and time-varying network delays. *IEEE Transactions on Automatic Control*, 58, 422–434.

Guan, X., Wu, H., and Bi, S. 2011. A game theory based obstacle avoidance routing protocol for wireless sensor networks. *Sensor*, 11, 9327–9343.

Halevi, Y. and Ray, A. 1988. Integrated communication and control systems: Part I—Analysis. *Journal of Dynamic Systems and Measure Control*, 110, 367–373.

Heemels, W.P.M.H., van de Wouw, A.R., and Nesic, D. 2010. Networked control systems with communication constraints: Tradeoff between transmission intervals, delays and performance. *IEEE Transactions on Automatic Control*, 55, 1781–1796.

Hespanha, J.P., Naghshtabri, J.P., and Xu, Y. 2007. A survey of recent results in networked control system. *Proceedings of the IEEE*, 95, 138–162.

Heydari, A. and Balakrishnan, S.N. 2011. Finite horizon input-constrainted nonlinear optimal control using single network adaptive critics. *Proceedings of the American Control Conference*, pp. 3047–3052.

Hu, S.S. and Zhu, Q.X. 2003. Stochastic optimal control and analysis of stability of networked control systems with long delay. *Automatica*, 39, 1877–1884.

Kumar, D. 2014. Performance analysis of energy efficient clustering protocols for maximising lifetime of wireless sensor networks. *IET Wireless Sensor Systems*, 4, 9–16.

Kumar, P.R. 2009. Fundamental issues in networked control system. *Proceedings of the 3rd WIDE Ph.D. School on Networked Control System*, Siena, Italy.

Lewis, F.L. and Syrmos, V. 1995. *Optimal Control*, 2nd edition, Wiley, New York.

Lian, F., Moyne, J., and Tilbury, D. 2003. Modeling and optimal controller design of networked control systems with multiple delays. *International Journal of Control*, 76, 591–606.

Lim, K. and Lee, Y.H. 1997. Optimal partitioning of heterogeneous traffic sources in mobile communication networks. *IEEE Transactions on Computers*, 46, 312–325.

Liou, L.W. and Ray, A. 1991. A stochastic regulator for integrated communication and control systems: Part I—Formulation of control law. *ASME Journal of Dynamic System Measurement and Control*, 4, 688–694.

Liu, X.H. and Goldsmith, A. 2004. Wireless medium access control in networked control system. *Proceedings of the American Control Conference*, Boston, MA, USA, pp. 688–694.

Nie, N. and Comaniciu, C. 2006. Adaptive channel allocation spectrum etiquette for cognitive radio networks. *Journal of Mobile Networks and Applications*, 11, 779–797.

Nilsson, J., Bernhardsson, B., and Wittenmark, B. 1998. Stochastic analysis and control of real-time systems with random time delays. *Automatica*, 34, 57–64.

Salem, T., El-kader, S.M., Ramadan, S.M., and Abdel-Mageed, M.Z. 2014. Opportunistic spectrum access in cognitive radio ad hoc networks. *International Journal of Computer Science*, 11, 41–50.

Schenato, L., Sinopoli, B., Franceschetti, M., Poolla, K., and Sastry, S. 2007. Foundations of control and estimation over lossy networks. *Proceedings of the IEEE*, 95, 163–187.

Song, J., Mok, A.K., Chen, D., and Nixon, M. 2006. Challenges of wireless control in process control. *Workshop on Research Directions for Security and Networking in Critical Real-Time and Embedded Systems*, Arlington, VA, USA, April 4, 2006.

Stengel, R.F. 1986. *Stochastic Optimal Control: Theory and Application*. Wiley-Interscience, New York.

Tabuada, P. 2007. Event-triggered real-time scheduling of stabilizing control tasks. *IEEE Transactions on Automatic Control*, 52, 1680–1685.

Walsh, G.C., Ye, H., and Bushnell, L. 1999. Stability analysis of networked control systems. *Proceedings of the American Control Conference*, San Diego, CA, USA, pp. 2876–2880.

Wang, F.Y., Jin, N., Liu, D., and Wei, Q.L. 2011. Adaptive dynamic programming for finite horizon optimal control of discrete-time nonlinear system with ε-error bound. *IEEE Transactions on Neural Networks*, 22, 24–36.

Werbos, P.J. 1991. Approximate dynamic programming for real-time control and neural modeling. In *Handbook of Intelligent Control*, D.A. White and D.A. Sorge, (eds.). Van Nostrand Reinhold, New York, pp. 493–525.

Wei, Y., Heidemann, J., and Estrin, D. 2002. An energy-efficient MAC protocol for wireless sensor network. *Proceedings of INFOCOM 2002*, New York, NY, USA, pp. 1567–1576.

Wu, J. and Chen, T. 2007. Design of networked control systems with packet dropouts. *IEEE Transactions on Automatic Control*, 52, 1314–1319.

Xia, F., Vinel, A., Gao, R., Wang, L., and Qiu, T. 2011. Evaluating IEEE 802.15.4 for cyber-physical systems. *EURASIP Journal on Wireless Communication and Networking*, 2011, 1–14.

Xu, H. and Jagannathan, S. 2012. Stochastic optimal controller design for unknown networked control systems under TCP. *Proceedings of the American Control Conference*, Montreal, Canada, pp. 6503–6508.

Zhang, H., Luo, Y., and Liu, D. 2011. Neural network based near optimal control for a class of discrete-time affine nonlinear system with control constraints. *IEEE Transactions on Neural Network*, 20, 1490–1530.

Zhang, W., Branicky, M.S., and Phillips, S. 2001. Stability of networked control systems. *IEEE Control System Magazine*, 21, 84–99.

Zhang, Z., Ma, M., and Yang, Y. 2008. Energy-efficient multihop polling in clusters of two-layered heterogeneous sensor networks. *IEEE Transactions on Computers*, 57, 231–245.

2 Background on Lyapunov Stability and Stochastic Optimal Control

In this chapter, we provide a brief background on dynamical systems and stochastic optimal control, mainly covering the topics that will be important for the development of an NCS. We will also cover the standard stochastic optimal control using the Riccati equation for discrete-time systems as it is needed for optimality. It is quite common for noncontrol engineers working in control applications to have little understanding of feedback control and dynamical systems. Many of the phenomena they observe are attributable of feedback control systems. For a detailed treatment of stochastic optimal control, refer to Åstrom (1970) and Stengel (1986). The neural network (NN) background material is taken from Sarangapani (2006). An incomplete understanding of any one of these can lead to incorrect conclusions being drawn, with inaccurate attributions of causes. Included in this chapter are discrete-time systems, computer simulation, norms, and stability first for determinstic systems, and then for traditional stochastic systems. One will notice that, in many cases, even for stochastic systems, the system dynamics are normally deterministic (Åstrom 1970; Stengel 1986), driven by a random disturbance or process noise with zero mean. Similarly, the output is acted upon by a sensor measurement noise with zero mean. In contrast, as derived in Chapter 3, the system dynamics will be a time-varying function of random variables for the NCS.

2.1 DETERMINISTIC DYNAMICAL SYSTEMS

This part of the chapter is from Sarangapani (2006). Many systems in nature, including neurobiological systems, are dynamical in nature, in the sense that they are acted upon by external inputs, have internal memory, and behave in certain ways that are captured by the notion of the development of activities through time. According to the notion of systems defined by Whitehead (1953), it is an entity distinct from its environment, whose interactions with the environment can be characterized through input and output signals. An intuitive feel for dynamic systems is provided by Luenberger (1979), which includes many examples.

2.1.1 Discrete-Time Systems

If the time index is an integer k instead of a real number t, the system is said to be of discrete time. A general class of discrete-time systems can be described by the nonlinear ordinary difference equation in discrete-time state space form

$$x(k+1) = f(x(k),u(k)), \quad y(k) = h(x(k),u(k)) \tag{2.1}$$

where $x(k) \in \Re^n$ is the internal state vector, $u(k) \in \Re^m$ is the contol input, and $y(k) \in \Re^p$ is the system output.

These equations may be derived directly from an analysis of the dynamical system or process being studied, or they may be sampled or discretized versions of continuous-time dynamics of a nonlinear system. Today, controllers are implemented in digital form by using *embedded hardware* making it necessary to have a discrete-time description of the controller. This may be determined by design, based on discrete-time system dynamics. Sampling of linear systems is well understood with many design techniques available. However, sampling of nonlinear systems is not an easy topic. In fact, the exact discretization of nonlinear continuous dynamics is based on the Lie derivatives and leads to an infinite series representation (see, e.g., Kalkkuhl and Hunt 1996). Various approximation and discretization techniques use truncated versions of the exact series.

2.1.2 Brunovsky Canonical Form

Letting $x(k) = [x_1(k)\ldots x_n(k)]^T$, a special form of nonlinear dynamics is given by the class of systems in the discrete Brunovsky canonical form:

$$\begin{aligned} x_1(k+1) &= x_2(k) \\ x_2(k+1) &= x_3(k) \\ &\vdots \\ x_n(k+1) &= f(x(k)) + g(x(k))u(k) \\ y(k) &= h(x(k)) \end{aligned} \tag{2.2}$$

As seen from Figure 2.1, this is a chain or cascade of unit delay elements z^{-1}, that is, a shift register. Each delay element stores information and requires an initial condition. The measured output $y(k)$ can be a general function of the states as shown, or can have more specialized forms such as

$$y(k) = h(x_1(k)) \tag{2.3}$$

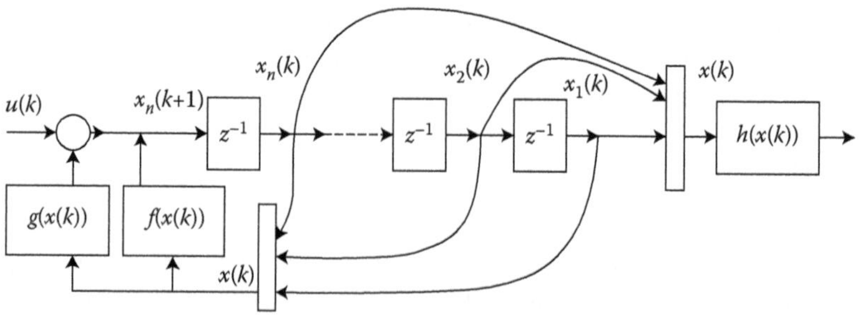

FIGURE 2.1 Discrete-time single-input Brunovsky form.

The discrete Brunovsky canonical form may equivalently be written as

$$x(k+1) = Ax(k) + bf(x(k)) + bg(x(k))u(k) \tag{2.4}$$

where

$$A = \begin{bmatrix} 0 & 1 & 0 & . & . & . & 0 \\ 0 & 0 & 1 & . & . & . & 0 \\ & & & . & & & \\ & & & . & & & \\ & & & . & & & \\ 0 & 0 & . & . & . & 1 & 0 \\ 0 & 0 & 0 & . & . & . & 0 \end{bmatrix}, \quad b = \begin{bmatrix} 0 \\ 0 \\ . \\ . \\ . \\ 0 \\ 1 \end{bmatrix} \tag{2.5}$$

A discrete-time form of the more general version may also be written. It is a system with m-parallel chains of delay elements of lengths n_1, n_2,... (e.g., m shift registers), each driven by one of the control inputs.

Many practical systems occur in the continuous-time Brunovsky form. However, if a system of the continuous Brunovsky form (Lewis et al. 1999) is sampled, the result is not the general form (2.2). Under certain conditions, general discrete-time systems of the form (2.1) can be converted to discrete Brunovsky canonical form systems (see, e.g., Kalkkuhl and Hunt 1996).

2.1.3 Linear Systems

A special and important class of dynamical systems is the discrete-time linear time-invariant (LTI) system

$$\begin{aligned} x(k+1) &= Ax(k) + Bu(k) \\ y(k) &= Cx(k) \end{aligned} \tag{2.6}$$

where A, B, C are constant matrices of general form (e.g., not restricted to Equation 2.5). An LTI is denoted by (A, B, C). Given an initial state $x(0)$, the solution to the LTI system can be explicitly written as

$$x(k) = A^k x(0) + \sum_{j=0}^{k-1} A^{k-j-1} Bu(j) \tag{2.7}$$

The next example shows the relevance of these solutions and demonstrates that the general discrete-time nonlinear systems are even easier to simulate on a computer than continuous-time systems, as no integration routine is needed.

EXAMPLE 2.1: DISCRETE-TIME SYSTEM—SAVINGS ACCOUNT (LEWIS ET AL. 1999)

Discrete-time descriptions can be derived from continuous-time systems by using Euler's approximation or system discretization theory. However, many phenomena are naturally modeled using discrete-time dynamics, including population growth/decline, epidemic spread, economic systems, and so on. The dynamics of the savings account using compound interest are given by the first-order system

$$x(k+1) = (1+i)x(k) + u(k)$$

where i represents the interest rate over each interval, k is the interval iteration number, and $u(k)$ is the amount of the deposit at the beginning of the kth period. The state $x(k)$ represents the account balance at the beginning of interval k.

2.1.3.1 Analysis

According to Equation 2.7, if equal annual deposits are made of $u(k) = d$, the account balance is

$$x(k) = (1+i)^k x(0) + \sum_{j=0}^{k-1} (1+i)^{k-j-1} d$$

with $x(0)$ being the initial amount in the account. Using the standard series summation formula

$$\sum_{j=0}^{k-1} a^j = \frac{1-a^k}{1-a}$$

one derives the standard formula for complex interest with constant annuities of d.

$$x(k) = (1+i)^k x(0) + d(1+i)^{k-1} \sum_{j=0}^{k-1} \frac{1}{(1+i)^j}$$

$$= (1+i)^k x(0) + d(1+i)^{k-1} \left[\frac{1-(1/(1+i)^k)}{1-(1/(1+i))} \right]$$

$$= (1+i)^k x(0) + d \left[\frac{(1+i)^{k-1} - 1}{i} \right]$$

2.1.3.2 Simulation

It is very easy to simulate a discrete-time system. No numerical integration driver program is needed in contrast to the continuous-time case. Instead, a simple "do loop" can be used. A complete MATLAB® program that simulates the compound interest dynamics is given by

```
%Discrete-Time Simulation program for Compound Interest Dynamics
d=100; i=0.08; % 8% interest rate
x(1) =1000;
for k=1:100
        x(k+1) = (1+i)*x(k)
end
k= [1:101];
plot(k,x);
```

2.2 MATHEMATICAL BACKGROUND

2.2.1 Vector and Matrix Norms

We assume the reader is familiar with norms, both vector and induced matrix norms (Lewis et al. 1993). We denote any suitable vector norm by $\|.\|$. When required to be specific, we denote the p-norm by $\|.\|_p$. Recall that for any vector $x \in \Re^n$

$$\| x \|_1 = \sum_{i=1}^{n} | x_i | \tag{2.8}$$

$$\| x \|_p = \left(\sum_{i=1}^{n} | x_i |^p \right)^{1/p} \tag{2.9}$$

$$| x |_\infty = \max_i | x_i | \tag{2.10}$$

The 2-norm is the standard Euclidean norm.

Given a matrix A, its induced p-norm is denoted by $\|A\|_p$. Letting $A = [a_{ij}]$, recall that the induced 1-norm is the maximum absolute column sum

$$\| A \|_1 = \sum_{i}^{\max} | a_{ij} | \tag{2.11}$$

and the induced ∞-norm is the maximum absolute row sum

$$\| A \|_\infty = \max_i \sum_i | a_{ij} | \tag{2.12}$$

The induced matrix p-norm satisfies the inequality, for any vector x,

$$\| A \|_p \leq \| A \|_p \| x \|_p \tag{2.13}$$

and for any two matrices A, B, one also has

$$\| AB \|_p \leq \| A \|_p \| B \|_p \tag{2.14}$$

Given a matrix $A = [a_{ij}]$, the Frobenius norm is defined as the root of the sum of the squares of all the elements:

$$\| A \|_F^2 \equiv \sum a_{ij}^2 = tr(A^T A) \tag{2.15}$$

where $tr(.)$ is the matrix trace (i.e., sum of diagonal elements). Though the Frobenius norm is not an induced norm, it is compatible with the vector 2-norm so that

$$\| Ax \|_2 \leq \| A \|_F \| x \|_2 \tag{2.16}$$

2.2.1.1 Singular Value Decomposition

The matrix norm $\|A\|_2$ induced by the vector 2-norm is the maximum singular value of A. For a general $m \times n$ matrix A, one may write the *singular value decomposition (SVD)*

$$A = U \Sigma V^T \tag{2.17}$$

where U is $m \times n$, V is $n \times n$, and both are orthogonal, that is

$$\begin{aligned} U^T U &= UU^T = I_m \\ V^T V &= VV^T = I_n \end{aligned} \tag{2.18}$$

where I_n is the $n \times n$ identity matrix. The $m \times n$ singular value matrix has the structure

$$\Sigma = diag\{\sigma_1, \sigma_2, \ldots, \sigma_r, 0, \ldots, 0\} \tag{2.19}$$

where r is the rank of A and σ_i are the singular values of A. It is conventional to arrange the singular values in a nonincreasing order, so that the largest singular value is $\sigma_{max}(A) = \sigma_1$. If A is full rank, then r is equal to either m or n, whichever is smaller. Then the minimum singular value is $\sigma_{min}(A) = \sigma_r$ (otherwise the minimum singular value is equal to zero).

The SVD generalizes the notion of eigenvalues to general nonsquare matrices. The singular values of A are the (positive) square roots of the nonzero eigenvalues of AA^T, or equivalently A^TA.

2.2.1.2 Quadratic Forms and Definiteness

Given an $n \times n$ matrix Q, the *quadratic form* x^TQx, with x an n-vector, will be important for stability analysis in this book. The quadratic form can in some cases have certain properties that are independent of the vector x selected. Four important definitions are

$$\begin{aligned}
&Q \text{ is positive definite, denoted } Q > 0, \text{ if } x^TQx > 0, \forall x \neq 0 \\
&Q \text{ is positive semidefinite, denoted } Q \geq 0, \text{ if } x^TQx \geq 0, \forall x \\
&Q \text{ is negative definite, denoted } Q < 0, \text{ if } x^TQx < 0, \forall x \neq 0 \\
&Q \text{ is negative semidefinite, denoted } Q \leq 0, \text{ if } x^TQx \leq 0, \forall x
\end{aligned} \tag{2.20}$$

If Q is symmetric, then it is positive definite if and only if all its eigenvalues are positive, and positive semidefinite if and only if all its eigenvalues are nonnegative. If Q is not symmetric, the tests are more complicated and involve determining the minors of the matrix. Tests for negative definiteness and semidefiniteness may be found by noting that Q is negative (semi) definite if and only if $-Q$ is positive (semi) definite.

If Q is a symmetric matrix, its singular values are the magnitudes of its eigenvalues. If Q is a symmetric positive semidefinite matrix, its singular values and its eigenvalues are the same. If Q is positive semidefinite, then, for any vector x, one has the useful inequality

$$\sigma_{\min}(Q) \| x \|^2 \leq x^TQx \leq \sigma_{\max}(Q) \| x \|^2 \tag{2.21}$$

2.2.2 Continuity and Function Norms

Given a subset $S \subset \Re^n$, a function $f(x): S \to \Re^m$ is *continuous* on $x_0 \in S$ if for every $\varepsilon > 0$ there exists a $\delta(\varepsilon, x_0) > 0$ such that $\|x - x_0\| < \delta(\varepsilon, x_0)$ implies that $\|f(x) - f(x_0)\| < \varepsilon$. If δ is independent of x_0, then the function is said to be *uniformly continuous*. Uniform continuity is often difficult to test. However, if $f(x)$ is continuous and its derivative $f'(x)$ is bounded, then it is uniformly continuous.

A function $f(x): \Re^n \to \Re^m$ is *differentiable* if its derivative $f'(x)$ exists. It is continuously differentiable if its derivative exists and is continuous. $f(x)$ is said to be locally Lipschitz if, for all $x, z \in S \subset \Re^n$, one has

$$\| f(x) - f(z) \| < L \| x - z \| \tag{2.22}$$

for some finite constant $L(S)$ where L is known as a *Lipschitz constant*. If $S = \Re^n$, then the function is globally Lipschitz.

If $f(x)$ is globally Lipschitz, then it is uniformly continuous. If it is continuously differentiable, it is locally Lipschitz. If it is differentiable, it is continuous.

For example, $f(x) = x^2$ is continuously differentiable. It is locally but not globally Lipschitz. It is continuous but not uniformly continuous.

Given a function $f(t):(0,\infty) \rightarrow \Re^n$, according to *Barbalat's lemma*, if

$$\int_0^\infty f(t)dt \leq \infty \tag{2.23}$$

and $f(t)$ is uniformly continuous, then $f(t) \rightarrow 0$ as $t \rightarrow \infty$. Given a function $f(t):(0,\infty) \rightarrow \Re^n$, its L_p (function) norm is given in terms of the vector norm $\|f(t)\|_p$ at each value of t by

$$\| f(\cdot) \|_p = \left(\int_0^\infty \| f(t) \|_p^p \, dt \right)^{1/p} \tag{2.24}$$

and if $p = \infty$

$$\| f(\cdot) \|_\infty = \sup_t \| f(t) \|_\infty \tag{2.25}$$

If the L_p norm is finite, we say $f(t) \in L_p$. Note that a function is in L_∞ if and only if it is bounded. For detailed treatment, refer to Lewis et al. (1993, 1999).

In the discrete-time case, let $Z_+ = \{0, 1, 2, \ldots\}$ be the set of natural numbers and $f(k): Z_+ \rightarrow \Re^n$. The l_p (function) norm is given in terms of the vector $\|f(k)\|_p$ at each value of k by

$$\| f(.) \|_p = \left(\sum_{k=0}^\infty \| f(k) \|_p^p \right)^{1/p} \tag{2.26}$$

and if $p = \infty$

$$\| f(.) \|_\infty = \sup_k \| f(k) \|_\infty \tag{2.27}$$

If the l_p norm is finite, we say $f(k) \in l_p$. Note that a function is in l_∞ if and only if it is bounded.

2.3 PROPERTIES OF DYNAMICAL SYSTEMS

In this section are discussed some properties of dynamical systems, including stability. For observability and controllability, please refer to Sarangapani (2006) and Goodwin and Sin (1984). If the original open-loop system is controllable and observable, then a feedback control system can be designed to meet the desired performance.

If the system has certain passivity properties, this design procedure is simplified and additional closed-loop properties such as robustness can be guaranteed. On the other hand, properties such as stability may not be present in the original open-loop system, but are design requirements for closed-loop performance.

Stability along with robustness (see Section 2.3.1) is a performance requirement for closed-loop systems. In other words, though the open-loop stability properties of the original system may not be satisfactory, it is desirable to design a feedback control system such that the closed-loop stability is adequate. We will discuss stability for discrete-time systems, but the same definitions also hold for continuous-time systems with obvious modifications.

Consider the dynamical system

$$x(k+1) = f(x(k), k) \tag{2.28}$$

where $x(k) \in \Re^n$, which might represent either an uncontrolled open-loop system or a closed-loop system after the control input $u(k)$ has been specified in terms of the state $x(k)$. Let the initial time be k_0, and the initial condition be $x(k_0) = x_0$. This system is said to be *nonautonomous* since the time k appears explicitly. If k does not appear explicitly in $f(.)$, then the system is autonomous. A primary cause of explicit time dependence in control systems is the presence of time-dependent disturbances $d(k)$.

A state x_e is an *equilibrium point* of the system $f(x_e, k) = 0$, $k \geq k_0$. If $x_0 = x_e$, so that the system starts out in the equilibrium state, then it will forever remain there. For linear systems, the only possible equilibrium point is $x_e = 0$; for nonlinear systems, x_e may be nonzero. In fact, there may even be an equilibrium set, such as a limit cycle.

2.3.1 Asymptotic Stability

An equilibrium point x_e is *locally asymptotically stable* (LAS) at k_0 if there exists a compact set $S \subset \Re^n$ such that for every initial condition $x_0 \in S$, one has $\|x(k) - x_e\| \to 0$ as $k \to \infty$. That is, the state $x(k)$ converges to x_e. If $S = \Re^n$ so that $x(k) \to x_e$ for all $x(k_0)$, then x_e is said to be *globally asymptotically stable* (GAS) at k_0. If the conditions hold for all k_0, the stability is said to be uniformly asymptotically stable or globally uniformly asymptotically stable (e.g., UAS, GUAS).

Asymptotic stability is a very strong property that is extremely difficult to achieve in closed-loop systems, even using advanced feedback controller design techniques. The primary reason is the presence of unknown but bounded system disturbances. A milder requirement is provided as follows.

2.3.2 Lyapunov Stability

An equilibrium point x_e *is stable in the sense of Lyapunov* (SISL) at k_0 if for every $\varepsilon > 0$ there exists $\delta(\varepsilon, x_0)$ such that $\|x_0 - x_e\| < \delta(\varepsilon, k_0)$ implies that $\|x(k) - x_e\| < \varepsilon$ for $k \geq k_0$. The stability is said to be *uniform* (e.g., uniformly SISL) if $\delta(.)$ is independent of k_0; that is the system is SISL for all k_0.

It is extremely interesting to compare these definitions to those of function continuity and uniform continuity. SISL is a notion of continuity for dynamical systems.

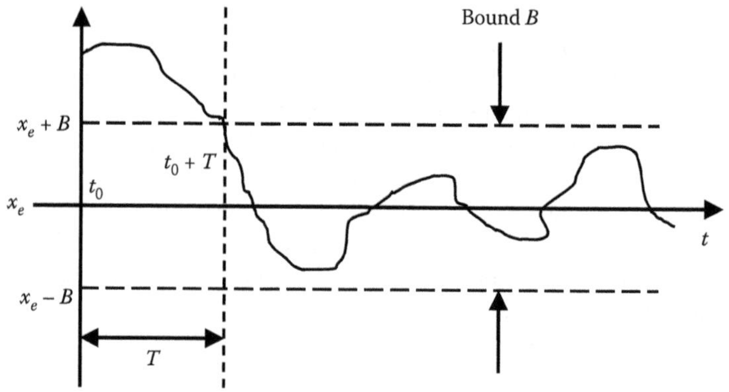

FIGURE 2.2 Illustration of uniform ultimate boundedness.

Note that for SISL there is a requirement that the state $x(k)$ be kept *arbitrarily close* to x_e by starting sufficiently close to it. This is still too strong a requirement for closed-loop control in the presence of unknown disturbances. Therefore, a practical definition of stability to be used as a performance objective for feedback controller design in this book is as follows.

2.3.3 Boundedness

This is illustrated in Figure 2.2. The equilibrium point x_e is said to be *uniformly ultimately bounded* (UUB) if there exists a compact set $S \subset \Re^n$ so that for all $x_0 \in S$ there exists a bound $\mu \geq 0$, and a number $N(\mu, x_0)$ such that $\|x(k)\| \leq \mu$ for all $k \geq k_0 + N$. The intent here is to capture the notion that for all initial states in the compact set S, the system trajectory eventually reaches, after a lapsed time of N, a bounded neighborhood of x_e.

The difference between UUB and SISL is that in UUB the bound μ cannot be made arbitrarily small by starting closer to x_e. In fact, the Vander Pol oscillator is UUB but not SISL. In practical closed-loop applications, μ depends on the disturbance magnitudes and other factors. If the controller is suitably designed, however, μ will be small enough for practical purposes. The term *uniform* indicates that N does not depend upon k_0. The term *ultimate* indicates that the boundedness property holds after a time lapse N. If $S = \Re^n$, the system is said to be *globally UUB* (GUUB).

2.3.4 A Note on Autonomous Systems and Linear Systems

If the system is autonomous so that

$$x(k+1) = f(x(k)) \tag{2.29}$$

where $f(x(k))$ is not an explicit function of time, the state trajectory is independent of the initial time. This means that if an equilibrium point is stable by any of the three definitions, the stability is automatically uniform. Nonuniformity is only a problem with nonautonomous systems.

If the system is linear so that

$$x(k+1) = A(k)x(k) \tag{2.30}$$

where $A(k)$ is an $n \times n$ matrix, then the only possible equilibrium point is the origin. For LTI systems, matrix A is time-invariant. Then the system poles are given by the roots of the characteristic equation

$$\Delta(z) = | zI - A | = 0 \tag{2.31}$$

where $|\cdot|$ is the matrix determinant and z is the Z transform variable. For LTI systems, AS corresponds to the requirement that all the system poles stay within the unit disk (i.e., none of them are allowed on the unit disk). SISL corresponds to marginal stability, that is, all the poles are within the unit disk and those on the unit disk are not repeated.

2.4 NONLINEAR STABILITY ANALYSIS AND CONTROLS DESIGN

For LTI systems, it is straightforward to investigate stability by examining the locations of the poles in the s-plane. However, for nonlinear or nonautonomous (e.g., time-varying) systems, there are no direct techniques. The (direct) Lyapunov approach provides methods for studying the stability of nonlinear systems and shows how to design control systems for such complex nonlinear systems. For more information, see Lewis et al. (1993), which deals with robot manipulator control, as well as Sarangapani (2006), Goodwin and Sin (1984), Landau (1979), Sastry and Bodson (1989), and Slotine and Li (1991), which have proofs and many excellent examples in continuous and discrete time.

2.4.1 Lyapunov Analysis for Autonomous Systems

The autonomous (time-invariant) dynamical system

$$x(k+1) = f(x(k)) \tag{2.32}$$

$x \in \Re^n$, could represent a closed-loop system after the controller has been designed. In Section 2.3.1, we defined several types of stability. We shall show here how to examine stability properties using a *generalized energy* approach. An isolated equilibrium point x_e can always be brought to the origin by redefinition of coordinates; therefore, let us assume without loss of generality that the origin is an equilibrium point. First, we give some definitions and results. Then some examples are presented to illustrate the power of the Lyapunov approach.

Let $L(x): \Re^n \to \Re$ be a scalar function such that $L(0) = 0$, and S be a compact subset of $\Re^n$. Then $L(x)$ is said to be

Locally positive definite if $L(x) > 0$ when $x \neq 0$, for all $x \in S$. (Denoted $L(x) > 0$.)

Locally positive semidefinite if $L(x) \geq 0$ when $x \neq 0$, for all $x \in S$. (Denoted $L(x) \geq 0$.)

Locally negative definite if $L(x) < 0$ when $x \neq 0$, for all $x \in S$. (Denoted $L(x) < 0$.)

Locally negative semidefinite if $L(x) \leq 0$ when $x \neq 0$, for all $x \in S$. (Denoted $L(x) \leq 0$.)

An example of a positive definite function is the quadratic form $L(x) = x^T P x$, where P is any matrix that is symmetric and positive definite. A definite function is allowed to be zero only when $x = 0$, a *semidefinite* function may vanish at points where $x \neq 0$. All these definitions are said to hold globally if $S = \Re^n$.

A function $L(x): \Re^n \to \Re$ with continuous partial differences (or derivatives) is said to be a *Lyapunov function* for the system (2.32), if, for some compact set $S \subset \Re^n$, one has locally

$$L(x) \text{ is positive definite}, L(x) > 0 \tag{2.33}$$

$$\Delta\, L(x) \text{ is negative semidefinite}, \Delta L(x) \leq 0 \tag{2.34}$$

where $\Delta L(x)$ is evaluated along the trajectories of (2.32) (as shown in an upcoming example). That is

$$\Delta L(x(k)) = L(x(k+1)) - L(x(k)) \tag{2.35}$$

Theorem 2.1: Lyapunov Stability

If there exists a Lyapunov function for a system (2.32), then the equilibrium point is SISL.

This powerful result allows one to analyze stability using a generalized notion of energy. The Lyapunov function performs the role of an energy function. If $L(x)$ is positive definite and its derivative is negative semidefinite, then $L(x)$ is nonincreasing, which implies that the state $x(t)$ is bounded. The next result shows what happens if the Lyapunov derivative is negative *definite*—then $L(x)$ continues to decrease until $\|x(k)\|$ vanishes.

Theorem 2.2: Asymptotic Stability

If there exists a Lyapunov function $L(x)$ for system (2.32) with the strengthened condition on its derivative $\Delta L(x)$ is negative definite,

$$\Delta L(x) < 0 \tag{2.36}$$

then the equilibrium point is AS.

To obtain global stability results, one needs to expand the set S to all of $\Re^n$, but also required is an additional *radial unboundedness property*.

Theorem 2.3: Global Stability

a. *Globally SISL*. If there exists a Lyapunov function $L(x)$ for the system (2.32) such that Equations 2.33 and 2.34 hold globally and

$$L(x) \to \infty \text{ as } \| x \| \to \infty \tag{2.37}$$

then the equilibrium point is globally SISL.

b. *Globally AS*. If there exists a Lyapunov function $L(x)$ for a system (2.32) such that Equations 2.33 and 2.36 hold globally and also the unboundedness condition (2.37) holds, then the equilibrium point is GAS.

The global nature of this result of course implies that the equilibrium point mentioned is the *only* equilibrium point.

The next examples show the utility of the Lyapunov approach and make several points. Among the points of emphasis are those stating that the Lyapunov function is intimately related to the *energy properties* of a system.

EXAMPLE 2.2: LOCAL AND GLOBAL STABILITY

a. Local stability
   Consider the system

$$x_1(k+1) = x_1(k)\left(\sqrt{x_1^2(k) + x_2^2(k)} - 2\right)$$

$$x_2(k+1) = x_2(k)\left(\sqrt{x_1^2(k) + x_2^2(k)} - 2\right)$$

Stability for nonlinear discrete-time systems can be examined by selecting the quadratic Lyapunov function candidate

$$L(x(k)) = x_1^2(k) + x_2^2(k)$$

which is a direct realization of an energy function and has first difference

$$\Delta L(x(k)) = x_1^2(k+1) - x_1^2(k) + x_2^2(k+1) - x_2^2(k)$$

Evaluating this along the system trajectories simply involves substituting the state differences from the dynamics to obtain, in this case,

$$\Delta L(x(k)) = -(x_1^2(k) + x_2^2(k))(3 - x_1^2(k) - x_2^2(k))$$

which is negative as long as

$$||x(k)|| = \sqrt{x_1^2(k) + x_2^2(k)} < \sqrt{3}$$

Therefore, $L(x(k))$ serves as a (local) Lyapunov function for the system, which is *locally AS*. The system is said to have a domain of attraction with a radius of *one*. Trajectories beginning outside $||x(k)|| = \sqrt{3}$ in the phase plane cannot be guaranteed to converge.

b. Global stability
Consider now the system

$$x_1(k+1) = x_1(k)x_2^2(k)$$

$$x_2(k+1) = x_2(k)x_1^2(k)$$

where the states satisfy $(x_1(k)x_2(k))^2 < 1$.
Selecting the Lyapunov function candidate

$$L(x(k)) = x_1^2(k) + x_2^2(k)$$

which is a direct realization of an energy function and has the first difference

$$\Delta L(x(k)) = x_1^2(k+1) - x_1^2(k) + x_2^2(k+1) - x_2^2(k)$$

Evaluating this along the system trajectories simply involves substituting the state differences from the dynamics to obtain, in this case,

$$\Delta L(x(k)) = -(x_1^2(k) + x_2^2(k))(1 - x_1^2(k)x_2^2(k))$$

Applying the constraint, the system is globally stable since the states are restricted.

EXAMPLE 2.3: LYAPUNOV STABILITY

Consider now the system

$$x_1(k+1) = x_1(k) - x_2(k)$$

$$x_1(k+1) = \sqrt{2x_1(k)x_2(k) - x_1^2(k)}$$

Selecting the Lyapunov function candidate

$$L(x(k)) = x_1^2(k) + x_2^2(k)$$

which is a direct realization of an energy function and has the first difference

$$\Delta L(x(k)) = x_1^2(k+1) - x_1^2(k) + x_2^2(k+1) - x_2^2(k)$$

Evaluating this along the system trajectories simply involves substituting the state differences from the dynamics to obtain, in this case,

$$\Delta L(x(k)) = -x_1^2(k)$$

This is only negative semidefinite (note that $\Delta L(x(k))$ can be zero when $x_2(k) \neq 0$). Therefore, $L(x(k))$ is a Lyapunov function, but the system is only shown by this method to be SISL and $\|x_1(k)\|$ and $\|x_2(k)\|$ are both bounded.

2.4.2 Controller Design Using Lyapunov Techniques

Though we have presented Lyapunov analysis only for unforced systems in the form (2.32), which have no control input, these techniques also provide a powerful set of tools for designing feedback control systems of the form

$$x(k+1) = f(x(k)) + g(x(k))u(k) \tag{2.38}$$

Thus, select a Lyapunov function candidate $L(x) = x^T(k)x(k) > 0$ and obtain the first difference of the Lyapunov function candidate along the system trajectories to obtain

$$\begin{aligned} \Delta L(x) &= L(x(k+1)) - L(x(k)) = x^T(k+1)x(k+1) - x^T(k)x(k) \\ &= (f(x(k)) + g(x(k))u(k))^T (f(x(k)) + g(x(k))u(k)) - x^T(k)x(k) \end{aligned} \tag{2.39}$$

Then, it is often possible to ensure that $\Delta L \leq 0$ by the appropriate selection of $u(k)$. When this is possible, it generally yields controllers in *state-feedback* form, that is, where $u(k)$ is a function of the states $x(k)$.

Practical systems with actuator limits and saturation often contain discontinuous functions, including the *signum function* defined for scalars $x \in \Re$ as

$$\operatorname{sgn}(x) = \begin{cases} 1, & x \geq 0 \\ -1, & x < 0 \end{cases} \tag{2.40}$$

shown in Figure 2.3, and for vectors $x = [x_1 \quad x_2 \quad \cdots \quad x_n]^T \in \Re^n$ as

$$\operatorname{sgn}(x) = [\operatorname{sgn}(x_i)] \tag{2.41}$$

where $[z_i]$ denotes a vector z with components z_i. The discontinuous nature of such functions often makes it impossible to apply i/o feedback linearization where

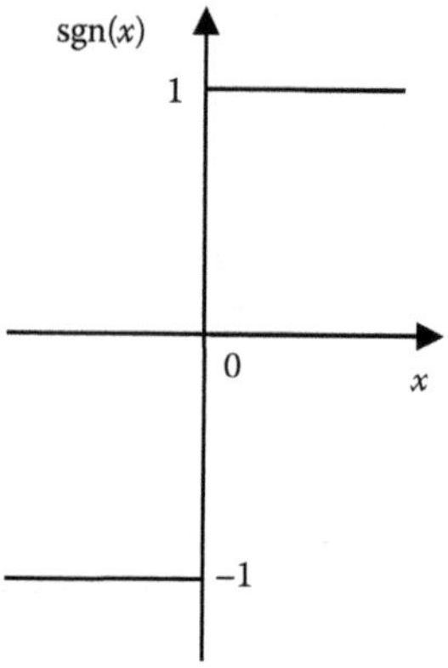

FIGURE 2.3 Signum function.

differentiation is required. In some cases, controller design can be carried out for systems containing discontinuities using Lyapunov techniques.

EXAMPLE 2.4: CONTROLLER DESIGN BY LYAPUNOV ANALYSIS

Consider the system

$$x_1(k+1) = x_2(k)\,\mathrm{sgn}(x_1(k))$$

$$x_2(k+1) = \sqrt{x_1(k)x_2(k) + u(k)}$$

having an actuator nonlinearity. A control input has to be designed using feedback linearization techniques (i.e., cancels all nonlinearities). A stabilizing controller can be easily designed using Lyapunov techniques.

Select the Lyapunov function candidate

$$L(x(k)) = x_1^2(k) + x_2^2(k)$$

and evaluate

$$\Delta L(x(k)) = x_1^2(k+1) - x_1^2(k) + x_2^2(k+1) - x_2^2(k)$$

Substituting the system dynamics in the above equation results in

$$\Delta L(x(k)) = x_2^2(k)\,\mathrm{sgn}^2(x_1(k)) - x_1^2(k) + (x_1(k)x_2(k) + u(k)) - x_2^2(k)$$

Now select the feedback control

$$u(k) = -x_2^2(k)\,\mathrm{sgn}^2(x_1(k)) + x_1^2(k) - x_1(k)x_2(k)$$

This yields

$$\Delta L(x(k)) = -x_2^2(k)$$

so that $L(x(k))$ is rendered a (closed-loop) Lyapunov function. Since $\Delta L(x(k))$ is negative semidefinite, the closed-loop system with this controller is SISL.

It is important to note that by slightly changing the controller, one can also show global asymptotic stability of the closed-loop system. Moreover, note that this controller has elements of feedback linearization (discussed in the next chapter), in that the control input $u(k)$ is selected to cancel nonlinearities. However, no difference of the right-hand side of the state equation is needed in the Lyapunov approach but the right-hand side becomes quadratic, which makes it hard to design controllers and show stability. This will be a problem for the discrete-time systems and we will be presenting how to select suitable Lyapunov function candidates for complex systems when standard adaptive control and NN-based controllers are deployed. Finally, there are some issues in this example, such as the selection of the discontinuous control signal, which could cause chattering. In practice, the system dynamics act as a low-pass filter so that the controllers work well.

2.4.2.1 Lyapunov Analysis and Controls Design for Linear Systems

For general nonlinear systems, it is not always easy to find a Lyapunov function. Thus, failure to find a Lyapunov function may be because the system is not stable, or because the designer simply lacks insight and experience. However, in the case of LTI systems

$$x(k+1) = Ax \tag{2.42}$$

Lyapunov analysis is simplified and a Lyapunov function is easy to find, if one exists.

2.4.2.2 Stability Analysis

Select as a Lyapunov function candidate the quadratic form

$$L(x(k)) = \frac{1}{2} x^T(k)Px(k) \tag{2.43}$$

where P is a constant symmetric positive definite matrix. Since $P > 0$, then x^TPx is a positive function. This function is a *generalized norm*, which serves as a system energy function. Then,

$$\Delta L(x(k)) = L(x(k+1)) - L(x(k)) = \frac{1}{2}[x^T(k+1)Px(k+1) - x^T(k)Px(k)] \tag{2.44}$$

$$= \frac{1}{2} x^T(k)[A^TPA - P]x(k) \tag{2.45}$$

For stability, one requires negative semidefiniteness. Thus, there must exist a symmetric positive semidefinite matrix Q such that

$$\Delta L(x) = -x^T(k)Qx(k) \tag{2.46}$$

This results in the next theorem.

Theorem 2.4: Lyapunov Theorem for Linear Systems

The system (2.40) is SISL, if there exist matrices $P > 0,\ Q \geq 0$ that satisfy the *Lyapunov equation*

$$A^T PA - P = -Q \tag{2.47}$$

If there exists a solution such that both P and Q are positive definite, the system is AS.

It can be shown that this theorem is both necessary and sufficient. That is, for LTI systems, if there is no Lyapunov function of the quadratic form (2.43), then there is no Lyapunov function. This result provides an alternative to examining the eigenvalues of the A matrix.

2.4.2.3 Lyapunov Design of LTI Feedback Controllers

These notions offer a valuable procedure for LTI control system design. Note that the closed-loop system with state feedback

$$x(k+1) = Ax(k) + Bu(k) \tag{2.48}$$

$$u = -Kx \tag{2.49}$$

is SISL if and only if there exist matrices $P > 0,\ Q \geq 0$ that satisfy the closed-loop Lyapunov equation

$$(A - BK)^T P(A - BK) - P = -Q \tag{2.50}$$

If there exists a solution such that both P and Q are *positive definite*, the system is AS.

Now suppose there exist $P > 0,\ Q \geq 0$ that satisfy the *Riccati equation*

$$P(k) = A^T P(k+1)(I + BR^{-1}B^T P(k+1))^{-1} A + Q \tag{2.51}$$

Now select the feedback gain as

$$K(k) = -(R + B^T P(k+1)B)^{-1} B^T P(k+1)A \tag{2.52}$$

and the control input as

$$u(k) = -K(k)x(k) \tag{2.53}$$

for some matrix $R > 0$.

These equations verify that this selection of the control input guarantees closed-loop asymptotic stability.

Note that the Riccati equation depends only on *known* matrices—the system (A, B) and two symmetric *design matrices* Q, R that need to be selected positive definite. There are many good routines that can find the solution P to this equation provided that (A, B) is *controllable* (e.g., MATLAB). Then, a stabilizing gain is given by Equation 2.52. If different design matrices Q, R are selected, different closed-loop poles will result. This approach goes far beyond classical frequency domain or root locus design techniques in that it allows the determination of stabilizing feedbacks for complex *multivariable* systems by simply solving a *matrix design equation.* For more details on this *linear quadratic (LQ) design technique*, see Lewis and Syrmos (1995).

2.4.3 Lyapunov Analysis for Nonautonomous Systems

We now consider nonautonomous (time-varying) dynamical systems of the form

$$x(k+1) = f(x(k),k), k \geq k_0 \tag{2.54}$$

$x \in \Re^n$. Assume again that the origin is an equilibrium point. For nonautonomous systems, the basic concepts just introduced still hold, but the explicit time dependence of the system must be taken into account. The basic issue is that the Lyapunov function may now depend on time. In this situation, the definitions of definiteness must be modified, and the notion of "decrescence" is needed.

Let $L(x(k),k) : \Re^n \times \Re \to \Re$ be a scalar function such that $L(0, k) = 0$, and S be a compact subset of $\Re^n$. Then, $L(x(k), k)$ is said to be

Locally positive definite if $L(x(k), k) \geq L_0(x(k))$ for some time-invariant positive definite $L_0(x(k))$, for all $k \geq 0$ and $x \in S$. (Denoted $L(x(k), k) > 0$.)

Locally positive semidefinite if $L(x(k), k) \geq L_0(x(k))$ for some time-invariant positive semidefinite $L_0(x(k))$, for all $k \geq 0$ and $x \in S$. (Denoted $L(x(k), k) \geq 0$.)

Locally negative definite if $L(x(k), k) \leq L_0(x(k))$ for some time-invariant negative definite $L_0(x(k))$, for all $k \geq 0$ and $x \in S$. (Denoted $L(x(k), k) < 0$.)

Locally negative semidefinite if $L(x(k), k) \leq L_0(x(k))$ for some time-invariant negative semidefinite $L_0(x(k))$, for all $k \geq 0$ and $x \in S$. (Denoted $L(x(k), k) \leq 0$.)

Thus, for definiteness of time-varying functions, a time-invariant definite function must be *dominated.* All these definitions are said to hold *globally* if $S \in \Re^n$.

A time-varying function $L(x(k),k) : \Re^n \times \Re \to \Re$ is said to be decrescent if $L(0, k) = 0$, and there exists a time-invariant positive definite function $L_1(x(k))$ such that

$$L(x(k),k) \leq L_1(x(k)), \forall k \geq 0 \tag{2.55}$$

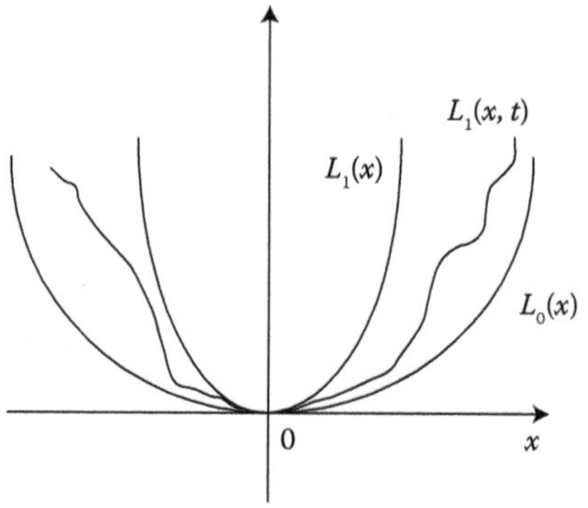

FIGURE 2.4 Time-varying function $L(x(k), k)$ that is positive definite $((L_0(x(k) < L(x(k), k))$ and decrescent $(L(x(k), k) \leq L_1(x(k))$.

The notions of decrescence and positive definiteness for time-varying functions are depicted in Figure 2.4.

EXAMPLE 2.5: DECRESCENT FUNCTION

Consider the time-varying function

$$L(x(k),k) = x_1^2(k) + \frac{x_2^2(k)}{3+\sin(kT)}$$

Note that $2 \leq 3 + \sin(kT) \leq 4$, so that

$$L(x(k),k) \geq L_0(x(k)) \equiv x_1^2(k) + \frac{x_2^2(k)}{4}$$

and $L(x(k), k)$ is globally positive definite. Also,

$$L(x(k),k) \leq L_1(x(k)) \equiv x_1^2(k) + x_2^2(k)$$

so that it is decrescent.

Theorem 2.5: Lyapunov Results for Nonautonomous Systems

a. *Lyapunov stability.* If, for system (2.54), there exists a function $L(x(k), k)$ with continuous partial derivatives (or differences), such that for x in a compact set $S \subset \Re^n$

$$L(x(k),k) \text{ is positive definite, } L(x(k),k) > 0 \tag{2.56}$$

$$\Delta L(x(k),k) \text{ is negative semidefinite, } \Delta L(x(k),k) \leq 0 \tag{2.57}$$

then the equilibrium point is SISL.

b. *Asymptotic stability*. If, furthermore, condition (2.57) is strengthened to

$$\Delta L(x(k),k) \text{ is negative definite}, \Delta L(x(k),k) < 0 \tag{2.58}$$

then the equilibrium point is AS.

c. *Global stability*. If the equilibrium point is SISL or AS, if $S = \Re^n$ and in addition the radial unboundedness condition holds:

$$L(x(k),k) \to \infty \text{ as } \| x(k) \| \to \infty, \forall k \tag{2.59}$$

then the stability is global.

d. *Uniform stability*. If the equilibrium point is SISL or AS, and in addition $L(x(k), k)$ is decrescent (e.g., (2.55) holds), then the stability is uniform (e.g., independent of k_0).

The equilibrium point may be both uniformly and globally stable—for example, if all the conditions of the theorem hold, then one has GUAS.

2.4.4 Extensions of Lyapunov Techniques and Bounded Stability

The Lyapunov results so far presented have allowed the determination of SISL. If there exists a function such that $L(x(k), k) > 0$, $\Delta L(x(k), k) \leq 0$, then the signal are bounded, if there exists a function such that $L(x(k), k) > 0$, $\Delta L(x(k), k) < 0$ then the system is AS. Various extensions of these results allow one to determine more about the stability properties by further examining the deeper structure of the system dynamics.

2.4.4.1 UUB Analysis and Controls Design

We have seen how to demonstrate that a system is SISL or AS using Lyapunov techniques. However, in practical applications, there are often unknown disturbances or modeling errors that make even SISL too much to expect in closed-loop systems. Typical examples are systems of the form

$$x(k+1) = f(x(k),k) + d(k) \tag{2.60}$$

with $d(k)$ an unknown but bounded disturbance. A more practical notion of stability is UUB. The next result shows that UUB is guaranteed if the Lyapunov derivative is negative outside some bounded region of $\Re^n$.

Theorem 2.6: UUB by Lyapunov Analysis

If, for system (2.60), there exists a function $L(x, k)$ with continuous partial differences such that for x in a compact set $S \subset \Re^n$

$$L(x(k),k) \text{ is positive definite}, L(x(k),k) > 0$$

$$\Delta L(x(k),k) < 0 \text{ for } \| x \| > R$$

for some $R > 0$ such that the ball of radius R is contained in S, then the system is UUB and the norm of the state is bounded to within a neighborhood of R. In this result, note that ΔL must be *strictly less* than zero outside the ball of radius R. If one only has $\Delta L(x(k), k) \leq 0$ for all $\|x\| > R$, then nothing may be concluded about the system stability.

For systems that satisfy the theorem, there may be some disturbance effects that push the state away from the equilibrium. However, if the state becomes too large, the dynamics tend to pull it back toward the equilibrium. Owing to these two opposing effects that balance when $\|x\| \approx R$, the time histories tend to remain in the vicinity of $\|x\| = R$. In effect, the norm of the state is effectively or *practically bounded* by R.

The notion of the ball outside which ΔL is negative should not be confused with that of *domain of attraction*—in Example 2.2a. It was shown there that the system is AS, provided $\|x\| < 1$, defining a domain of attraction of radius one.

The next examples show how to use this result. They make the point that it can also be used as a *control design technique* where the control input is selected to guarantee that the conditions of the theorem hold.

EXAMPLE 2.6: UUB OF LINEAR SYSTEMS WITH DISTURBANCE

It is common in practical systems to have unknown disturbances, which are often bounded by some known amount. Such disturbances result in UUB and require the UUB extension for analysis. Suppose the system

$$x(k+1) = Ax(k) + d(k)$$

has A stable and a disturbance $d(k)$ that is unknown but bounded so that $\|d(k)\| < d_M$, with the bound d_M known.

Select the Lyapunov function candidate

$$L(x(k)) = x^T(k)Px(k)$$

and evaluate

$$\begin{aligned}\Delta L(x(k)) &= x^T(k+1)Px(k+1) - x^T(k)Px(k)\\ &= x^T(k)(A^TPA - P)x(k) + 2x^T(k)A^TPd(k) + d^T(k)Pd(k)\\ &= -x^T(k)Qx(k) + 2x^T(k)A^TPd(k) + d^T(k)Pd(k)\end{aligned}$$

where (P, Q) satisfy the Lyapunov equation

$$A^TPA - P = -Q$$

One may now use the norm equalities to write

$$\Delta L(x(k)) \leq -[\sigma_{\min}(Q)\,||\,x(k)\,||^2 - 2\,||\,x(k)\,||\,\sigma_{\max}(A^T P)\,||\,d(k)\,|| - \sigma_{\max}(P)\,||\,d(k)\,||^2]$$

which is negative as long as

$$||\,x(k)\,|| \geq \frac{\sigma_{\max}(A^T P)d_M + \sqrt{\sigma_{\max}^2(A^T P)d_M^2 + \sigma_{\min}(Q)\sigma_{\max}(P)d_M^2}}{\sigma_{\min}(Q)}$$

Thus, if the disturbance magnitude bound increases, the norm of the state will also increase.

EXAMPLE 2.7: UUB OF CLOSED-LOOP SYSTEM

The UUB extension can be utilized to design stable closed-loop systems. The system described by

$$x(k+1) = x^2(k) - 10x(k)\sin x(k) + d(k) + u(k)$$

is excited by an unknown disturbance whose magnitude is bounded so that $||d(k)|| < d_M$. To find a control that stabilizes the system and mitigates the effect of disturbances, select the control input as

$$u(k) = -x^2(k) + 10x(k)\sin x(k) + k_v x(k)$$

This helps cancel the sinusoidal nonlinearity and provides a stabilizing term yielding the closed-loop system

$$x(k+1) = k_v x(k) + d(k)$$

Select the Lyapunov function candidate

$$L(x(k)) = x^2(k)$$

whose first difference is given by

$$\Delta L(x(k)) = x^2(k+1) - x^2(k)$$

Evaluating the first difference along the closed-loop system trajectories yields

$$\Delta L(x(k)) \leq -x^2(k)(1 - k_{v\max}^2) - 2x(k)k_v d(k) + d^2(k)$$

which is negative as long as

$$\| x(k) \| > \frac{k_{v\max} d_M + \sqrt{k_{v\max}^2 d_M^2 + (1-k_{v\max}^2) d_M^2}}{(1-k_{v\max}^2)}$$

which after simplification results in

$$\| x(k) \| > \frac{(1+k_{v\max})}{(1-k_{v\max}^2)} d_M$$

The UUB bound can be made smaller by moving the closed-loop poles near the origin. Placing the poles at the origin will result in a deadbeat controller and it should be avoided under all circumstances.

2.5 STOCHASTIC DISCRETE-TIME CONTROL

Stochastic stability along with robustness is a performance requirement for closed-loop stochastic systems similar to the case of stability for deterministic systems. In other words, though the open-loop stability properties of the original stochastic system may not be satisfactory, it is desirable to design a feedback control system such that the closed-loop stability is adequate. We will discuss stability for stochastic discrete-time systems. Refer to Khasminskii and Milstein (2011), Stengel (1986), Åstrom (1970), and Kumar and Varaiya (1986) for details. Normally, the stability of stochastic systems should be shown using probability. The criteria for convergence with probability one is often difficult to establish. We will therefore use convergence in the mean square (Åstrom 1970) or convergence in the mean for the simple reason that it leads to simple analysis. The definitions are made simple for a common engineer to interpret and apply the results.

2.5.1 STOCHASTIC LYAPUNOV STABILITY

Consider the dynamical system

$$x(k+1) = f(x(k),k) + w(k) \tag{2.61}$$

where $x(k) \in \Re^n$, which might represent either an uncontrolled open-loop system or a closed-loop system after the control input $u(k)$ has been specified in terms of the state $x(k)$, and $w(k) \in \Re^n$ is the Gaussian process random variable representing the disturbance (or process noise). Let the initial time be k_0, and the initial condition be $E(x(k_0)) = x_0$ or initial state is deterministic where $E(\cdot)$ is the expectation operator. This system is said to be a *nonautonomous* stochastic system since the time k appears explicitly. If k does not appear explicitly in $f(\cdot)$, then the system is an autonomous stochastic system. A primary cause of explicit time dependence in control systems is the presence of time-dependent disturbances $d(k)$.

A state x_e is an *equilibrium point* of the system $f(x_e, k) = 0$, $k \geq k_0$ if one can show with high probability (Kumar and Varaiya 1986). If $x_0 = x_e$, so that the system starts out in the equilibrium state, then it will forever remain there with probability. For linear stochastic systems, the only possible equilibrium point $x_e = 0$ with high probability; for nonlinear systems, x_e may be nonzero. In fact, there may even be an equilibrium set, such as a limit cycle.

2.5.1.1 Asymptotic Stable in the Mean Square

An equilibrium point x_e is *locally AS* in the mean square (or probability) at k_0, if there exists a set $S \subset \Re^n$ such that, for every initial condition $x_0 \in S$, one has $E\{\|x(k) - x_e\|^2\} \to 0$ as $k \to \infty$ where $E\{\cdot\}$ is expect operator. That is, the state $E\{x(k)\}$ converges to $E\{x_e\}$. If $S = \Re^n$ so that $E\{x(k) \to E\{x_e\}$ for all $x(k_0)$, then $E\{x_e\}$ is said to be GAS at k_0.

2.5.1.2 Lyapunov Stable in the Mean Square

An equilibrium point x_e *is* SISL in the mean square (or probability) at k_0 if for every $\varepsilon > 0$ there exists $\delta(\varepsilon, x_0)$ such that $E\{\|x_0 - x_e\|^2\} < \delta(\varepsilon, k_0)$ implies that $E\{\|x(k) - x_e\|^2\} < \varepsilon$ for $k \geq k_0$.

2.5.1.3 Bounded in the Mean Square

The equilibrium point x_e is said to be *bounded* in the mean square (or probability) if there exists a set $S \subset \Re^n$ so that for all $x_0 \in S$ there exists a bound $\mu \geq 0$, and a number $N(\mu, x_0)$ such that $E\{\|x(k)\|^2\} \leq \mu$ for all $k \geq k_0 + N$. The intent here is to capture the notion that for all initial states in the set S, the system trajectory eventually reaches, after a lapsed time of N, a bounded neighborhood of x_e.

2.5.1.4 Bounded in the Mean

An equilibrium point x_e is said to be bounded in the mean if there exists a compact set $\Omega_x \subset \Re^n$ so that for all initial values of $x_0 \in \Omega_x$, there exists a bound B and a time $T(B, x_0)$ such that $E\|x_k - x_e\| \leq B$ for all $k \geq k_0 + T$.

2.5.2 Stochastic Linear Discrete-Time Optimal Control

First it is important to note that there is no single optimal trajectory (Stengel 1986) and it is therefore necessary to base the stochastic optimization of linear/nonlinear systems on the principle of optimality that is utilized for determinitic systems. Before presenting a general case, we will show the optimal control of a stochastic linear system and contrast it with its deterministic scenario when the system matrices are time-invariant.

Consider the deterministic linear dynamical system over finite time horizon described by

$$x(k+1) = Ax(k) + Bu(k) + w(k), \quad k = 0, \ldots, N-1$$

where $x(0), \ldots, x(N)$, $u(0), \ldots, u(N-1)$ and w_k are random variables, A and B denote deterministic system matrices, w_k being the zero mean Gaussian process noise or disturbance at time k with $E(w(k)) = 0$, $E(w(k)w^T(k)) = W$, $x(0)$ is the initial value of the state independent of $w(k)$, with $E(x(0)) = 0$, $E(x(0)x^T(0)) = X$. Here, the state vector is assumed to be perfectly measured without any measurement noise. Let $u(k) = \phi(x(k))$, $k = 0, \ldots, N-1$ where $\phi(.) : \mathbb{R}^n \rightarrow \mathbb{R}^m$ represents the control policy at the time instant k. By using the control policy in the above equation, we obtain

$$x(k+1) = Ax(k) + B\phi(x(k)) + w(k), \quad k = 0, \ldots, N-1$$

Next, the object is to generate the control policies by minimizing the performance index J, as

$$J = E\left(\sum_{k=0}^{N-1}(x^T(k)Qx(k) + u^T(k)Ru(k)) + x^T(N)Q_f x(N)\right)$$

with Q, $Q_f \geq 0$, $R > 0$ representing the positive semidefinite and definite matrices, respectively, and $V_N(x(N)) = x^T(N)Q_f x(N)$ denoting the terminal cost. The solution can be obtained by using dynamic programming (DP). Let $V(x(k))$ be the optimal value function of the objective, from time instant k onward

$$V(x(k)) = \min_{\phi_0 \ldots, \phi_{N-1}} E\left(\sum_{\tau=k}^{N-1}(x^T(k)Qx(k) + u^T(k)Ru(k)) + x^T(N)Q_f x(N)\right)$$

subject to the constraint $x(k+1) = Ax(k) + Bu(k) + w(k)$, $u(k) = \phi_k(x(k))$. The minimal cost is given by $J^* = E(V(x(0)))$. The value function at any given time instant $V(x(k))$ can be found by backward recursion: for $k = 0, \ldots, N-1$

$$V(x(k)) = x^T(N)Q_f x(N) + \min_{u(k)}\{u^T(k)Ru(k) + E(V(Ax(k) + Bu(k) + w(k)))\}$$

The optimal policies have the following form:

$$\phi^*(.) = \arg\min_{u(k)}\{\mathrm{u}^T(k)Ru(k) + E(V(Ax(k) + Bu(k) + w(k)))\}$$

Next, let us show by taking a quadratic value function for this linear system of the form (Stengel 1986)

$$V(x(k)) = x(k)P_k x(k) + q_k, \quad k = 0, \ldots, N-1$$

where $P_k \geq 0$ is a positive definite kernel matrix with $P_N = Q_N$, $q_N = 0$. From the above equation, it follows that

$$V(x(k+1)) = x^T(k+1)P_{k+1}x(k+1) + q_{k+1}$$

The Bellman recursion is given by

$$V(x(k)) = x^T(N)Q_f x(N) + \min_{u(k)}\{u^T(k)Ru(k) + \mathrm{E}((Ax(k) + Bu(k) + w(k))^T$$

$$P_{k+1}(Ax(k) + Bu(k) + w(k)) + q_{k+1})\}$$

$$= x^T(N)Q_f x(N) + tr(WP_{k+1}) + q_{k+1} + \min_{u(k)}\{u^T(k)Ru(k)$$

$$+(Ax(k) + Bu(k))^T P_{k+1}(Ax(k) + Bu(k))\}$$

where $E(w^T(k)P_{k+1}\, w(k)) = tr(WP_{k+1})$ with $tr(\cdot)$ as the trace operator. This recursion is the same as that of a deterministic case (Lewis and Syrmos 1995) with an added term. The optimal policy is obtained by using the stationarity condition (Stengel 1986) (see below for the generalized case) as a linear state feedback $\phi^*(x(k)) = K_k x(k)$, where the Kalman gain, K_k, is represented by

$$K_k = -(B^T P_{k+1} B + R)^{-1} B^T P_{k+1} A$$

Note that the Kalman gain is exactly same as that of the deterministic case. By substituting the optimal policy and value function in the Bellman recursion, we obtain the Riccati equation and an additional recursion given by

$$P_k = A^T P_{k+1} A - A^T P_{k+1} B(B^T P_{k+1} B + R)^{-1} B^T P_{k+1} A + Q$$

$$q_k = q_{k+1} + tr(WP_{k+1})$$

Note that the first recursion is known as the certainty-equivalent Riccati equation and it is the same as that of a linear quadratic regulator (LQR) for the deterministic case while the second term is an additional term in the value function, which is referred to as *value function increment* for the stochastic case due to process noise. However, the optimal policy for the stochastic control is the same as that of the deterministic case and it is independent of X and W.

Moving on, the optimal cost is given by

$$J^* = E(V(x(0))) = tr(XP_0) + q_0 = tr(XP_0) + \sum_{k=1}^{N} tr(WP_k)$$

Strangely, the first term becomes $x^T(0)P_0x(0)$ when the process noise is absent, $w(0) = \cdots = w(N-1) = 0$, which is the optimal cost for the deterministic case of LQR, while the other terms are the average optimal costs due to the presence of Gaussian process noise.

For the infinite-horizon case, select the control policies to minimize average cost as

$$J = \lim_{N\to\infty} \frac{1}{N} E \sum_{k=0}^{N-1} (x^T(k)Qx(k) + u^T(k)Ru(k))$$

with the optimal average stage cost given by

$$J^* = tr(WP_\infty)$$

where P_∞ is the steady-state value satisfying the algebraic Riccati equation (ARE)

$$P_\infty = Q + A^T P_\infty A - A^T P_\infty B(R + B^T P_\infty B)^{-1} B^T P_\infty A$$

For the infinite horizon, the cost function does not include the terminal weighting term. In addition, here the optimal average cost does not depend on X. The suboptimal (or optimal for infinite horizon) policy is a constant linear state feedback given by (Stengel 1986; Lewis and Syrmos 1995)

$$u_k = K_\infty x_k$$

where the steady-state Kalman gain sequence is given by

$$K_\infty = -(R + B^T P_\infty B)^{-1} B^T P_\infty A$$

Note that the steady-state gain K_∞ does not depend upon X and/or W. This derivation presents two interesting aspects: (1) a stochastic system is traditionally obtained from a deterministic system driven by process noise or disturbance and (2) difference between traditional LQR and its stochastic scenario. These derivations will help with the optimal controller design for an NCS though there are differences. In the case of an NCS, as mentioned in Chapters 1 and 3, after incorporating the network imperfections, one obtains a time-varying stochastic system even when there is no process noise.

Now we formally generalize the above derivation for linear time-varying systems driven by Gaussian process noise. To move on, consider a stochastic linear discrete-time system given by

$$x(k+1) = A(k)x(k) + B(k)u(k) + w(k) \tag{2.62}$$

with $x(k)$, $u(k)$, $w(k)$ representing system state, control input, and disturbance (or white noise process), respectively, and disturbance belonging to some probability space with a known probability density function. Here, $A(k) \in \Re^{n\times n}$, $B(k) \in \Re^{n\times m}$ denote deterministic system matrices with the mean value of the disturbance signal considered zero. Under certain mild assumptions and using standard stochastic optimal control theory (Åstrom 1970; Stengel 1986), the *certainty-equivalent* stochastic optimal value function can be written as

$$\begin{aligned} V^*(x(k)) &= \min_{u(k)} E[x^T(N)S(N)x(N) + \sum_{l=k}^{N-1} r(x(l),u(l))] \\ &= \min_{u(k)} E\left[x^T(N)S(N)x(N) + \sum_{l=k}^{N-1} (x^T(l)Q_d x(l) + u^T(l)R_d u(l)) \right] \quad \forall k = 0, N-1 \end{aligned} \tag{2.63}$$

where $E(\cdot)$ is the expectation operator, cost-to-go is defined as $r(x(k), u(k)) = x^T(k) Q_d x(k) + u^T(k)R_d u(k)$, and Q_d, R_d, S_N are symmetric positive semidefinite and positive definite matrices, respectively.

Next, according to DP, the stochastic version of the Bellman equation in discrete time can be expressed as

$$\left\{0 = \min_{u_{d,k}} E(r(x(k),u(k)) + V^*(x(k+1)) - V^*(x(k)))\right. \tag{2.64}$$

$$\left\{0 = E[x^T(\mathrm{N})S(N)x(\mathrm{N}) - V^*(x(\mathrm{N}))]\right. \tag{2.65}$$

Assuming that a minimum on the right-hand side of Equation 2.64 exists and is unique, then the optimal control can be derived as (Åstrom 1970; Stengel 1986)

$$E[u^*(k)] = -\frac{1}{2}E\left[R_d^{-1}B(k)\frac{\partial V^*(x(k+1))}{\partial x(k+1)}\right] \quad \forall k = 0,1,\ldots,N-1 \tag{2.66}$$

Substituting the optimal control input (2.66) into the Bellman equation (2.64), the certainty-equivalent version of the stochastic HJB equation in discrete time can be represented as

$$\begin{cases} 0 = E\left[x^T(k)Q_d x(k) + \dfrac{1}{4}\dfrac{\partial V^{*T}(x(k+1))}{\partial x(k+1)}B^T(k)R_d^{-1}B(k)\right. \\ \qquad \left.\dfrac{\partial V^*(x(k+1))}{\partial x(k+1)} + V^*(x(k+1)) - V^*(x(k))\right] \\ 0 = E\left[x^T(N)S(N)x(N) - V^*(x(N))\right] \end{cases} \tag{2.67}$$

According to Åstrom (1970) and Stengel (1986) for linear systems, the certainty-equivalence stochastic value function (2.63) can be formulated as a quadratic function of the system state vector and thus can be expressed as

$$\begin{cases} V^*(x(k)) = E(x^T(k)P(k)x(k)) \qquad \forall k = 0,1,\ldots,N-1 \\ V^*(x(N)) = E(x^T(N)P(N)x(N)) \end{cases} \tag{2.68}$$

where P_k, $\forall k = 0, 1, \ldots, N$ is a positive-definite kernel matrix. Substituting Equation 2.68 into Equation 2.66, discrete-time HJB becomes a well-known certainty-equivalent stochastic Riccati equation (SRE), which is given by

$$\begin{cases} 0 = E[A^T(k)[P(k) - P(k)B(k)(B^T(k)P(k)B(k) + R_d)^{-1}B^T(k)P(k)] \\ \qquad \times A(k) + Q_d - P(k)] \\ E(P(N)) = E(S(N)) \end{cases} \tag{2.69}$$

with $S(N)$ being the terminal weighting matrix. Moreover, the optimal control policy can be represented in terms of the SRE solution, $P(k)$, as

$$E[u^*(k)] = -E[(B^T(k)P(k)B(k)+R_d)^{-1}B^T(k)P(k)A(k)x(k)] \tag{2.70}$$

From Equation 2.70, it is important to note that the stochastic optimal control policy generation requires the system matrices. However, an NCS will be obtained by incorporating the network imperfections with deterministic continuous or discrete-time systems leading to an uncertain stochastic time-varying system. The resulting closed-loop stochastic system for an NCS may not have a random disturbance or process noise. Therefore, the certainty-equivalent version of the stochastic value function and the Riccati equation will be utilized in the presence of perfect state measurements. Otherwise, for the traditional stochastic system given by Equations 2.28 and 2.62, one has to include the *stochastic value function increment* (Stengel 1986) in the presence of white process and measurement noise signals. Nevertheless, the traditional methods function backward-in-time and require system dynamics. In order to deal with uncertainties in system dynamics and function forward-in-time, one has to resort to Q-learning for linear systems and NNs for nonlinear systems combined with approximate or adaptive DP. In the next section, the notation for the system dynamics is slightly altered.

2.5.3 Stochastic Q-Learning

In this section, a stochastic Q-function approach is introduced for designing a certainty-equivalence controller. Consequently, a stochastic adaptive estimator (AE) is proposed to learn this $Q(\cdot)$ function online. Consider the stochastic system with uncertain dynamics described as

$$z_{k+1} = A_{zk}z_k + B_{zk}u_k + w_k$$

with w_k being the white noise process. Given the unique equilibrium point at $z = 0$ for the stochastic system on a set Ω, assume that the states are considered perfectly measurable (Stengel 1986) with the linear system dynamics A_{zk}, B_{zk} being unknown. The stochastic optimal control can still be obtained in terms of estimated Q-function in a forward-in-time manner without using value and policy iterations in contrast with existing Q-function-based DP schemes where value and policy iterations are needed. Next, the Q-function setup is described.

2.5.3.1 Q-Function Setup

According to these conditions, the stochastic optimal control input that minimizes the cost function J_k for stochastic system can be derived as $u_k^* = -K_k z_k$ with K_k being the optimal gain and u_k^* being the control input. According to the optimal control theory (Stengel 1986), the certainty-equivalence stochastic cost function can be represented as

$$J_k = \underset{\tau,\gamma}{E}(z_k^T P_k z_k) \tag{2.71}$$

where $P_k \geq 0$ is the solution to the certainty-equivalence part of the SRE (Stengel 1986). The optimal action-dependent value function or simply Q-function denoted as $Q(\cdot)$ of the stochastic system is defined in terms of expected value as

$$Q(z_k,u_k) = \underset{\tau,\gamma}{E}\{[r(z_k,u_k)+J_{k+1}]\} = \underset{\tau,\gamma}{E}\{[z_k^T u_k^T]H_k[z_k^T u_k^T]^T\} \tag{2.72}$$

where $r(z_k,u_k) = z_k^T S_z z_k + u_k^T R_z u_k$. It is important to note that the matrix H_k is time-varying. Since the stochastic optimal control, u_k^*, is dependent on state z_k, which is known at time k, Q-function can be expressed as $Q(z_k,u_k) = [z_k^T u(z_k)^T]\underset{\tau,\gamma}{E}(H_k)[z_k^T u(z_k)^T]^T$. Then, using the Bellman equation (Stengel 1986) and certainty-equivalent stochastic cost function definition, the following equation can be formulated by applying Q-function (2.72) as

$$\begin{aligned}
\begin{bmatrix} z_k \\ u_k \end{bmatrix}^T \underset{\tau,\gamma}{E}(H_k)\begin{bmatrix} z_k \\ u_k \end{bmatrix} &= \underset{\tau,\gamma}{E}\{[r(z_k,u_k)+J_{k+1}]\} \\
&= z_k^T S_z z_k + u_k^T R_z u_k + \underset{\tau,\gamma}{E}(z_{k+1}^T P_{k+1} z_{k+1}) \\
&= \begin{bmatrix} z_k \\ u_k \end{bmatrix}^T \begin{bmatrix} S_z & 0 \\ 0 & R_z \end{bmatrix}\begin{bmatrix} z_k \\ u_k \end{bmatrix} \\
&\quad + \underset{\tau,\gamma}{E}\left\{\left(\begin{bmatrix} z_k \\ u_k \end{bmatrix}^T \begin{bmatrix} A_{zk}^T \\ B_{zk}^T \end{bmatrix}^T P_{k+1} \begin{bmatrix} A_{zk}^T \\ B_{zk}^T \end{bmatrix}\begin{bmatrix} z_k \\ u_k \end{bmatrix}\right)\right\} \\
&= \begin{bmatrix} z_k \\ u_k \end{bmatrix}^T \begin{bmatrix} S_z + \underset{\tau,\gamma}{E}(A_{zk}^T P_{k+1} A_{zk}) & \underset{\tau,\gamma}{E}(A_{zk}^T P_{k+1} B_{zk}) \\ \underset{\tau,\gamma}{E}(B_{zk}^T P_{k+1} A_{zk}) & R_z + \underset{\tau,\gamma}{E}(B_{zk}^T P_{k+1} B_{zk}) \end{bmatrix}\begin{bmatrix} z_k \\ u_k \end{bmatrix}
\end{aligned} \tag{2.73}$$

Therefore, $E_{\tau,\gamma}(H_k)$ can be written in terms of system matrices and solution to the SRE as

$$\bar{H}_k = \underset{\tau,\gamma}{E}(H_k) = \begin{bmatrix} \bar{H}_k^{zz} & \bar{H}_k^{zu} \\ \bar{H}_k^{uz} & \bar{H}_k^{uu} \end{bmatrix} = \begin{bmatrix} S_z + \underset{\tau,\gamma}{E}(A_{zk}^T P_{k+1} A_{zk}) & \underset{\tau,\gamma}{E}(A_{zk}^T P_{k+1} B_{zk}) \\ \underset{\tau,\gamma}{E}(B_{zk}^T P_{k+1} A_{zk}) & R_z + \underset{\tau,\gamma}{E}(B_{zk}^T P_{k+1} B_{zk}) \end{bmatrix} \tag{2.74}$$

The optimal action-dependent value function $Q(z_k, u_k)$ is equal to stochastic cost function J_k. Therefore, we have

$$J_k = Q(z_k,u_k) \tag{2.75}$$

Then using Equation 2.74 and stochastic control theory (Stengel 1986), the optimal time-varying gain can be expressed in terms of $\bar{H}_k$ as

$$\underset{\tau,\gamma}{E}(K_k) = [R_z + \underset{\tau,\gamma}{E}(B_{zk}^T P_{k+1} B_{zk})]^{-1} \underset{\tau,\gamma}{E}(B_{zk}^T P_{k+1} A_{zk}) = (\bar{H}_k^{uu})^{-1} \bar{H}_k^{uz} \tag{2.76}$$

Remark 2.1

According to Equation 2.76, if the solution to the SRE, P_k+1 is known, then the time-varying system matrices A_{zk}, B_{zk} are still required to compute the controller gains in a backward-in-time manner. On the other hand, if time-varying matrix $\bar{H}_k$ can be learned online at time k without the knowledge of linear time-varying system dynamics, certainty-equivalent optimal controller gain can be solved not only without NCS system matrices but also forward-in-time.

2.5.3.2 Model-Free Online Tuning Based on Adaptive Estimator and Q-Learning

The proposed online tuning approach entails one AE, which is used to learn the Q-function. Since the Q-function includes the $\bar{H}_k$ matrix, this matrix can be solved online and the control signal can be obtained using Equation 2.74. We make the following assumption (Kreisselmeier 1986) since the stochastic system is linear and slowly time-varying.

Assumption 2.1

The Q-function, $Q(z_k, u_k)$, can be expressed as the linear in the unknown parameters (LIP).

By using the stochastic adaptive control theory (Åstrom 1970) and the definition of Q-function (2.72), $Q(z_k, u_k)$ can be represented in vector form similar to the AE representation as

$$Q(z_k, u_k) = w_k^T \bar{H}_k w_k = \bar{h}_k^T \bar{w}_k \tag{2.77}$$

where$\bar{h}_k = vec(\bar{H}_k), w_k = [z_k^T\ u^T(z_k)]^T, w_k \in \mathbb{R}^{n+(\bar{d}+1)m=l}$,and$\bar{w}_k = (w_{k1}^2, \ldots, w_{k1}w_{kl}, w_{k2}^2, \ldots, w_{kl-1}w_{kl}, w_{kl}^2)$ is the Kronecker product quadratic polynomial basis vector and $\bar{h}_k = \text{vec}(\bar{H}_k)$ with the vector function acting on $l \times l$ matrices, thus yielding an $l(l+1)/2 \times 1$ column vector (Note: the vec(⋅) function is constructed by stacking the columns of the matrix into one column vector with the off-diagonal elements, which can be combined as $H_{mn} + H_{nm}$).

The time-varying matrix $\bar{H}_k$ can be considered as slowly varying consistent with standard adaptive control (Goodwin and Sin 1984; Kreisselmeier 1986; Narendra and Annaswamy 1989; Sastry and Bodson 1989; Chen and Guo 1991). Then Q-function can be expressed as an unknown time-varying target parameter vector and the regression function $\bar{w}_k$. Next, the Q-function $Q(z_k, u_k)$ estimation will be considered.

EXAMPLE 2.8

The networked control of inverted pendulum system with random delays and packet losses is given

$$\begin{bmatrix} \dot{x} \\ \ddot{x} \\ \dot{\phi} \\ \ddot{\phi} \end{bmatrix} = \begin{bmatrix} 0 & 1 & 0 & 0 \\ 0 & -0.1818 & 2.6727 & 0 \\ 0 & 0 & 0 & 1 \\ 0 & -0.4545 & 31.1818 & 0 \end{bmatrix} \begin{bmatrix} x \\ \dot{x} \\ \phi \\ \dot{\phi} \end{bmatrix} + \begin{bmatrix} 0 \\ 1.8182 \\ 0 \\ 4.5455 \end{bmatrix} u + w \tag{2.78}$$

where white noise is given as $w \sim N(0, 1)$.

SOLUTION

The proposed Q-learning-based adaptive optimal controller is implemented on the system with unknown system dynamics. After incorporating random disturbance, the resulting system is the stochastic time-varying system. Assume the state measurements are perfect.

The system state z_k is generated as $z_k = [x_k\ \Delta x_k\ \phi_k\ \Delta\phi_k]^T \in \Re^{4\times1}, \forall k$ and $w = [z\, u] \in \Re^{5\times1}$ the initial stabilizing policy for the algorithm was selected as $u_0(z_k) = [-70.71\ \ -37.83\ \ 105.53\ \ 20.92]z_k$ while the regression function for the Q-function was generated as $\{w_1^2, w_1w_2, w_1w_3, \ldots, w_2^2, \ldots, w_5^2, \ldots, w_6^2\}$.

The design parameter for the Q-function $Q(z_k, u_k)$ was selected as $\alpha_h = 10^{-6}$ while the initial parameters for the AE were set to zero at the beginning of the simulation. The initial parameters of the Q-function estimator were chosen to reflect the initial stabilizing control. The simulation was run for 500 times steps, and for the first 250 times steps, exploration noise with mean zero and variance 0.06 was added to the system in order to ensure the persistency of excitation (PE) condition holds.

In Figure 2.5, the proposed Q-learning-based stochastic optimal controller makes the NCS state tracking errors converge to zero even when the NCS dynamics are unknown. According to the above results, the proposed Q-learning-based adaptive optimal control algorithm will have nearly the same performance as the NCS with unknown dynamics as that of an optimal controller for the NCS when the system dynamics, delays, and packet losses are known.

2.5.4 Stochastic Nonlinear Discrete-Time Optimal Control

In this section, a stochastic *cost* approach is introduced for designing a certainty-equivalent controller for stochastic nonlinear systems. If a dynamic system is driven by uncertain disturbance inputs, the stochastic optimal trajectory must be generated using measurements. If the measurements contain random errors due to measurement white noise process, then the feedback control responds not only to the effects of random inputs but to the measurement errors as well. In the presence of measurement noise process, a neighboring control technique (Stengel 1986) needs to be designed. In contrast, in this book, the optimal feedback control is considered when the state is measured perfectly and when there is no process noise. However, for the purpose of background, we consider a deterministic system driven by a white process noise.

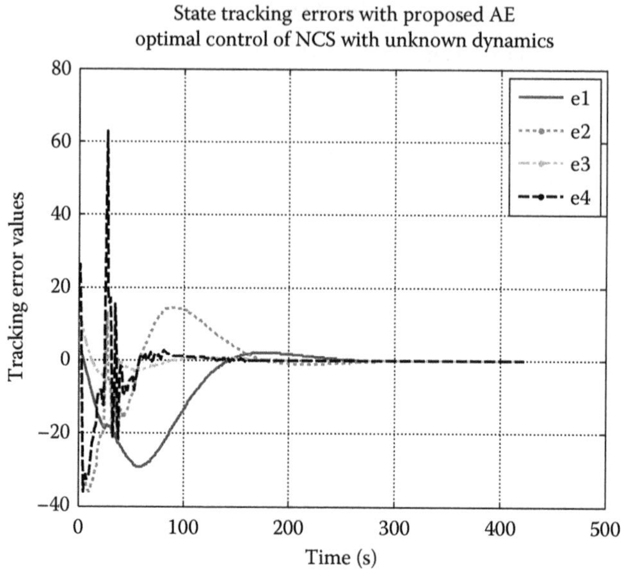

FIGURE 2.5 System response.

Consider the affine nonlinear discrete-time system given by

$$x(k+1) = f(x(k)) + g(x(k))u(k) + w(\mathrm{k}) \tag{2.79}$$

where $x(k)$, $u(k)$, and $w(k)$ represent the system state, control input, and white process noise while $f(x(k))$ and $g(x(k))$ denote the internal dynamics and control coefficient matrix, respectively. According to Stengel (1986), the certainty-equivalent optimal control input is derived by minimizing the certainty-equivalent stochastic value function expressed as

$$\begin{cases} V_k(x,k) = E\left[\phi_N(x(N)) + \sum_{l=k}^{N-1} r(x(l),u(l))\right] \\ \qquad = E\left[\phi_N(x(N)) + \sum_{l=k}^{N-1} (Q_d(x(l)) + u^T(l)R_d u(l))\right] \forall k = 0,1,\ldots,N-1 \\ V_N(x,N) = E[\phi_N(x(N))] \end{cases} \tag{2.80}$$

where the cost-to-go is denoted as $r(x(l), u(l)) = Q_d(x(l)) + u^T(l)R_d u(l)$, $\forall k = 0, 1, \ldots,$ $N-1$, NT_s is the final time instant, $Q_d(x) \geq 0$, $\phi_N(x) \geq 0$, and R_d being the symmetric positive definite matrix, and $E(\cdot)$ is the expectation operator (the mean value). Here, the terminal constraint $\phi_N(x)$ needs to be satisfied in the finite-horizon optimal control design. Equation 2.80 can also be rewritten as

$$V_k(x,k) = E\left[r(x(k),u(k)) + \phi_N(x(N)) + \sum_{l=k}^{N-1} r(x(l),u(l))\right] \tag{2.81}$$

$$= E\left[(Q_d(x(k)) + u^T(k)R_d u(k)) + V_{k+1}(x(k+1))\right], k = 0,\ldots,N-1$$

According to the observability condition (Chen and Guo 1991), when $x = 0$, $V_k(x) = 0$, the value function $V_k(x)$ serves as a Lyapunov function (Sarangapani 2006). Based on the Bellman principle of optimality (Stengel 1986; Lewis and Syrmos 1995), the certainty-equivalence stochastic optimal value function also satisfies the stochastic HJB equation in discrete time and is given by

$$\begin{cases} V^*(x,k) = \min\limits_{u_{d,k}}(V_k(x,k) = \min\limits_{u(k)}\left[E[Q_d(x(k)) + u^T(k)R_d u(k) + V^*(x(k+1))]\right], \\ \forall k = 0,\ldots,N-1 \\ V^*(x_{d,N},N) = E\left[\phi_N(x(k))\right] \end{cases} \tag{2.82}$$

Differentiating Equation 2.81, the certainty-equivalent optimal control $u^*(k)$ is obtained as

$$E\left[\frac{\partial(Q_d(x(k)) + u^T(k)R_d u(k))}{\partial u(k)} + \frac{\partial x^T(k+1)}{\partial u(k)}\frac{\partial V^*(x(k+1))}{\partial x(k+1)}\right] = 0 \tag{2.83}$$

In other words

$$\begin{cases} E\left[u^*(x(k))\right] = -\dfrac{1}{2}E\left[R_d^{-1}g^T(x(k))\dfrac{\partial V^*(x(k+1))}{\partial x(k+1)}\right], \quad k = 0,\ldots,N-1 \\ E\left[u^*(x(N-1))\right] = -\dfrac{1}{2}E\left[R_d^{-1}g^T(x(N-1))\dfrac{\partial \phi_N(x(N))}{\partial x(N)}\right] \end{cases} \tag{2.84}$$

Substituting Equation 2.81 into Equation 2.80, the discrete-time HJB Equation 2.80 becomes

$$V^*(x,k) = E\left[Q_d(x(k)) + \frac{1}{4}\frac{\partial V^{*T}(x(k+1))}{\partial x(k+1)}g(x(k))R_d^{-1}\right.$$

$$\left.\times g(x(k))\frac{\partial V^{*T}(x(k+1))}{\partial x(k+1)}] + V^*(x(k+1))\right] \quad \forall k = 0,\ldots,N-1$$

$$V^*(x,N-1)=E\left[Q_d(x(N-1))+\frac{1}{4}\frac{\partial\phi_N^T(x(N))}{\partial x(N)}g(x(N-1))R_d^{-1}\right.$$
$$\left.\times g(x(N-1))\frac{\partial\phi_N(x(N))}{\partial x(N)}+\phi_N(x(N))\right] \quad (2.85)$$

EXAMPLE 2.9

The continuous-time version of original nonlinear affine system is given by $\dot{x}=f(x)+g(x)u+w, x\in\Re^2, y=Cx$, where $f(x)=\begin{bmatrix}-x_1+x_2\\ -0.5x_1-0.5x_2(1-(\cos(2x_1)+2)^2\end{bmatrix}$, $g(x)=\begin{bmatrix}0\\ \cos(2x_1)+2\end{bmatrix}$, and $C=\begin{bmatrix}0 & 1\\ 2 & 0\end{bmatrix}$, and random disturbance is $w\sim N(0,\ 1)$. Obtain discrete-time version of the system. Apply the proposed stochastic Q-learning and plot the system response.

Solution

The proposed stochastic optimal control is implemented for the NNCS with unknown system dynamics in the presence of random delays and packet losses after discretization as shown in Chapter 3 (Section 3.11). After discretization, the system becomes the nonlinear stochastic time-varying system. We will assume that there is no disturbance and measurement noise.

The system state y_k^o is generated as $y_k\in\Re^{2\times1},\forall k$, and the initial stabilizing policy for the proposed algorithm was selected as $u_o(y_k)=[-2\ -5]\ y_k$, while the activation functions for NN-based identifier were generated as $\tanh\{(y_1)^2,y_1y_2,(y_2)^2,(y_1)^4,(y_1)^3y_2^o,\ldots,(y_2)^6\}$, critic NN activation function were selected as the sigmoid of six-order polynomial $\{(y_1)^2,y_1y_2,(y_2)^2,(y_1)^4,(y_1)^3y_2^o,\ldots,(y_2)^6\}$, and action NN activation function were generated from the gradient of critic NN activation function. The design parameters for NN-based identifier, critic NN, and action NN were selected as $\alpha_C=0.002$, $\alpha_V=10^{-4}$, and $\alpha_u=0.005$ while the NN-based identifier and critic NN weights are set to zero at the beginning of the simulation. The initial weights of the action NN are chosen to reflect the initial stabilizing control. The simulation was run for 20 s, and for the first 10 s, exploration noise with mean zero and variance 0.06 was added to the system in order to ensure the PE condition.

The performance of the proposed stochastic optimal controller is evaluated. This stochastic optimal controller can make the state regulation errors converge to zero even when the system dynamics are unknown as shown in Figure 2.6.

2.5.5 Background on Neural Networks

A brief background and NN function approximation is dealt with in this part of the chapter (Sarangapani 2006) as NN will be utilized to approve the value function and control policy for nonlinear systems. Artificial NNs are modeled on biological processes for information processing, including specifically the nervous system and its basic unit, the neuron. Signals are propagated in the form of potential differences between the inside and outside of cells. Dendrites bring signals from other neurons

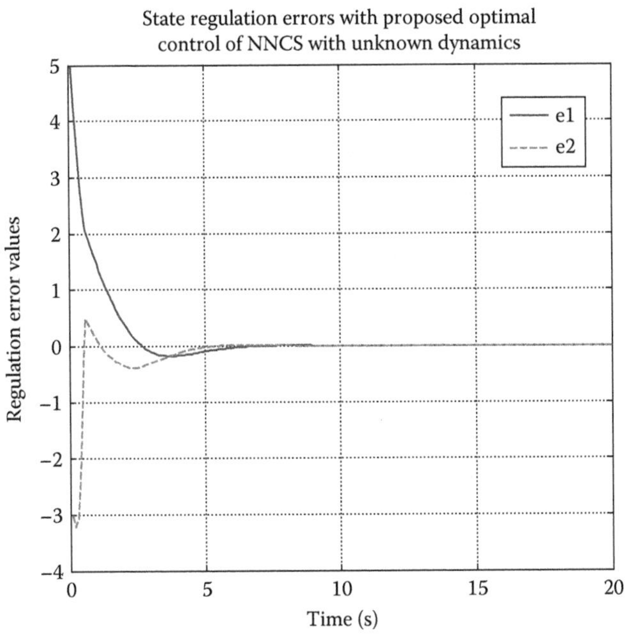

FIGURE 2.6 Performance of stochastic optimal controller with unknown network imperfections.

into the cell body or soma, possibly multiplying each incoming signal by a transfer weighting coefficient. In the soma, cell capacitance integrates the signals, which collect in the axon hillock. Once the combined signal exceeds a certain cell threshold, a signal, the action potential, is transmitted through the axon. Cell nonlinearities make the composite action potential a nonlinear function of the combination of arriving signals. The axon connects through the synapses with the dendrites of subsequent neurons. The synapses operate through the discharge of neurotransmitter chemicals across intercellular gaps, and can be either excitatory (tending to fire the next neuron) or inhibitory (tending to prevent the firing of the next neuron).

2.5.6 Two-Layer Neural Networks

A two-layer NN, which has two layers of neurons, with one layer having L neurons feeding a second layer having m neurons, is depicted in Figure 2.7. The first layer is known as the *hidden layer*, with L being the *number of hidden-layer neurons*; the second layer is known as the *output layer*. An NN with multiple layers is called a *multilayer perceptron*; its computing power is significantly enhanced over the one-layer NN. With a one-layer NN, it is possible to implement digital operations such as AND, OR, and COMPLEMENT (see the problems section). However, research in NN was stopped many years ago when it was shown that the one-layer NN is incapable of performing the EXCLUSIVE OR operation, which is a basic problem in

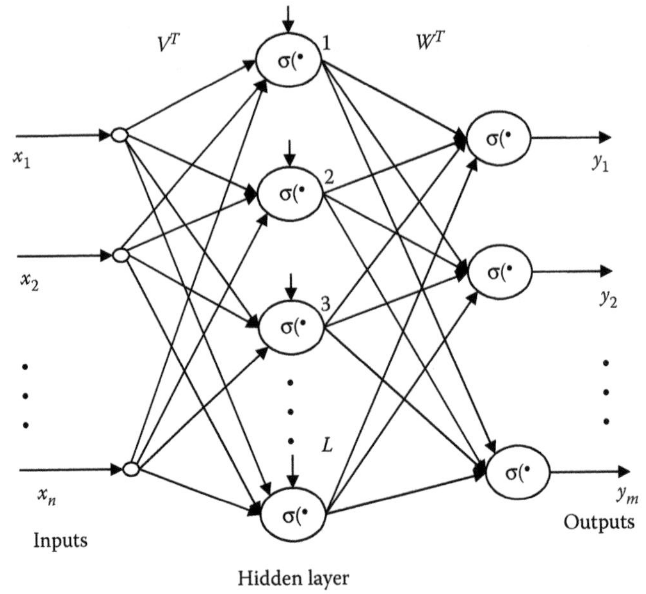

FIGURE 2.7 A two-layer neural network.

digital logic design. It was later demonstrated that the two-layer NN can implement the EXCLUSIVE OR (X-OR) and this again accelerated NN research in the early 1980s. Several researchers in the literature presented solutions to the X-OR operation by using sigmoid activation functions.

The output of the two-layer NN is given by the recall equation

$$y_i = \sigma\left(\sum_{l=1}^{L} w_{il}\sigma\left(\sum_{j=1}^{n} v_{lj}x_j + v_{l0}\right) + w_{i0}\right);\ i = 1,2,\ldots,m \tag{2.86}$$

Defining the hidden-layer outputs z_l allows one to write

$$\begin{aligned} z_l &= \sigma\left(\sum_{j=1}^{n} v_{lj}x_j + v_{l0}\right);\ l = 1,2,\ldots,L \\ y_i &= \sigma\left(\sum_{l=1}^{L} w_{il}z_l + w_{i0}\right);\ \mathrm{i} = 1,2,\ldots,m \end{aligned} \tag{2.87}$$

Defining first-layer weight matrices $\bar{V}$ and V as in the previous section, and second-layer weight matrices as

$$\overline{W}^T \equiv \begin{bmatrix} w_{11} & w_{12} & \cdots & w_{1L} \\ w_{21} & w_{22} & \cdots & w_{2L} \\ \vdots & \vdots & & \vdots \\ w_{m1} & w_{m2} & \cdots & w_{mL} \end{bmatrix}, \; b_w = \begin{bmatrix} w_{10} \\ w_{20} \\ \vdots \\ w_{m0} \end{bmatrix}$$

$$W^T \equiv \begin{bmatrix} w_{10} \; w_{11} & w_{12} & \cdots & w_{1L} \\ w_{20} \; w_{21} & w_{22} & \cdots & w_{2L} \\ \vdots & \vdots & & \vdots \\ w_{m0} \; w_{m1} & w_{m2} & \cdots & w_{mL} \end{bmatrix} \tag{2.88}$$

one may write the NN output as

$$y = \overline{\sigma}(\overline{W}^T \overline{\sigma}(\overline{V}^T \overline{x} + b_v) + b_w) \tag{2.89}$$

or, in streamlined form as

$$y = \overline{\sigma}(W^T \sigma(V^T x)) \tag{2.90}$$

In these equations, the notation $\overline{\sigma}$ means the vector defined in accordance with Equation 2.90. In Equation 2.89, it is necessary to use the augmented vector

$$\sigma(w) \equiv [1 \; \overline{\sigma}(w)^T]^T = [1 \; \sigma(w_1) \; \sigma(w_2) \ldots \sigma(w_L)]^T \tag{2.91}$$

where a "1" is placed as the first entry to allow the incorporation of the thresholds w_{i0} as the first column of W^T. In terms of the hidden-layer output vector $z \in \Re^L$, one may write

$$\overline{z} = \sigma(V^T x) \tag{2.92}$$

$$y = \sigma(W^T z) \tag{2.93}$$

where $z \equiv [1 \; \overline{z}^T]^T$.

In the remainder of this book, we shall not show the overbar on vectors—the reader will be able to determine by the context whether the leading "1" is required. We shall generally be concerned in the later chapters with two-layer NN with linear activation functions in the output layer, so that

$$y(k) = W^T \sigma(V^T x(k)) \tag{2.94}$$

It is important to mention that the input to the hidden-layer weights will be selected randomly and held fixed whereas those hidden to the output-layer weights will be tuned. This will minimize the computational complexity associated with using NN in feedback control applications while ensuring that one can use NN in control.

EXAMPLE 2.10: OUTPUT SURFACE FOR TWO-LAYER NEURAL NETWORK (LEWIS ET AL. 1999)

A two-layer NN with two inputs and one output is given by the equation below:

$$y = \bar{W}^T \bar{\sigma}(\bar{V}^T \bar{x} + b_v) + b_w \equiv w\sigma(vx + b_v) + b_w$$

with weight matrices and thresholds given by

$$v = \bar{V}^T = \begin{bmatrix} -2.69 & -2.80 \\ -3.39 & -4.56 \end{bmatrix}, \; b_v = \begin{bmatrix} -2.21 \\ 4.76 \end{bmatrix}$$

$$w = \bar{W}^T = \begin{bmatrix} -4.91 & 4.95 \end{bmatrix}, \; b_w = \begin{bmatrix} -2.28 \end{bmatrix}$$

Plots of the NN output surface y as a function of the inputs x_1, x_2 over the grid $[-2, 2] \times [-2, 2]$ can be generated. Different outputs can be illustrated corresponding to the use of different activation functions. To make the plot in Figure 2.8, the MATLAB NN Toolbox 4.0 was used with the sequence of commands given in Example 2.10.

```
% Example 2.10: Output surface of two-layer NN
% Set up NN weights
v= [-2.69 -2.80; -3.39 -4.56];
```

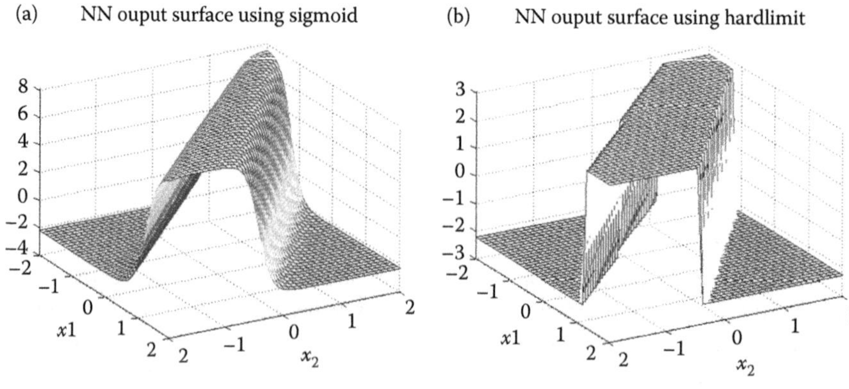

FIGURE 2.8 Output surface of a two-layer NN. (a) Using sigmoid activation function. (b) Using hard limit activation function.

```
bv= [-2.21; 4.76];
w= [-4.91 4.95];
bw= [-2.28];

% Set up plotting grid for sampling x
[x1,x2] =meshgrid(-2:0.1:2);

% Compute NN input vectors p and simulate NN using sigmoid
p1=x1(:);
p2=x2(:);
p= [p1'; p2'];

net=nnt2ff(minmax(p),{v,w},{bv,bw},{'hardlim','purelin'});
a=sim(net,p);

% Format results for using 'mesh' or 'surfl' plot routines:
a1=eye(41);
a1(:) =a';
mesh(x1,x2,a1);
AZ=60, EL=30;
view(AZ,EL);
xlabel('x1');
ylabel('x2');
%title('NN output surface using sigmoid');
title('NN output surface using hardlimit');
```

Plotting the NN output surface over a region of values for x reveals graphically the decision boundaries of the network and aids in visualization.

2.5.7 NN Function Approximation

If the first-layer weights and the thresholds V in Equation 2.73 are predetermined by some *a priori* method, then only the second-layer weights and thresholds W are considered to define the NN, so that the NN has only one layer of weights. One may then define the fixed function $\phi(x) = \sigma(V^T x)$ so that such a one-layer NN has the recall equation

$$y = W^T \phi(x) \tag{2.95}$$

where $x \in \Re^n$ (recall that technically x is augmented by "1"), $y \in \Re^m$, $\phi(\cdot): \Re \to \Re^L$, and L is the number of hidden-layer neurons. This NN is *linear* in the NN parameters W. This will make it easier for us to deal with such networks in subsequent chapters. Specifically, it is easier to train the NN by tuning the weights. This one-layer having only output-layer weights W should be contrasted with the one-layer NN discussed in Equation 2.73, which had only input-layer weights V.

More generality is gained if $\sigma(\cdot)$ is not diagonal, for example, as defined in Equation 2.93, but $\phi(\cdot)$ is allowed to be a general function from $\Re^n$ to $\Re^L$. This is called a *functional link neural net (FLNN)*. Some special FLNN are now discussed. We often use $\sigma(\cdot)$ in place of $\phi(\cdot)$, with the understanding that, for LIP nets, this activation function vector is not diagonal, but is a general function from $\Re^n$ to $\Re^L$.

Of fundamental importance in NN closed-loop control applications is the *universal function approximation property* of NN having at least two layers. (One-layer NNs do not generally have a universal approximation capability.) The approximation capabilities of NN have been studied by many researchers.

The basic universal approximation result says that any smooth function $f(x)$ can be approximated arbitrarily closely on a compact set using a two-layer NN with appropriate weights. This result has been shown using sigmoid activations, hard limit activations, and others. Specifically, let $f(x):\mathfrak{R}^n \rightarrow \mathfrak{R}^m$ be a smooth function. Then, given a compact set $S \in \mathfrak{R}^n$ and a positive number ε_N, there exists a two-layer NN such that

$$f(x) = W^T\sigma(V^T x) + \varepsilon \tag{2.96}$$

with $\|\varepsilon\| = \varepsilon_N$ for all $x \in S$, for some sufficiently large value L of hidden-layer neurons. The value ε (generally a function of x) is called the NN function approximation error, and it decreases as the hidden-layer size L increases. We say that, on the compact set S, as S becomes larger, the required L generally increases correspondingly. Approximation results have also been shown for smooth functions with a finite number of discontinuities.

Even though the result says that there exists an NN that approximates $f(x)$, it should be noted that it does not show how to determine the required weights. It is in fact not an easy task to determine the weights so that an NN does indeed approximate a given function $f(x)$ closely enough. In the next section, we shall show how to accomplish this using backpropagation tuning. If the function approximation is to be carried out in the context of a dynamic closed-loop feedback control scheme, the issue is thornier and is solved in subsequent chapters.

2.5.7.1 Functional Link Neural Networks

Consider a special class of a one-layer NN known as FLNN written as

$$y = W^T\phi(x) \tag{2.97}$$

with W the NN output weights (including the thresholds) and $\phi(\cdot)$ a general function from $\mathfrak{R}^n$ to $\mathfrak{R}^L$. In subsequent chapters, these NN have a great advantage in that they are easier to train than a general two-layer NN since they are *linear in the tunable parameters* (LIP). Unfortunately, for an LIP NN, the functional approximation property does not generally hold. However, an FLNN can still approximate functions as long as the activation functions $\phi(\cdot)$ are selected as a *basis*, which must satisfy the following two requirements on a compact simply connected set S of $\mathfrak{R}^n$:

1. A constant function on S can be expressed as Equation 2.97 for a finite number L of hidden-layer functions.
2. The functional range of Equation 2.97 is dense in the space of continuous functions from S to $\mathfrak{R}^m$ for a countable L.

If $\phi(\cdot)$provides a basis, then a smooth function $f(x):\Re^n \rightarrow \Re^m$ can be approximated on a compact set S of $\Re^n$, by

$$f(x) = W^T\phi(x) + \varepsilon \tag{2.98}$$

for some ideal weights and thresholds W and some number of hidden-layer neurons L. In fact, for any choice of a positive number ε_N, one can find a feedforward NN such that $\|\varepsilon\| < \varepsilon_N$ for all x in S.

Barron (1993) has shown that for all LIP approximators, there is a fundamental lower bound, so that ε is bounded below by terms on the order of $1/L^{2/n}$. Thus, as the number of NN inputs "n" increases, increasing L to improve the approximation accuracy becomes less effective. This lower-bound problem does not occur in the multilayer nonlinear-in-the-parameters network.

PROBLEMS

Section 2.1

Problem 2.1.1: *Simulation of Compound Interest System.* Simulate the system of Example 2.1 and plot the state with respect to time.

Problem 2.1.2: *Genetics.* Many congenital diseases can be explained as a result of both genes at a single location being the same recessive gene (Luenberger 1979). Under some assumptions, the frequency of the recessive gene at generation k is given by the recursion

$$x(k+1) = \frac{x(k)}{1+x(k)}$$

Simulate in MATLAB using $x(0) = 75$. Observe that $x(k)$ converges to zero, but very slowly. This explains why deadly genetic diseases can remain active for a very long time. Simulate the system starting for a small negative value of $x(0)$ and observe that it tends away from zero.

Problem 2.1.3: *Discrete-Time System.* Simulate the system

$$x_1(k+1) = \frac{x_2(k)}{1+x_2(k)}$$

$$x_2(k+1) = \frac{x_1(k)}{1+x_2(k)}$$

using MATLAB. Plot the phase-plane plot.

Section 2.4

Problem 2.4.1: *Lyapunov Stability Analysis.* Using the Lyapunov stability analysis, examine the stability for the following systems. Plot time histories to substantiate your claims.

a. $x_1(k+1) = x_1(k)\sqrt{(x_1^2(k)+x_2^2(k))}$

$x_2(k+1) = x_2(k)\sqrt{(x_1^2(k)+x_2^2(k))}$

b. $x(k+1) = -x^2(k) - 10x(k)\sin x(k)$

Problem 2.4.2: *Lyapunov Control Design.* Useing Lyapunov techniques, design controllers to stabilize the following system. Plot time histories of the states to verify your design. Verify passivity and dissipativity of the systems.

a. $x(k+1) = -x^2(k)\cos x(k) - 10x(k)\sin x(k) + u(k)$

$x_1(k+1) = x_1(k)x_2(k)$

b. $x_2(k+1) = x_1^2(k) - \sin x_2(k) + u(k)$

Problem 2.4.3: *Stability Improvement Using Feedback.* The system

$$x(k+1) = Ax(k) + Bu(k) + d(k)$$

has a disturbance $d(k)$ that is unknown but bounded so that $\|d(k)\| < d_M$, with the bounding constant known. In Example 2.7, the system with no control input, $B = 0$, and A stable was shown to be UUB. Show that by selecting the control input as $u(k) = -Kx(k)$ it is possible to improve the UUB stability properties of the system by making the bound on $\|x(k)\|$ smaller. In fact, if feedback is allowed, the initial system matrix A need not be stable as long as (A, B) is *stabilizable*.

Section 2.5

Problem 2.5.1: Consider the linearized version of the power system dynamics given by

$$\dot{x} = \begin{bmatrix} -0.665 & 11.5 & 0 & 0 \\ 0 & -2.5 & 2.5 & 0 \\ -9.5 & 0 & -13.736 & -13.736 \\ 0.6 & 0 & 0 & 0 \end{bmatrix} x + \begin{bmatrix} 0 \\ 0 \\ 13.136 \\ 0 \end{bmatrix} u$$

Implement the linear stochastic optimal control policy by assuming that the dynamics are uncertain.

REFERENCES

Åstrom, K.J. 1970. *Introduction to Stochastic Control Theory.* Academic Press, New York.

Barron, A.R. 1993. Universal approximation bounds for superpositions of a sigmoidal function. *IEEE Transactions on Information Theory*, 39(3), 930–945.

Chen, H.F. and Guo, L. 1991. *Identification and Stochastic Adaptive Control.* Cambridge Press, Cambridge, MA.
Goodwin, C.G. and Sin, K.S. 1984. *Adaptive Filtering, Prediction, and Control.* Prentice-Hall, Englewood Cliffs, NJ.
Kalkkuhl, J.C. and Hunt, K.J. 1996. Discrete-time neural model structures for continuous-time nonlinear systems. *Neural Adaptive Control Technology*, Zbikowski, R. and Hunt, K.J. (eds.), Chapter 1, World Scientific, Singapore.
Khasminskii, R. and Milstein, G.N. 2011. *Stochastic Stability of Differential Equations*, 2nd edition, Springer, Berlin.
Kreisselmeier, G. 1986. Adaptive control of a class of slowly time-varying plants. *System and Control Letters*, 8, 97–103.
Kumar, P.R. and Varaiya, P. 1986, *Stochastic Systems: Estimation, Identification and Control.* Prentice-Hall, Englewood Cliffs, NJ.
Landau, Y.D. 1979. *Adaptive Control: The Model Reference Approach.* Marcel Dekker, Inc., Basel.
Lewis, F.L., Abdallah, C.T., and Dawson, D.M. 1993. *Control of Robot Manipulators.* Macmillan, New York.
Lewis, F.L. and Syrmos, V.L. 1995. *Optimal Control*, 2nd edition. John Wiley & Sons, New York.
Lewis, F.L., Jagannathan, S., and Yesiderek, A. 1999. *Neural Network Control of Robot Manipulators and Nonlinear Systems.* Taylor & Francis, London, UK.
Luenberger, D.G. 1979. *Introduction to Dynamic Systems.* Wiley, New York.
Narendra, K.S. and Annaswamy, A.M. 1989. *Stable Adaptive Systems.* Prentice-Hall, Englewood Cliffs, NJ.
Sarangapani, J. 2006. *Neural Network Control of Nonlinear Discrete-Time Systems.* Taylor & Francis (CRC Press), Boca Raton, FL.
Sastry, S. and Bodson, M. 1989. *Adaptive Control.* Prentice-Hall, Englewood Cliffs, NJ.
Slotine, J.-J.E. and Li, W. 1991. *Applied Nonlinear Control.* Prentice-Hall, Englewood Cliffs, NJ.
Stengel, R.F. 1986. *Stochastic Optimal Control: Theory and Application.* Wiley-Interscience, New York.
Whitehead, A.N. 1953. *Science and the Modern World*, Lowell Lectures (1925). Macmillan, New York.

3 Optimal Adaptive Control of Uncertain Linear Network Control Systems

An NCS that uses a real-time communication network in its feedback control loop has been considered as the next-generation control system. However, as observed in Chapter 1, inserting a network into the feedback loop brings many challenging issues due to network imperfections such as network-induced delays and packet losses that occur during exchanging data among devices. Moreover, these network imperfections can degrade the control system performance significantly, causing instability.

Therefore, Halevi and Ray (1988) and Zhang et al. (2001) analyzed and proposed the stability region of an NCS with network-induced delays and packet losses, respectively. On the other hand, by using stochastic optimal control theory (Åstrom 1970; Stengel 1986), Nilsson et al. (1998) derived infinite-horizon stochastic optimal control of an NCS with network imperfections and the solution is presented backward-in-time with known NCS system dynamics by assuming that the network imperfections are known *a priori*. However, NCS system dynamics after incorporating network imperfections are not typically known beforehand due to random delays and packet losses. Moreover, the finite-horizon optimal scheme for an uncertain NCS is not considered (Halevi and Ray 1988; Nilsson et al. 1998; Zhang et al. 2001).

The adaptive dynamics programming (ADP) techniques by Werbos (1983) and Barto et al. (1983), on the other hand, have been developed to obtain optimal controller design for uncertain linear/nonlinear systems in the forward-in-time manner. In ADP, by using policy or value iterations (VIs), a reinforcement learning scheme is combined with dynamic programming to solve optimal adaptive control. In Wang et al. (2011) and Heydari and Balakrishnan (2011), a finite-horizon optimal control scheme by using the iterative ADP approach is derived for an affine nonlinear system.

However, for achieving optimality, iteration-based ADP methods (Barto et al. 1983; Werbos 1983; Wang et al. 2011) may need a significant number of iterations within a sample interval, which can become a bottleneck for hardware implementation as demonstrated in Dierks and Jagannathan (2012). Therefore, Dierks and Jagannathan (2012) proposed a time-based ADP approach to derive infinite-horizon optimal control of an affine nonlinear discrete-time system in a forward-in-time manner. In this scheme, the past history of system states and cost function estimates have been utilized instead of iteration-based approximated optimal design.

However, existing ADP approaches (Barto et al. 1983; Werbos 1983; Wang et al. 2011; Dierks and Jagannathan 2012) are not applicable for an NCS since (a) network imperfections are ignored; (b) in many cases infinite-horizon-based optimal control is considered; and finally (c) an iterative scheme is utilized. Network imperfections cause stochastic uncertainty in system dynamics and traditional controllers can become unstable in the presence of random delays and packet losses.

In the recent literature, in Xu et al. (2012), a novel infinite-horizon stochastic optimal control has been proposed for linear NCS (LNCS) in the presence of unknown system dynamics and network imperfections. The finite-horizon case has not been addressed so far in the literature for such uncertain stochastic dynamic systems in a forward-in-time manner until very recently (Xu and Jagannathan 2013). Compared with infinite-horizon optimal control, a terminal constraint has to be satisfied in finite-horizon optimal design, which is more difficult and challenging (Stengel 1986). In addition, demonstrating convergence of closed-loop signals to optimal value is quite difficult with finite-horizon optimal control.

In this chapter, optimal adaptive control scheme using ADP is undertaken to generate finite-horizon stochastic optimal regulation of an LNCS with uncertain system dynamics (Xu and Jagannathan 2013) resulting from unknown network imperfections such as network-induced delays and packet losses. First, with an initial admissible control, a novel adaptive estimator (AE) (Sarangapani 2006), which is tuned online, is proposed and updated forward-in-time to learn the certainty-equivalent stochastic value function by using the Bellman equation (Stengel 1986) given terminal constraint and perfect state measurements.

Next, certainty-equivalent stochastic optimal adaptive control is generated by optimizing the stochastic value function while satisfying the terminal constraint via the AE. In contrast with the traditional finite-horizon stochastic optimal regulator, which needs the full knowledge of system dynamics to solve the SRE, this certainty-equivalent AE-based optimal adaptive control scheme relaxes the requirement on system dynamics as well as value or policy iterations for an LNCS. In other words, the parameters of the AE are updated once a sampling instant is consistent with the traditional adaptive control. The case of infinite time horizon is also deduced. Convergence analysis is also included in this chapter.

In the derivation of stochastic optimal control, first, the deterministic linear time-invariant continuous-time system is modeled as a linear time-varying stochastic system due to the presence of a communication network in contrast with standard stochastic optimal control where random disturbance (or white process noise) and measurement noise are utilized. Here, the stochastic certainty-equivalent value function and control policy is derived and the stochastic increment utilized in traditional stochastic optimal control (Stengel 1986) is not considered. If the random disturbance inputs and measurement noise signals have to be added with an NCS, the approach introduced in this chapter has to be rederived (Xu et al. 2014).

Note that the control of an LNCS with network imperfections is different from the control of time-delay systems with known deterministic delays (Luck and Ray 1990; Mahmoud and Ismail 2005) since the random delays and packet losses cannot be easily handled by time-delay systems. Therefore, these approaches from Mahmoud and Ismail (2005) and Luck and Ray (1990) may not perform well with an LNCS.

First, Section 3.1 presents the traditional control design by using the Riccati equation-based solution. Subsequently, Section 3.2 develops the finite-horizon optimal adaptive control for an LNCS (Xu and Jagannathan 2013). Eventually, Section 3.3 extends the results of Section 3.2 to the infinite-horizon case.

3.1 TRADITIONAL CONTROL DESIGN AND STOCHASTIC RICCATI EQUATION-BASED SOLUTION

Consider a stochastic linear discrete-time system given by

$$x_{d,k+1} = A_{d,k}x_{d,k} + B_{d,k}u_{d,k} + w_k \tag{3.1}$$

where $x_{d,k}$, $u_{d,k}$, w_k represent system state, control input, and white process noise with zero mean, respectively, and $A_{d,k} \in \Re^{n\times n}, B_{d,k} \in \Re^{n\times m}$ denote system matrices. Using standard stochastic optimal control theory (Åstrom 1970; Stengel 1986), the certainty-equivalent optimal value function in the presence of perfect state measurements can be written as

$$\begin{aligned} V^*(x_{d,k}) &= \min_{u_{d,l}} E\left[x_{d,N}^T S_N x_{d,N} + \sum_{l=k}^{N-1} r(x_{d,l}, u_{d,l}) \right] \\ &= \min_{u_{d,l}} E\left[x_{d,N}^T S_N x_{d,N} + \sum_{l=k}^{N-1} (x_{d,l}^T Q_d x_{d,l} + u_{d,l}^T R_d u_{d,l}) \right] \quad \forall k = 0,\dots,N-1 \end{aligned} \tag{3.2}$$

where $E(\cdot)$ is the expectation operator, cost-to-go is defined as $r(x_{d,l}, u_{d,l}) = x_{d,l}^T Q_d x_{d,l} + u_{d,l}^T R_d u_{d,l}$, and Q_d, R_d, S_N are symmetric positive semidefinite and positive definite matrices, respectively.

Next, according to dynamic programming (DP), the Bellman equation in discrete time can be expressed as

$$\left\{ 0 = \min_{u_{d,k}} E(r(x_{d,k}, u_{d,k}) + V^*(x_{d,k+1}) - V^*(x_{d,k})) \right. \tag{3.3}$$

with the boundary condition

$$\left\{ 0 = E\left[x_{d,N}^T S_N x_{d,N} - V^*(x_{d,N}) \right] \right. \tag{3.4}$$

Assuming that the minimum on the right-hand side of Equation 3.3 exists and is unique, the optimal control can be derived as (Lewis and Syrmos 1995)

$$u_{d,k}^* = -\frac{1}{2} E\left[R_d^{-1} B_{d,k} \frac{\partial V^*(x_{d,k+1})}{\partial x_{d,k+1}} \right] \quad \forall k = 0,1,\dots,N-1 \tag{3.5}$$

Substituting the optimal control input (3.5) into Bellman Equation 3.3, the discrete-time HJB equation can be represented as

$$\begin{cases} 0 = E\left[x_{d,k}^T Q_d x_{d,k} + \dfrac{1}{4}\dfrac{\partial V^{*T}(x_{d,k+1})}{\partial x_{d,k+1}} B_d^T R_d^{-1} B_d \dfrac{\partial V^*(x_{d,k+1})}{\partial x_{d,k+1}} \right. \\ \left. \quad + V^*(x_{d,k+1}) - V^*(x_{d,k+1}) \right] \\ 0 = E\left[x_{d,N}^T S_N x_{d,N} - V^*(x_{d,N}) \right] \end{cases} \tag{3.6}$$

According to Stengel (1986) for linear systems, certainty-equivalent stochastic value function (3.2) can be formulated as a quadratic function of the system state vector and thus can be expressed as

$$\begin{cases} V^*(x_{d,k}) = E\left(x_{d,k}^T P_k x_{d,k} \right) & \forall k = 0,1,\ldots,N-1 \\ V^*(x_{d,N}) = E\left(x_{d,N}^T P_N x_{d,N} \right) \end{cases} \tag{3.7}$$

where P_k, $\forall k = 0,1,\ldots,N$ is a positive definite kernel matrix. Substituting Equation 3.7 into Equation 3.5, discrete-time HJB becomes a well-known certainty-equivalent SRE, which is given by

$$\begin{cases} 0 = E\left[A_{d,k}^T \left[P_k - P_k B_{d,k} \left(B_{d,k}^T P_k B_{d,k} + R_d \right)^{-1} B_{d,k}^T P_k \right] A_{d,k} + Q_d - P_k \right] \\ E(P_N) = E(S_N) \end{cases} \tag{3.8}$$

with S_N being the terminal weighting matrix. Moreover, the certainty-equivalent optimal control input can be represented in terms of the SRE solution, P_k, as

$$u_k^* = -E\left[\left(B_{d,k}^T P_k B_{d,k} + R_d \right)^{-1} B_{d,k}^T P_k A_{d,k} x_{d,k} \right] \quad \forall k = 0,1,\ldots,N-1 \tag{3.9}$$

The solution of the certainty-equivalent SRE equation and the control input evolve in a backward-in-time manner provided the system matrices are known beforehand. For the finite horizon, the control gain becomes time-varying. When the system matrices are uncertain, the SRE solution cannot be found. Additionally, generating the control input in a forward-in-time manner has significant practical value for hardware implementation, which is not possible with traditional optimal control techniques. For an LNCS, the systems matrices are time-varying and uncertain due to network imperfections necessitating an optimal adaptive approach. In the presence of bounded deterministic disturbances, the optimal control can be obtained as described in the next chapter while the disturbances are considered to be absent in this chapter.

3.2 FINITE-HORIZON OPTIMAL ADAPTIVE CONTROL

In this section, novel ADP techniques are used to obtain stochastic optimal control of an LNCS (Xu and Jagannathan 2013) with uncertain dynamics. First, the background of the LNCS is introduced. Then, a novel AE is proposed to obtain the unknown value function online for the LNCS. Next, a model-free online tuning of the parameters of the value function estimator by using the ADP method is introduced in order to relax the need for the SRE solution. Eventually, the convergence proof is given.

3.2.1 Background

The basic structure of an LNCS is shown in Figure 3.1 where a communication network is used to close the feedback control loop. The LNCS incorporates the network-induced delays and packet losses, which includes (1) $\tau_{sc}(t)$: sensor-to-controller delay, (2) $\tau_{ca}(t)$: controller-to-actuator delay, and (3) $\gamma(t)$: indicator of network-induced packet losses.

According to the NCS literatures (Nilsson et al. 1998; Xu and Jagannathan 2012) with standard communication network protocol, the following assumptions (Goldsmith 2003; Hu and Zhu 2003) are needed for the LNCS stochastic optimal design.

Assumption 3.1

(a) For a wide area network, two types of networked-induced delays are considered independent, ergodic and unknown, whereas their probability distribution functions are considered known. The sensor-to-controller delay is kept less than one sampling interval. (b) The sum of two delays is bounded while the initial state of the system is deterministic (Hu and Zhu 2003).

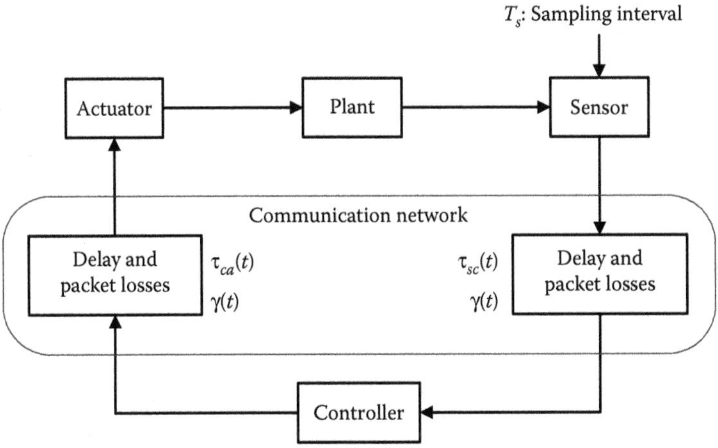

FIGURE 3.1 Linear networked control system.

Considering the network-induced delays and packet losses, the original time-invariant plant, $\dot{x}(t) = Ax(t) + Bu(t)$, can be represented as

$$\dot{x}(t) = Ax(t) + B\gamma(t)u(t - \tau(t)) \tag{3.10}$$

where

$$\gamma(t) = \begin{cases} \mathbf{I}^{n\times n} & \text{if packet has been received at time } t \\ \mathbf{0}^{n\times n} & \text{if packet has been lost } x(t) \text{ at time } t \end{cases}$$

and $x(t) \in \Re^n, u(t) \in \Re^m$ represent the system state and control inputs of the LNCS, respectively, $A \in \Re^{n\times n}, B \in \Re^{n\times m}$ denote the system matrices, and $\tau(t)$ denotes networked-induced delay.

By using Xu et al. (2012), and after integrating Equation 3.10 over a sampling interval $[kT_s, (k+1)T_s)$, the LNCS can be expressed as

$$x_{k+1} = A_s x_k + B_k^1 \gamma_{k-1} u_{k-1} + \cdots + B_k^{\bar{d}} \gamma_{k-\bar{d}} u_{k-\bar{d}} + B_k^0 \gamma_k u_k \tag{3.11}$$

where $\bar{d}T_s$ is the upper bound of network-induced delay, A_s, $B_k^0, B_k^1, \ldots, B_k^{\bar{d}}$ and γ_k are defined similar to that by Xu et al. (2012). To simplify the LNCS representation (3.11), we define a new augment state that incorporates current state and previous control inputs (i.e., $z_k = \left[x_k^T \; u_{k-1}^T \; u_{k-2}^T \ldots u_{k-\bar{d}}^T \right]^T \in \Re^{n+\bar{d}*m}$). Then, Equation 3.11 can be expressed as a stochastic linear time-varying system given by

$$z_{k+1} = A_{zk} z_k + B_{zk} u_k \quad \forall k = 0,1,2,\ldots N \tag{3.12}$$

with time-varying system matrices expressed as

$$A_{zk} = \begin{bmatrix} A_s & \gamma_{k-1}B_k^1 & \cdots & \cdots & \gamma_{k-\bar{d}+1}B_k^{\bar{d}-1} & \gamma_{k-\bar{d}}B_k^{\bar{d}} \\ 0 & 0 & \cdots & \cdots & 0 & 0 \\ 0 & I_m & \cdots & \cdots & 0 & 0 \\ \vdots & \vdots & \ddots & & \vdots & \vdots \\ \vdots & \vdots & & \ddots & \vdots & \vdots \\ 0 & 0 & \cdots & \cdots & I_m & 0 \end{bmatrix}, \quad B_{zk} = \begin{bmatrix} \gamma_k B_k^0 \\ I_m \\ 0 \\ 0 \\ \vdots \\ 0 \end{bmatrix}$$

It is clear from Equation 3.12 that system matrices are time-varying and stochastic due to network imperfections. Owing to unknown delays and packet losses, the system dynamics become uncertain. Before designing the controller, we need to ensure that Equation 3.12 is controllable. It was shown in Xu et al. (2012) that if the original time-invariant system is controllable, then Equation 3.12 is also controllable.

Next, according to the LNCS representation (3.12), the stochastic optimal control policy can be developed by minimizing the value function over a finite horizon, which is given as

$$V(z_k,k) = \underset{\tau,\gamma}{E}\left[z_N^T S_{z,N} z_N + \sum_{i=k}^{N-1}\left(z_i^T Q_z z_i + u_i^T R_z u_i\right)\right], \quad \forall k = 0,\ldots,N-1 \tag{3.13}$$

where NT_s is the final time instant, $S_{z,N}$, Q_z, and R_z are symmetric positive semi-definite and positive definite constant matrices, respectively, and $\underset{\tau,\gamma}{E}(\cdot)$ is the expectation operator (i.e., mean value) of $z_N^T S_{z,N} z_N + \sum_{i=k}^{N-1}\left(z_i^T Q_z z_i + u_i^T R_z u_i\right)$ in terms of network-induced delay τ and packet losses γ.

Remark 3.1

For the proposed stochastic scheme, there are two important things to note: (1) compared with the traditional linear time-invariant system, due to network imperfections, LNCS become time-varying (Chen 1999) and of higher dimension due to augmented system states resulting from past control inputs; and (2) traditional control designs developed for deterministic time-delay systems (Mahmoud and Ismail 2005) are not suitable for LNCS since delay randomness cannot be eliminated. In this chapter, stochastic analysis will be utilized for LNCS with random network imperfections.

Next, before we present the optimal control design, a useful lemma about utilizing geometric sequence analysis in Lyapunov stability theory is introduced.

Lemma 3.1

If there exists a positive definite Lyapunov function $L(x_k)$ for the LNCS (3.12) with the strengthened condition on its first difference

$$\Delta L(x_k) = L(x_{k+1}) - L(x_k) \le -\alpha L(x_k) < 0 \tag{3.14}$$

where $0 < \alpha < 1$ and $\Delta L(x_k)$ is negative definite. Then, at any time instant (i.e., $k = 1,2,\ldots$), Lyapunov function $L(x_k)$ can be expressed in terms of bounded initial Lyapunov function $L(x_0)$ and constant α. In addition, the first difference can be proven to be uniformly ultimately bounded with positive ultimate bound $B_{L,k}$ given by

$$L(x_k) \le (1-\alpha)^k L(x_0) \equiv B_{L,k} \tag{3.15}$$

Moreover, $L(x_k) \to 0$ and $B_{L,k} \to 0$ as $k \to \infty$.
Proof: Refer to Appendix 3A.

3.2.2 Stochastic Value Function

Consider an LNCS with network imperfection represented as Equation 3.12. Given the LNCS with a unique equilibrium point, $z = 0$, on a set **S**, stochastic optimal control signal $u_k^* = -L_k z_k \quad \forall k = 0,1,\ldots,N-1$ (*Note*: L_k is the optimal control gain) can be obtained by minimizing stochastic value function $V(z_k)$ given by Equation 3.13. The stochastic value function (Åstrom 1970; Stengel 1986) is represented in the quadratic form as

$$V(z_k,k) = \begin{cases} \underset{\tau,\gamma}{E}\left(z_k^T P_{z,k} z_k, k\right) & \forall k = 0,1,\ldots,N-1 \\ \underset{\tau,\gamma}{E}\left(z_N^T S_{z,N} z_N, N\right) \end{cases} \tag{3.16}$$

where $P_{z,k} \geq 0$ is the solution to the SRE (Åstrom 1970; Stengel 1986). Then, the Bellman equation can be expressed in terms of the expected value as

$$V^*(z_k,k) = \underset{\tau,\gamma}{E}[r(z_k,u_k) + V^*(z_{k+1},k+1)] = \underset{\tau,\gamma}{E}\left\{\left[z_k^T\ u_k^T\right]\Psi_k\left[z_k^T\ u_k^T\right]^T\right\} \tag{3.17}$$

where $r(z_k,u_k) = z_k^T Q_z z_k + u_k^T R_z u_k$. In contrast to Xu and Jagannathan (2012), for the finite-horizon case, substituting Equation 3.16 into the Bellman Equation 3.17 during $[0, (N-1)T_s]$ yields

$$\begin{bmatrix} z_k \\ u_k \end{bmatrix}^T \underset{\tau,\gamma}{E}(\Psi_k)\begin{bmatrix} z_k \\ u_k \end{bmatrix} = \underset{\tau,\gamma}{E}[r(z_k,u_k) + V^*(z_{k+1},k+1)] \tag{3.18}$$

$$= \begin{bmatrix} z_k \\ u_k \end{bmatrix}^T \begin{bmatrix} Q_z + \underset{\tau,\gamma}{E}\left(A_{zk}^T P_{z,k+1} A_{zk}\right) & \underset{\tau,\gamma}{E}\left(\left(A_{zk}^T P_{z,k+1} B_{zk}\right)\right. \\ \underset{\tau,\gamma}{E}\left(\left(B_{zk}^T P_{z,k+1} A_{zk}\right)\right. & R_z + \underset{\tau,\gamma}{E}\left(\left(B_{zk}^T P_{z,k+1} B_{zk}\right)\right. \end{bmatrix} \begin{bmatrix} z_k \\ u_k \end{bmatrix}, \quad \forall k = 0,1,\ldots,N-1$$

Next, after incorporating the terminal constraint in the value function and Equation 3.18, $\bar{\Psi}_k = \underset{\tau,\gamma}{E}(\Psi_k)$ can be expressed as

$$\bar{\Psi}_k = \begin{bmatrix} \bar{\Psi}_k^{zz} & \bar{\Psi}_k^{zu} \\ \bar{\Psi}_k^{uz} & \bar{\Psi}_k^{uu} \end{bmatrix} = \begin{bmatrix} Q_z + \underset{\tau,\gamma}{E}\left(A_{zk}^T P_{z,k+1} A_{zk}\right) & \underset{\tau,\gamma}{E}\left(\left(A_{zk}^T P_{z,k+1} B_{zk}\right)\right. \\ \underset{\tau,\gamma}{E}\left(\left(B_{zk}^T P_{z,k+1} A_{zk}\right)\right. & R_z + \underset{\tau,\gamma}{E}\left(\left(B_{zk}^T P_{z,k+1} B_{zk}\right)\right. \end{bmatrix},$$

$$\forall k = 0,\ldots,N-1 \tag{3.19}$$

and

$$\bar{\Psi}_N = \begin{bmatrix} S_{z,N} & 0 \\ 0 & 0 \end{bmatrix}$$

Next, the stochastic optimal control gain (3.19) (Stengel 1986) for LNCS can be expressed in terms of value function parameters, $\bar{\Psi}_k$, $\forall k = 0,1,\ldots,N-1$, as

$$\begin{aligned} L_k &= \left[R_z + \underset{\tau,\gamma}{E}\left(B_{zk}^T P_{k+1} B_{zk}\right)\right]^{-1} \underset{\tau,\gamma}{E}\left(B_{zk}^T P_{k+1} A_{zk}\right) \\ &= \left(\bar{\Psi}_k^{uu}\right)^{-1} \bar{\Psi}_k^{uz} \quad \forall k = 0,1,\ldots,N-1 \end{aligned} \tag{3.20}$$

It is important to note that solving the SRE solution P_k, $k = 0,\ldots,N$ requires system matrices. However, according to Equation 3.20, system dynamics are not needed to obtain the optimal control gain matrix if parameter vector $\bar{\Psi}_k$ $k = 0,1,\ldots,N-1$ can be estimated online. In addition, the terminal constraint is also accommodated. This approach is utilized in this chapter. Next, online tuning of the parameters of the AE is introduced.

3.2.3 Model-Free Online Tuning of Adaptive Estimator

In this section, an AE has been proposed to estimate the value function (3.16) and corresponding time-dependent parameter $\bar{\Psi}_k$ $\forall k = 0,1,\ldots,N-1$. Moreover, using Equation 3.20, stochastic optimal control gain, $L_k \forall k = 0,1,\ldots,N-1$, can be obtained by using estimated $\bar{\Psi}_k$, $\forall k = 0,1,\ldots,N-1$ even without the knowledge of LNCS system dynamics.

Assuming the value function can be represented as the LIP and based on Equation 3.16, the value function can be expressed as

$$V(z_k,k) = \varphi_k^T \bar{\Psi}_k \varphi_k = \bar{\theta}_k \bar{\varphi}_k = \underset{\tau,\gamma}{E}[W^T \sigma(N-k)\bar{\varphi}_k, k], \quad \forall k = 0,\ldots,N \tag{3.21}$$

where

$$\bar{\theta}_k = \text{vec}(\bar{\Psi}_k) = W^T \sigma(N-k), \quad \varphi_k = \underset{\tau,\gamma}{E}\left\{\left[z_k^T \ u\left(z_k^T\right)\right]^T\right\}$$

and

$$\bar{\varphi}_k = \left(\varphi_{k1}^2, \ldots, \varphi_{k1}\varphi_{kl}, \varphi_{k2}^2, \ldots, \varphi_{kl-1}\varphi_{kl}, \varphi_{kl}^2\right)$$

is the Kronecker product quadratic polynomial basis vector (Sarangapani 2006), including augment state and control input, and vec(·) function is defined similar to Sarangapani (2006). Moreover, $\sigma(\cdot)$ is the time-dependent regression function for the value function parameter vector $\bar{\theta}_k$. It is important to note that $\bar{\theta}_N = \text{vec}(\bar{\Psi}_N)$ (3.19) is considered a known terminal constraint in the LNCS stochastic optimal problem. Therefore, it is obvious that target W and regression function $\sigma(\cdot)$ should satisfy $\bar{\theta}_N = \underset{\tau,\gamma}{E}[W^T\sigma(0)]$.

Next, the stochastic value function (3.21) can be approximated by using AE as

$$\hat{V}(z_k,k) = \hat{\bar{\theta}}_k\bar{\phi}_k = \underset{\tau,\gamma}{E}\left[\hat{W}_k^T\sigma(N-k)\bar{\phi}_k,k\right] \quad \forall k = 0,\ldots,N \tag{3.22}$$

where $\underset{\tau,\gamma}{E}(\hat{W}_k)$ is the estimated parameter at time kT_s. It is important to note that value function (3.22) is a nonautonomous function, which is different from the infinite-horizon case (Stengel 1986; Lewis and Syrmos 1995). On the other hand, since time is finite, the time-dependent function can be bounded as $\sigma_{\min} \le \sigma(k) \le \sigma_{\max}, \forall k = 0,1,\ldots,N$ with $\sigma_{\min}$, $\sigma_{\max}$ being positive constants. Moreover, the value function can also be bounded as $\hat{V}_{\min}(z_k) \le \hat{V}(z_k,k) \le \hat{V}_{\max}(z_k)$, $k = 0,\ldots,N$, where $\hat{V}_{\min}(z_k) = \underset{\tau,\gamma}{E}\left[\hat{W}_k^T\sigma_{\min}\bar{\varphi}_k\right]$ and $\hat{V}_{\max}(z_k) = \underset{\tau,\gamma}{E}\left[\hat{W}_k^T\sigma_{\max}\bar{\varphi}_k\right]$. Note that, the bounded functions $\hat{V}_{\min}(z_k), \hat{V}_{\max}(z_k)$ are positive definite time-independent functions.

Then, substituting Equation 3.22 into the Bellman Equation 3.17 may result in residual errors. By using delayed values for convenience (Xu and Jagannathan 2012), the residual error associated with Equation 3.22 can be expressed as

$$\begin{aligned}\underset{\tau,\gamma}{E}(e_{FTBE,k}) &= \underset{\tau,\gamma}{E}[\hat{V}(z_k,k) - \hat{V}(z_{k-1},k-1) + r(z_{k-1},u_{k-1})] \\ &= \underset{\tau,\gamma}{E}[r(z_{k-1},u_{k-1}) + \hat{W}_k^T\Delta\phi(z_{k-1},k-1)] \quad k = 0,1,\ldots,N\end{aligned} \tag{3.23}$$

where $\Delta\phi(x_{k-1},k-1) = \sigma(N-k)\bar{\varphi}_k - \sigma(N-k+1)\bar{\varphi}_{k-1}$ which is bounded as $\Delta\phi_{\min}(z_{k-1}) \le \Delta\phi(z_{k-1},k-1) \le \Delta\phi_{\max}(z_{k-1}), \Delta\phi_{\max}(z_{k-1}) = \sigma_{\max}\bar{\varphi}_k - \sigma_{\min}\bar{\varphi}_{k-1}$ and $\Delta\phi_{\min}(z_{k-1}) = \sigma_{\min}\bar{\varphi}_k - \sigma_{\max}\bar{\varphi}_{k-1}$, $\underset{\tau,\gamma}{E}(e_{FTBE,k})$ is the Bellman equation residual error for the finite-horizon scenario. Further, the dynamics of Equation 3.23 can be expressed as

$$\underset{\tau,\gamma}{E}(e_{FTBE,k+1}) = \underset{\tau,\gamma}{E}\left[r(z_k,u_k) + \hat{W}_{k+1}^T\Delta\phi(x_k,k)\right] \quad \forall k = 0,\ldots,N$$

Besides considering $\underset{\tau,\gamma}{E}(e_{FTBE,k})$, the estimation error $\underset{\tau,\gamma}{E}(e_{FC,k})$ due to the terminal constraint needs to be considered and therefore given by

$$\underset{\tau,\gamma}{E}(e_{FC,k}) = \underset{\tau,\gamma}{E}\left[\bar{\theta}_N - \hat{W}_k^T\sigma(0)\right] \tag{3.24}$$

Next, auxiliary residual error vector and terminal constraint estimation error vector can be defined as

$$\begin{aligned}\underset{\tau,\gamma}{E}(\Xi_{FTBE,k}) &= \underset{\tau,\gamma}{E}\left(\Gamma_{k-1} + \hat{W}_k^T\Delta\Phi(z,k-1)\right) \\ \underset{\tau,\gamma}{E}(\Xi_{FC,k}) &= \underset{\tau,\gamma}{E}\left[\bar{\theta}_N\Lambda - \hat{W}_k^T\sigma(0)\Lambda\right]\end{aligned} \tag{3.25}$$

with

$$\Gamma_{k-1} = [r(z_{k-1},u_{k-1})r(z_{k-2},u_{k-2})\dots r(z_{k-1-j},u_{k-1-j})], \quad \Delta\Phi(z,k-1)$$
$$= [\Delta\phi(x_{k-1},k-1)\dots\Delta\phi(x_{k-1-i},k-1-i)], \quad 0<j<k-1\in\mathbb{N}$$

with $\mathbb{N}$ being the set of positive natural numbers, and Λ a constant normalized matrix to make the dimension of $\underset{\tau,\gamma}{E}(\Xi_{FC,k})$ match the dimension of $\underset{\tau,\gamma}{E}(\Xi_{FTBE,k})\in\Re^{1\times(1+i)}$. Since $\Delta\phi(z_k,k)$ is bounded, the term $\Delta\Phi(z,k-1)$ is also bounded such that

$$\Delta\Phi_{\min}(z)\le\Delta\Phi(z,k-1)\le\Delta\Phi_{\max}(z)$$

where $\Delta\Phi_{\min}(z)$ and $\Delta\Phi_{\max}(z)$ are positive time-independent functions. The historical residual error vector $\underset{\tau,\gamma}{E}(\Xi_{FTBE,k})$ has been recalculated by using the most current $\underset{\tau,\gamma}{E}\left(\hat{W}_k\right)$.

Then, the dynamics of auxiliary error vectors (3.25) can be derived as

$$\begin{aligned}\underset{\tau,\gamma}{E}(\Xi_{FTBE,k+1}) &= \underset{\tau,\gamma}{E}\left(\Gamma_k+\hat{W}_{k+1}^T\Delta\Phi(z,k)\right)\\ \underset{\tau,\gamma}{E}(\Xi_{FC,k+1}) &= \underset{\tau,\gamma}{E}\left[\bar{\theta}_N\Lambda-\hat{W}_{k+1}^T\sigma(0)\Lambda\right]\end{aligned} \tag{3.26}$$

For forcing both the Bellman equation and terminal constraint estimation errors convergence to zero, the update law for the AE can be derived as

$$\begin{aligned}\underset{\tau,\gamma}{E}(\hat{W}_{k+1}) = \underset{\tau,\gamma}{E}[&\Theta(z,k)(\Theta^T(z,k)\Theta(z,k))^{-1}(\alpha_W\Xi_{FTBE,k}^T\\ &+\alpha_W\Xi_{FC,k}^T-\Gamma_k^T-\bar{\theta}_N\Lambda)], \quad \forall k=0,1,\dots,N-1\end{aligned} \tag{3.27}$$

where $\Theta(z,k)=\Delta\Phi(z,k)-\sigma(0)\Lambda$ and $0<\alpha_W<1$. It is important to note that the term $\Theta(x,k)$ is bounded as $\Theta_{\min}(z)\le\Theta(z,k)\le\Theta_{\max}(z)$ with $\Theta_{\min}(z)=\Delta\Phi_{\min}(z)-\sigma(0)\Lambda$ and $\Theta_{\max}(z)=\Delta\Phi_{\max}(z)-\sigma(0)\Lambda$ where $\Theta_{\min}(z)$ and $\Theta_{\max}(z)$ are positive time-independent functions. Substituting Equation 3.27 into Equation 3.26 yields

$$\underset{\tau,\gamma}{E}(\Xi_{FTBE,k+1}+\Xi_{FC,k+1})=\alpha_W\underset{\tau,\gamma}{E}(\Xi_{FTBE,k}+\Xi_{FC,k}) \tag{3.28}$$

Moreover, the Bellman Equation 3.17 can be rewritten by using the AE as

$$\underset{\tau,\gamma}{E}[W^T\sigma(N-k)\bar{\phi}_k]=\underset{\tau,\gamma}{E}[r(z_k,u_k)+W^T\sigma(N-k-1)\bar{\phi}_{k+1}] \tag{3.29}$$

whereas the cost-to-go term $r(z_k,u_k)$ can be represented as

$$\underset{\tau,\gamma}{E}[r(z_k,u_k)] = \underset{\tau,\gamma}{E}[W^T\sigma(N-k)\bar{\phi}_k - W^T\sigma(N-k-1)\bar{\phi}_{k+1}]$$
$$= \underset{\tau,\gamma}{E}(W^T\Delta\varphi(x_k,k)) \tag{3.30}$$

Define the parameter estimation error as $\underset{\tau,\gamma}{E}\left(\tilde{W}_k\right) = \underset{\tau,\gamma}{E}\left(W - \hat{W}_k\right)$ and utilize Equation 3.25 with $\underset{\tau,\gamma}{E}(e_{FTBE,k+1} + e_{FC,k+1}) = \alpha_W \underset{\tau,\gamma}{E}(e_{FTBE,k+1} + e_{FC,k+1})$ from Equation 3.28 to yield

$$\underset{\tau,\gamma}{E}\{\tilde{W}_{k+1}^T[\Delta\varphi(x_k,k) - \sigma(0)]\} = -\alpha_W \underset{\tau,\gamma}{E}\{r(z_{k-1},u_{k-1}) - \bar{\theta}_N - \hat{W}_k^T[\Delta\varphi(x_{k-1},k-1) - \sigma(0)]\},$$
$$\forall k = 0,1,\ldots,N \tag{3.31}$$

Since $\underset{\tau,\gamma}{E}[r(z_{k-1},u_{k-1})] = \underset{\tau,\gamma}{E}(W^T\Delta\phi_k)$ and $\underset{\tau,\gamma}{E}(\bar{\theta}_N) = \underset{\tau,\gamma}{E}\left[\hat{W}^T\sigma(0)\right]$, Equation 3.31 can be rewritten as

$$\underset{\tau,\gamma}{E}\left\{\tilde{W}_{k+1}^T[\Delta\varphi_k - \sigma(0)]\right\} = -\alpha_W \underset{\tau,\gamma}{E}\left\{\tilde{W}_k^T[\Delta\varphi_{k-1} - \sigma(0)]\right\} \tag{3.32}$$

Next, the convergence of stochastic value function and terminal constraint estimation errors, along with AE parameter estimation error $\underset{\tau,\gamma}{E}\left(\tilde{W}_k\right)$ given as Equation 3.32 will be analyzed. Theorem 3.1 will be used to prove the overall closed-loop system stability in Theorem 3.2.

Theorem 3.1: Boundedness in the Mean of the AE Errors

Given u_{0k}, an initial admissible control policy for an LNCS (3.11) with final time NT_s, consider the initial parameter vector W_0 of the AE is bounded in the set Ω. Let the AE parameter update law be given as Equation 3.27. Then, during finite horizon (i.e., $t \in [0,NT_s]$), there exists a positive constant α_W with $0 < \alpha_W < 1$ such that AE parameter estimation errors are *bounded* in the mean (Xu et al. 2012). The ultimate bound is dependent upon the final time value (i.e., NT_s) and initial AE estimation error bound $B_{V,0}$.

Proof: Refer to Appendix 3B.

Next, using standard Lyapunov theory (Sarangapani 2006), during finite horizon, the parameter estimation and value function estimation errors can be proven to be *UUB* in the mean (Xu et al. 2012) where ultimate bounds are dependent on initial conditions and final time NT_s. The details are given as follows:

Assume the AE estimation error initially is bounded such that $\underset{\tau,\gamma}{E}\left\|\tilde{W}_0^T\Theta(z,-1)\right\| < B_{V,0}$ with $B_{V,0}$ being the positive bound constant. According to Lyapunov theory (Sarangapani 2006), $\underset{\tau,\gamma}{E}\left\|\tilde{W}_k^T\Theta(z,k-1)\right\| \quad \forall k = 1,2,\ldots,N$ can be expressed as

$$\underset{\tau,\gamma}{E}\left\|\tilde{W}_k^T\Theta(z,k-1)\right\| = \sqrt{\begin{array}{l}\underset{\tau,\gamma}{E}\left\|\tilde{W}_k^T\Theta(z,k-1)\right\|^2 + \underbrace{\left(\underset{\tau,\gamma}{E}\left\|\tilde{W}_0^T\Theta(z,0)\right\|^2 - \underset{\tau,\gamma}{E}\left\|\tilde{W}_0^T\Theta(z,-1)\right\|^2\right)}_{=0} \\ +\cdots+\underbrace{\left(\underset{\tau,\gamma}{E}\left\|\tilde{W}_{k-1}^T\Theta(z,k-2)\right\|^2 - \underset{\tau,\gamma}{E}\left\|\tilde{W}_{k-1}^T\Theta(z,k-2)\right\|^2\right)}_{=0}\end{array}} \tag{3.33}$$

It is important to note though adding and substituting terms $\underset{\tau,\gamma}{E}\left\|\tilde{W}_{k-1}^T\Theta(z,k-2)\right\|^2,\ldots,\underset{\tau,\gamma}{E}\left\|\tilde{W}_0^T\Theta(z,-1)\right\|^2$, Equation 3.33 holds. Next, by changing the sequence of terms, Equation 3.33 can be rewritten as

$$\begin{aligned}\underset{\tau,\gamma}{E}\left\|\tilde{W}_k^T\Theta(z,k-1)\right\| &= \sqrt{\begin{array}{l}\underset{\tau,\gamma}{E}\left\|\tilde{W}_0^T\Theta(z,-1)\right\|^2 + \underbrace{\left(\underset{\tau,\gamma}{E}\left\|\tilde{W}_1^T\Theta(z,0)\right\|^2 - \underset{\tau,\gamma}{E}\left\|\tilde{W}_0^T\Theta(z,-1)\right\|^2\right)}_{\Delta L_V(\tilde{W}_0)} \\ +\cdots+\underbrace{\left(\underset{\tau,\gamma}{E}\left\|\tilde{W}_k^T\Theta(z,k-1)\right\|^2 - \underset{\tau,\gamma}{E}\left\|\tilde{W}_{k-1}^T\Theta(z,k-2)\right\|^2\right)}_{\Delta L_V(\tilde{W}_{k-1})}\end{array}} \\ &\le \sqrt{B_{V,0}^2 + \Delta L_V(\tilde{W}_0,0)+\cdots+\Delta L_V(\tilde{W}_{k-1},k-1)} \\ &\le \sqrt{B_{V,0}^2 + \sum_{i=0}^{k-1}\Delta L_V(\tilde{W}_i,i)} \quad \forall k=1,2,\ldots,N\end{aligned} \tag{3.34}$$

Using Equation 3C.2 and geometric sequence analysis, ultimate bound $B_{V,k}$ can be expressed as

$$\begin{aligned}\underset{\tau,\gamma}{E}\left\|\tilde{W}_k^T\Theta(z,k-1)\right\| &\le \sqrt{B_{V,0}^2 + \sum_{i=0}^{k-1}\Delta L_V(\tilde{W}_i,i)} \\ &\le \sqrt{B_{V,0}^2 - \sum_{i=0}^{k-1}\left[(1-\alpha_W^2)\underset{\tau,\gamma}{E}\left\|\tilde{W}_i^T\Theta(z,i-1)\right\|^2\right]} \\ &\le \sqrt{B_{V,0}^2 - (1-\alpha_W^2)B_{V,0}^2\sum_{i=0}^{k-1}(\alpha_W^2)^i} \\ &\le \sqrt{1-(1-\alpha_W^{2k})}B_{V,0} \\ &\le \alpha_W^k B_{V,0} \equiv B_{V,k} \quad \forall k=1,2,\ldots,N\end{aligned} \tag{3.35}$$

where $B_{V,k}$ is bounded value for $k = 1,2,\ldots,N$. Since initial AE estimation error $B_{V,0}$ is a bounded, tuning parameter $0 < \alpha_W < 1$ and α_W^k will decrease while k increases, the ultimate bound $B_{V,k}$ (3.35) will also decrease when k increases. Further, when final time NT_s increases, the AE errors and value function estimation errors will remain *bounded* in the mean, but also the bound will decrease with the elongated time horizon.

Remark 3.2

Observe that since initial AE estimation error $B_{V,0}$ is bounded constant and $0 < \alpha_W < 1$, α_W^k bound $B_{V,k} = \alpha_W^k B_{V,0}$ will converge to zero *exponentially* when time goes to infinity (i.e., $B_{V,k} \to 0$ as $k \to \infty$). Therefore, the parameter estimation error and value function estimation error should be all asymptotically stable in the mean (Xu and Jagannathan 2013) for the infinite-horizon case (i.e., $k \to \infty, \underset{\tau,\gamma}{E}\left(\tilde{W}_k\right) \to 0$ and $\hat{V}(z_k,k) \to V(z_k,k)$).

According to Equations 3.9 and 3.20, the stochastic optimal control inputs based on the estimated matrix can be expressed as

$$\hat{u}_k = -\hat{L}_k z_k = \left(\hat{\bar{\Psi}}_k^{uu}\right)^{-1} \hat{\bar{\Psi}}_k^{uz} z_k \quad \forall k = 0,1,\ldots,N-1 \tag{3.36}$$

where

$$\hat{\bar{\Psi}}_k = \text{vec}^{-1}\left(\hat{\bar{\theta}}_k\right) = \text{vec}^{-1}\left[\underset{\tau,\gamma}{E}\left[\hat{W}_k^T \sigma(N-k)\right]\right] \quad \forall k = 0,\ldots,N-1$$

$\text{vec}^{-1}(\cdot)$ is the inverse operation of $\text{vec}(\cdot)$ given in Xu and Jagannathan (2012).

3.2.4 Closed-Loop System Stability

In this section, the value function estimation error and AE parameter estimation errors will be proven to be *bounded* in the mean (Xu et al. 2012). Further, within the finite time horizon, closed-loop LNCS system states and estimated stochastic optimal control will converge to an ultimate bound near-optimal system states and control inputs, respectively. It is important to note that these ultimate bounds are dependent upon initial conditions such as initial system states, value function estimation errors, and time horizon NT_s. When the duration of the horizon increases, these bounds will decrease. When the time horizon goes to infinity such as $N \to \infty$, the proposed design will become the infinite-horizon optimal solution for an LNCS. The flowchart of the proposed stochastic optimal regulator of an LNCS is shown in Figure 3.2.

The initial LNCS system states are assumed to reside within the set Ω stabilized by using the initial admissible control input u_0 (Xu et al. 2012). Then, the value

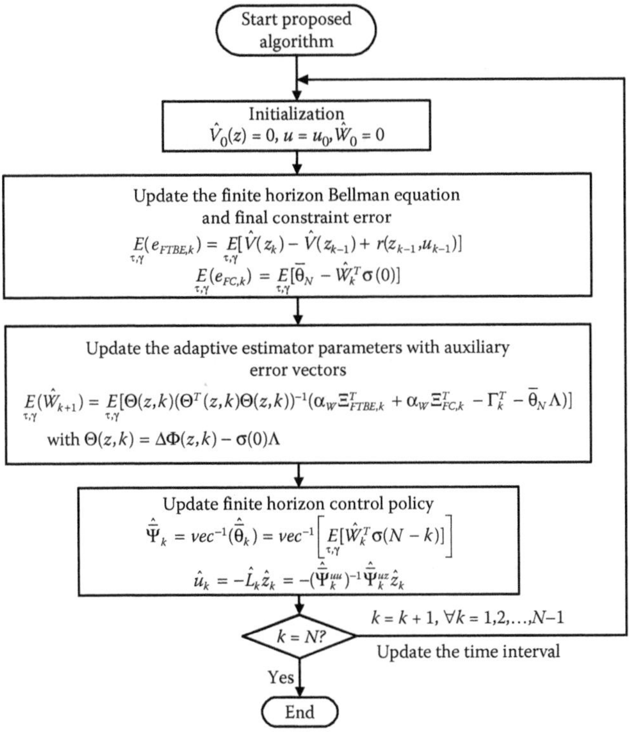

FIGURE 3.2 Finite-horizon stochastic optimal regulator for an LNCS.

function estimator tuning gain α_W can be derived to maintain that all the future signals are *bounded* in the mean. Finally, the proposed control design can be shown to converge to the finite-horizon stochastic optimal control design within ultimate bounds in the mean. Before proceeding further, the following lemma is needed for deriving the bounds of the stochastic optimal closed-loop dynamics when the stochastic optimal control is applied to the LNCS.

Lemma 3.2

Consider an LNCS with network imperfections and a sequence of stochastic optimal control policies such that the following inequality is satisfied:

$$\left\| \underset{\tau,\gamma}{E}(A_{zk} z_k + B_{zk} u_k^*) \right\|^2 \le l_o \left\| \underset{\tau,\gamma}{E}(z_k) \right\|^2 \quad \forall k = 0,1,\ldots,N-1 \tag{3.37}$$

where u_k^* is the optimal control policy and $0 < l_o < (1/2)$ is a constant.

Proof: The proof follows similar to Xu et al. (2012) and is therefore omitted here. Next, the main result is stated.

Theorem 3.2: Convergence of Optimal Control Signals

Given the initial system state z_0 and initial value function estimator $\underset{\tau,\gamma}{E}(\hat{W}_0)$ are bounded in the set Ω, let u_0 be any initial admissible control policy for an LNCS with network imperfections satisfying the bounds provided by Equation 3.37. During the time horizon (i.e., $t \in [0,NT_s]$), let the value function estimator be tuned and estimated control policy be given by Equations 3.27 and 3.36, respectively. Then, there exist positive constants l_o given by Lemma 3.2 and α_W provided by Theorem 3.1 such that the system state z_k and corresponding parameter estimation error vector $\underset{\tau,\gamma}{E}(\tilde{W}_k)$ of the value function estimator are all *bounded* in the mean (Xu et al. 2012) within the finite time horizon. Moreover, the bounds are dependent upon final time (i.e., NT_s), bounded initial value function estimation error $B_{V,0}$, and bounded initial LNCS system state $B_{z,0}$.

Proof: Refer to Appendix 3C.

Next, using standard Lyapunov theory (Sarangapani 2006), during the finite horizon, all the signals can be proven *bounded* in the mean (Xu and Jagannathan 2012), where bounds are dependent on initial conditions and final time NT_s. The details are given below.

Assume an LNCS system state is initiated as a positive bound constant B_z,0 (i.e., $\underset{\tau,\gamma}{E}\|z_k\|^2 = B_{z,0}$) and value function estimation error from Theorem 3.1 as B_V,0. Using standard Lyapunov theory and similar to Equations 3.33 and 3.34, $\underset{\tau,\gamma}{E}\|z_k\|$ and $\underset{\tau,\gamma}{E}\left\|\tilde{W}_k^T\Theta(z,k-1)\right\|$ for $k = 1,2,\ldots,N$ can be represented as

$$
\begin{aligned}
&\|\Pi\| \underset{\tau,\gamma}{E}\|\Pi\|\|z_k\|^2 + \underset{\tau,\gamma}{E}\left\|\tilde{W}_k^T\Theta(z,k-1)\right\|^2 = \|\Pi\| \underset{\tau,\gamma}{E}\|z_0\|^2 + \underset{\tau,\gamma}{E}\left\|\tilde{W}_0^T\Theta(z,-1)\right\|^2 \\
&\quad + \|\Pi\| \underbrace{\left(\underset{\tau,\gamma}{E}\|z_1\|^2 - \underset{\tau,\gamma}{E}\|z_0\|^2\right)}_{\Delta L_z(z_0)} + \underbrace{\left(\underset{\tau,\gamma}{E}\left\|\tilde{W}_1^T\Theta(z,0)\right\|^2 - \underset{\tau,\gamma}{E}\left\|\tilde{W}_0^T\Theta(z,-1)\right\|^2\right)}_{\Delta L_V(\tilde{W}_0)} + \cdots \\
&\quad + \|\Pi\| \underbrace{\left(\underset{\tau,\gamma}{E}\|z_k\|^2 - \underset{\tau,\gamma}{E}\|z_{k-1}\|^2\right)}_{\Delta L_z(z_{k-1})} + \underbrace{\left(\underset{\tau,\gamma}{E}\left\|\tilde{W}_k^T\Theta(z,k-1)\right\|^2 - \underset{\tau,\gamma}{E}\left\|\tilde{W}_{k-1}^T\Theta(z,k-2)\right\|^2\right)}_{\Delta L_V(\tilde{W}_0)} \\
&\le \|\Pi\| B_{z,0}^2 + B_{V,0}^2 + \underbrace{\Delta L_z(z_0) + \Delta L_V(\tilde{W}_0,0)}_{\Delta L_0} + \cdots + \underbrace{\Delta L_z(z_{k-1}) + \Delta L_V(\tilde{W}_{k-1},k-1)}_{\Delta L_{k-1}} \\
&\le \|\Pi\| B_{z,0}^2 + B_{V,0}^2 + \sum_{i=0}^{k-1}\Delta L_i, \quad \forall k = 1,2,\ldots,N
\end{aligned}
\tag{3.38}
$$

Using Equation 3C.3 and the property of geometric sequence, Equation 3.38 can be rewritten as

$$\|\Pi\| \underset{\tau,\gamma}{E}\|z_k\|^2 + \underset{\tau,\gamma}{E}\left\|\tilde{W}_k^T\Theta(x,k-1)\right\|^2 \le \|\Pi\| B_{z,0}^2 + B_{V,0}^2 + \sum_{i=0}^{k-1}\Delta L_i$$

$$\le \|\Pi\| B_{z,0}^2 + B_{V,0}^2 - \sum_{i=0}^{k-1}\left[(1-2l_o)\|\Pi\| \underset{\tau,\gamma}{E}\|z_i\|^2\right] - \sum_{i=0}^{k-1}\left[\frac{1}{2}(1-\alpha_W^2)\underset{\tau,\gamma}{E}\left\|\tilde{W}_i\Theta(z,i-1)\right\|^2\right]$$

$$\le \|\Pi\| B_{z,0}^2 + B_{V,0}^2 - (1-2l_o)\|\Pi\| B_{z,0}^2 \sum_{i=0}^{k-1}(2l_o^i) - \frac{(1-\alpha_W^2)B_{V,0}^2}{2}\sum_{i=0}^{k-1}\frac{(1+\alpha_W^2)^i}{2^i}$$

$$\le \|\Pi\| B_{z,0}^2 + B_{V,0}^2 - [1-(2l_o)^k]\|\Pi\| B_{z,0}^2 - \left[1-\frac{(1+\alpha_W^2)^k}{2^k}\right]B_{V,0}^2$$

$$\le (2l_o)^k\|\Pi\| B_{z,0}^2 + \left[\frac{(1+\alpha_W^2)}{2}\right]^k B_{V,0}^2 \quad \forall k = 1,2,\ldots,N \tag{3.39}$$

Therefore, the bounds for the system state and value function estimation error can be expressed as

$$\underset{\tau,\gamma}{E}\|z_k\| \le \sqrt{(2l_o)^k B_{z,0}^2 + \left[\frac{(1+\alpha_W^2)}{2}\right]^k \frac{B_{V,0}^2}{\|\Pi\|}} \equiv B_{z,k}^{CL} \tag{3.40}$$

or

$$\underset{\tau,\gamma}{E}\left\|\tilde{W}_k^T\Theta(z,k-1)\right\| \le \sqrt{(2l_o)^k\|\Pi\| B_{z,0}^2 + \left[\frac{(1+\alpha_W^2)}{2}\right]^k B_{V,0}^2} \equiv B_{V,k}^{CL}, \quad \forall k = 1,2,\ldots,N \tag{3.41}$$

where $B_{z,k}^{CL}$ and $B_{V,k}^{CL}$ are the values of bounds for $k = 1,2,\ldots,N$. According to the representation of Equations 3.40 and 3.41 and values of l_o, α_W, it is important to note that since tuning parameter $0 < l_o < (1/2)$ and $0 < \alpha_W < 1$ given in Theorems 3.1 and 3.2, $0 < 2l_o < 1, 0 < ((1+\alpha_W^2)/2) < 1$ and terms $(2l_o)^k$, $((1+\alpha_W^2)/2)^k$ will decrease while k increases. Moreover, since the initial value function estimation error $B_{V,0}$ and the LNCS system state $B_{z,0}$ are positive bounded constants, the closed-loop bounds $B_{z,k}^{CL}$ (3.40) and $B_{V,k}^{CL}$ (3.41) decrease when k increases. Further, when the final time NT_s increases, the system state and value function estimation errors will remain bounded and decrease with time (Xu et al. 2012).

Remark 3.3

Note since $0<2l_o<1, 0<((1+\alpha_W^2)/2)<1<1$, the initial value function estimation error $B_{V,0}$ and the LNCS system state $B_{z,0}$ are positive bounded constants, both terms $(2l_o)^k$, $((1+\alpha_W^2)/2)^k$ and the closed-loop bounds $B_{z,k}^{CL}$ (3.40) and $B_{V,k}^{CL}$ (3.41) will converge to zero *exponentially* when time goes to infinity, that is, $B_{z,k}^{CL} \to 0$, $B_{V,k}^{CL} \to 0$ as $k \to \infty$. Therefore, while time goes to infinity (i.e., $k \to \infty$), the LNCS system states and value function estimation error are all asymptotically stable in the mean (Xu et al. 2012), that is, $\underset{\tau,\gamma}{E}(z_k) \to 0$ and $\hat{V}(z_k,k) \to V(z_k,k)$ as $k \to \infty$ and proposed finite-horizon stochastic optimal will become infinite-horizon optimal control, Table 3.1 presents the proposed stochastic optimal design.

3.2.5 Simulation Results

The performance of the proposed stochastic finite-horizon optimal control design for an LNCS with unknown system dynamics and network imperfections is evaluated in this section. Before demonstrating the results, the simulation example is first given as follows

TABLE 3.1
Finite-Horizon Stochastic Optimal Design

Compute Finite-Horizon Control Input

$$\hat{u}_k = -\hat{L}_k z_k = \left(\hat{\bar{\Psi}}_k^{uu}\right)^{-1} \hat{\bar{\Psi}}_k^{uz} z_k \quad \forall k = 0,1,\ldots,N-1$$

where

$$\hat{\bar{\Psi}}_k = \text{vec}^{-1}\left(\hat{\bar{\theta}}_k\right) = \text{vec}^{-1}\left[\underset{\tau,\gamma}{E}\left[\hat{W}_k^T \sigma(N-k)\right]\right] \quad \forall k = 0,\ldots,N-1$$

Compute the Bellman Equation Residual Error and Terminal Constraint Estimation Error

Bellman Equation Residual Error

$$\underset{\tau,\gamma}{E}(\Xi_{FTBE,k+1}) = \underset{\tau,\gamma}{E}(\Gamma_k + \hat{W}_{k+1}^T \Delta\Phi(z,k))$$

Terminal Constraint Estimation Error

$$\underset{\tau,\gamma}{E}(\Xi_{FC,k}) = \underset{\tau,\gamma}{E}\left[\bar{\theta}_N \Lambda - \hat{W}_k^T \sigma(0)\Lambda\right]$$

Parameter Update

$$\underset{\tau,\gamma}{E}(\hat{W}_{k+1}) = \underset{\tau,\gamma}{E}\left[\Theta(z,k)(\Theta^T(z,k)\Theta(z,k))^{-1}\left(\alpha_W \Xi_{FTBE,k}^T + \alpha_W \Xi_{FC,k}^T - \Gamma_k^T - \bar{\theta}_N \Lambda\right)\right],$$

$$\forall k = 0,1,\ldots,N-1$$

where α_W is the tuning parameter.

EXAMPLE 3.1

The continuous-time version of a batch reactor system dynamics is shown as (Carnevale et al. 2007)

$$\dot{x} = \begin{bmatrix} 1.38 & -0.2077 & 6.715 & -5.676 \\ -0.5814 & -4.29 & 0 & 0.675 \\ 1.067 & 4.273 & -6.654 & 5.893 \\ 0.048 & 4.273 & 1.343 & -2.104 \end{bmatrix} x + \begin{bmatrix} 0 & 0 \\ 5.679 & 0 \\ 1.136 & -3.146 \\ 1.136 & 0 \end{bmatrix} u \quad (3.42)$$

Then, LNCS parameters can be selected as (Xu and Jagannathan 2012): (1) LNCS sampling time: $T_s = 50$ ms; (2) upper bound of network-induced delay is defined as two, that is, $\bar{d} = 2$; (3) the network-induced delays: $E(\tau_{sc}) = 35$ ms and $E(\tau) = 75$ ms; (4) network-induced packet losses follow Bernoulli distribution with $\bar{\gamma} = 0.3$; (5) final time: $t_f = NT_s = 10$ s with simulation time steps $N = 200$; and (6) to obtain ergodic performance, Monte Carlo simulations with 1000 iterations were run.

3.2.5.1 LNCS State Regulation Error and Performance

First, the proposed stochastic optimal control is implemented into an LNCS in the presence of unknown network imperfections. Then, terminal constraint $S_N = \mathbf{I}$ and the LNCS augment state z_k can be represented as $z_k = [x_k \; u_{k-1} \; u_{k-2}]^T \in \Re^{8\times 1}$ and $\phi = [\hat{z} \; u]^T \in \Re^{10\times 1}$. Also, the initial admissible policy can be designed as

$$u_0 = -\begin{bmatrix} 0.87 & 0.85 & -0.1 & 1.24 & 0.03 & 0 & 0.13 & 0.01 \\ -1.51 & 0.09 & -2.55 & 2.47 & 0 & 0.08 & -0.05 & 0.52 \end{bmatrix} \hat{z}_k$$

Moreover, the state-dependent part of the regression functions for stochastic value function estimation is generated as $\{\varphi_1^2, \varphi_1\varphi_2, \varphi_1\varphi_3, \ldots, \varphi_2^2, \ldots, \varphi_{10}^2\}$ and the time-dependent part of the regression function (i.e., $\sigma(\cdot)$) is selected as saturation polynomial time function as $sat\{(N-k)^{10}, (N-k)^9, \ldots, 1\}$. It is important to note the saturation operator for time function is to ensure the magnitude of the time function stays inside a reasonable range such that the AE weights are computable. The learning rate for the stochastic value function estimator is selected as $\alpha_W = 10^{-3}$ when initial parameters are taken as zero.

As shown in Figures 3.3 and 3.4, the proposed stochastic optimal control design can force the LNCS system states and control inputs converge close to zero within a finite time horizon (i.e., $t \in [0, NT_s]$) even when system dynamics and network imperfections are unknown. It is important to note that (1) since the proposed scheme needs to tune the value function estimator, there is a slight overshoot at the beginning of the simulation; (2) due to the tuning parameter, $\alpha_W = 10^{-3}$ is small and according to Equations 3.40 and 3.41, the bounds for both LNCS the system states and value function estimation errors are close to zero after 10 s.

3.2.5.2 Bellman Equation and Terminal Constraint Errors

To evaluate the performance of the proposed scheme further, the Bellman equation and terminal constraint errors are considered. As shown in Figures 3.5 and 3.6, both

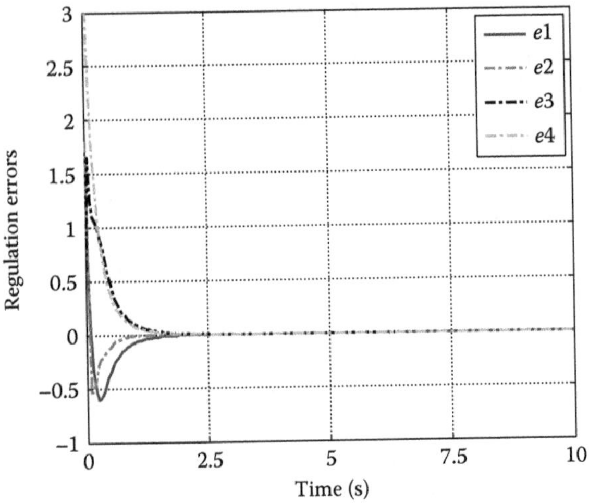

FIGURE 3.3 State regulation error for an NCS.

the Bellman equation and terminal constraint errors converge close to zero during the finite horizon, that is, $t \in [0, NT_s]$, which indicates that the proposed design can converge close to the stochastic LNCS optimal design while satisfying the terminal constraint. Note that convergence is dependent upon the tuning rate, which is given in Theorem 3.2.

Moreover, according to Theorems 3.1 and 3.2, when the final time step N increases, the upper bound on the proposed optimal design estimation error (i.e., $e_{total} = e_{FTBE} + e_{FC}$) will decrease. As shown in Figure 3.7, the proposed optimal estimation error keeps decreasing with time (i.e., NT_s). It is obvious that when the final

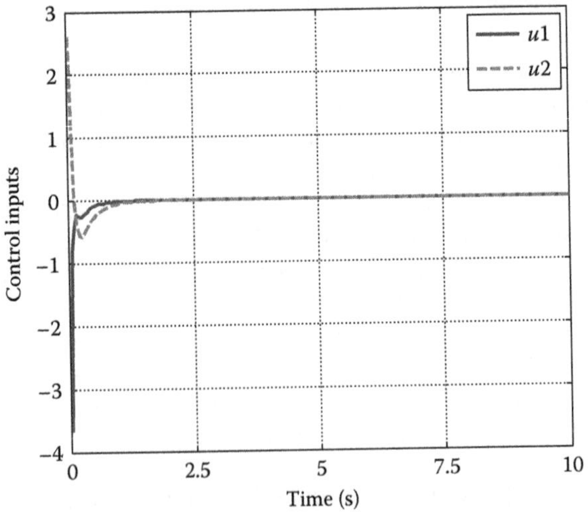

FIGURE 3.4 Estimated LNCS control signals.

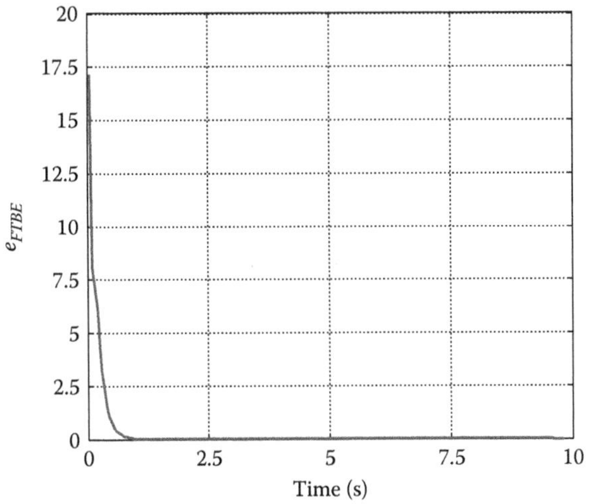

FIGURE 3.5 Bellman error.

time goes to infinity, that is, $N \to \infty$, the optimal estimation error will converge to zero, that is, $e_{total} \to 0$ and the proposed design will converge to the desired infinite-horizon optimal design.

3.2.5.3 Optimality Analysis of the Proposed Scheme

For comparison, a traditional optimal design for an NCS, which solves the SRE (Åstrom 1970; Stengel 1986) backward-in-time with known system dynamics is considered and network imperfections have been incorporated. As shown in Figure 3.8, although the value function of the proposed finite-horizon optimal design is slightly

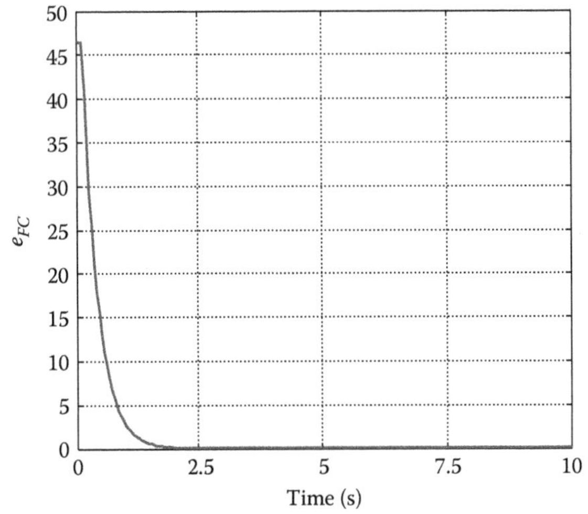

FIGURE 3.6 Terminal constraint error.

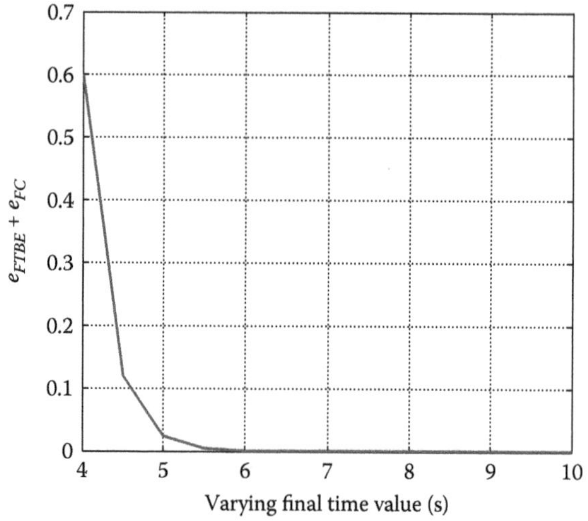

FIGURE 3.7 Effect of final time step N on optimal estimation error.

higher than the traditional optimal design, this result confirms the validity of the proposed finite-horizon optimal adaptive controller design.

Next, according to optimal control theory (Stengel 1986), the value of the state weighting matrix (i.e., Q_z, R_z) will affect the system performance. In Figure 3.9, three different Q_z values have been considered (i.e., $Q_z = 0.1$, $Q_z = 1$, and $Q_z = 5$). When the value of the weighting matrix Q_z increases, the proposed design can force the LNCS system state converge to zero quicker. It is because the large system state weighting

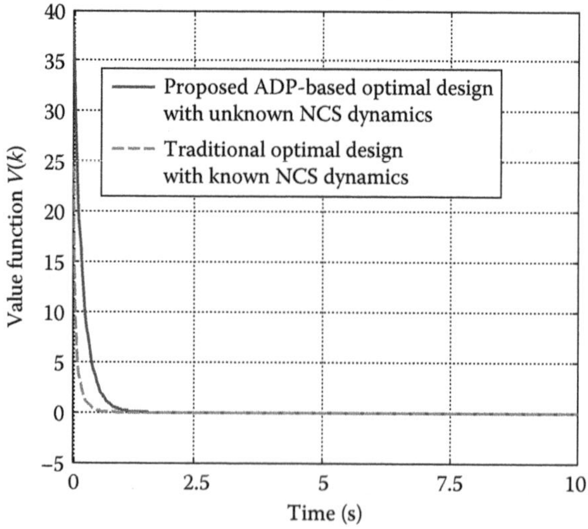

FIGURE 3.8 Comparison of value function between finite-horizon optimal design with unknown NCS dynamics and traditional optimal design with known dynamics.

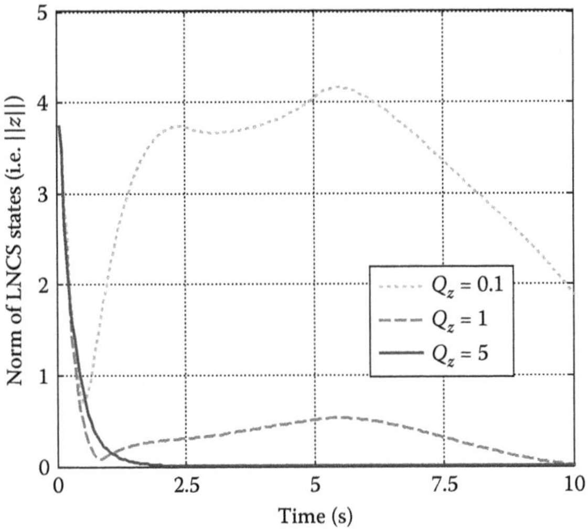

FIGURE 3.9 Effect of weighting matrix Q_z.

matrix Q_z will weigh the system state heavily in the value function. When the proposed stochastic optimal scheme minimizes the value function, the intermediate system state will be smaller due to the larger state weighting matrix Q_z. Moreover, three different R_z values have been considered (i.e., $R_z = 0.1$, $R_z = 1$, and $R_z = 5$) in Figure 3.10. Since the large control input weighting matrix R_z could weigh the control input heavily in the value function, the intermediate control input would be smaller while weighing matrix R_z increases.

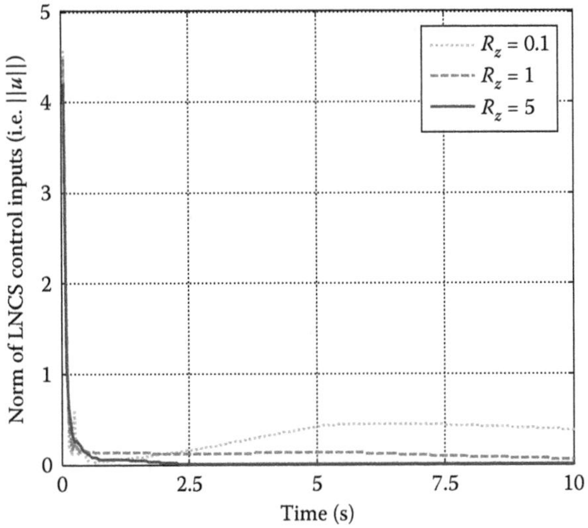

FIGURE 3.10 Effect of weighting matrix R_z.

According to the results shown from Figures 3.3 through 3.10, after a short initial learning time, the proposed stochastic value function estimator-based stochastic optimal control of an LNCS with unknown system dynamics resulting from network imperfections could have nearly the same performance as that of the conventional stochastic optimal control for an LNCS in the presence of known system dynamics and network imperfections.

3.3 EXTENSIONS TO INFINITE HORIZON

It is important to note that infinite-horizon optimal design can be derived when the terminal constraint is ignored and the final time is stretched to infinity, that is, $N \to \infty$. The details of infinite-horizon optimal adaptive design are given as follows.

3.3.1 Adaptive Estimation for Optimal Regulator Design

According to the definition of value function, the relationship between Ψ_k matrix in Equation 3.19 and the stochastic cost function is given as

$$V(x_k) = \underset{\tau,\gamma}{E}\left(\varphi_k^T \overline{\Psi}_k \varphi_k\right) = \underset{\tau,\gamma}{E}\left(\overline{\theta}_k^T \overline{\varphi}_k\right) \tag{3.43}$$

Then, the value function $V(x_k)$ can be approximated by using an AE as

$$\hat{V}(x_k) = \underset{\tau,\gamma}{E}\left(\hat{\overline{\theta}}_k^T \overline{\varphi}_k\right) \tag{3.44}$$

where $\hat{\overline{\theta}}_k$ is the estimate value of the target parameter vector $\hat{\overline{\theta}}_k$ with regressor satisfying $\left\|\overline{\varphi}_k\right\| = 0$ for $\left\|z_k\right\| = 0$.

It is observed that the Bellman equation can be rewritten as $V(x_{k+1}) - V(x_k) + \underset{\tau,\gamma}{E}\left[r(z_k, u_k)\right] = 0$. This relationship, however, is not guaranteed to hold when the estimated matrix $\hat{\overline{\Psi}}_k$ is applied. Hence, using delayed values for convenience, the residual error, e_{hk}, associated with Equation 3.15 can be expressed as $\hat{V}(x_k) - \hat{V}(x_{k-1}) + \underset{\tau,\gamma}{E}\left[r(z_{k-1}, u_{k-1})\right] = \underset{\tau,\gamma}{E}(e_{hk})$, that is,

$$\begin{aligned}
\underset{\tau,\gamma}{E}(e_{hk}) &= \underset{\tau,\gamma}{E}\left[r(z_{k-1}, u_{k-1}) + \hat{\overline{\theta}}_k^T \overline{\varphi}_k - \hat{\overline{\theta}}_k^T \overline{\varphi}_{k-1}\right] \\
&= \underset{\tau,\gamma}{E}\left[r(z_{k-1}, u_{k-1}) + \hat{\overline{\theta}}_k^T (\overline{\varphi}_k - \overline{\varphi}_{k-1})\right] \\
&= \underset{\tau,\gamma}{E}\left[r(z_{k-1}, u_{k-1}) + \hat{\overline{\theta}}_k^T \Delta\varphi_{k-1}\right]
\end{aligned} \tag{3.45}$$

where $\Delta\varphi_{k-1} = \overline{\varphi}_k - \overline{\varphi}_{k-1}$.

The residual dynamics in Equation 3.45 are then rewritten as

$$\underset{\tau,\gamma}{E}(e_{hk+1}) = \underset{\tau,\gamma}{E}\left[r(z_k,u_k)+\hat{\bar{\theta}}_{k+1}^T\Delta\varphi_k\right] \tag{3.46}$$

Next, we define an auxiliary residual error vector as

$$\underset{\tau,\gamma}{E}(\Xi_{hk}) = \underset{\tau,\gamma}{E}\left(\Gamma_{k-1}+\hat{\bar{\theta}}_k^T\Omega_{k-1}\right)\in\Re^{1\times(i+1)} \tag{3.47}$$

where

$$\Gamma_{k-1} = [r(z_{k-1},u_{k-1}) \quad r(z_{k-2},u_{k-2}) \quad \dots \quad r(z_{k-1-i},u_{k-1-j})]$$

and

$$\Omega_{k-1} = [\Delta\varphi_{k-1} \quad \Delta\varphi_{k-2} \quad \dots \quad \Delta\varphi_{k-1-j}],\quad 0<j<k-1\in\mathbb{N}$$

with $\mathbb{N}$ being the set of positive natural numbers. It is important to note that Equation 3.47 indicates a time history of the previous $j+1$ residual errors (3.45) recalculated by using the most recent $\hat{\bar{\theta}}_k$.

The dynamics of the auxiliary vector (3.47) are generated similar to Equation 3.46 and revealed to be

$$\underset{\tau,\gamma}{E}\left(\Xi_{hk+1}\right) = \underset{\tau,\gamma}{E}\left(\Gamma_k+\hat{\bar{\theta}}_{k+1}^T\Omega_k\right) \tag{3.48}$$

Now define the update law of the time-varying matrix $\bar{\Psi}_k$ as

$$\underset{\tau,\gamma}{E}\left(\hat{\bar{\theta}}_{k+1}\right) = \underset{\tau,\gamma}{E}\left[\Omega_k\left(\Omega_k^T\Omega_k\right)^{-1}\left(\alpha_h\Xi_{hk}^T-\Gamma_k^T\right)\right] \tag{3.49}$$

where $0<\alpha_h<1$. Substituting Equation 3.49 into Equation 3.48 results

$$\underset{\tau,\gamma}{E}(\Xi_{hk+1}) = \underset{\tau,\gamma}{E}(\alpha_h\Xi_{hk}) \tag{3.50}$$

It is observed that the stochastic value function $V(x_k)$ and AE (3.44) will become zero only when $z_k = 0$. Hence, when the system states have converged to zero, the value function approximation is no longer updated. It can be seen as a PE requirement for the inputs to the value function estimator wherein the system states must be persistently exciting long enough for the AE to learn the stochastic cost function.

Now define the parameter estimation error to be $\tilde{\bar{\theta}}_k = \bar{\theta}_k - \hat{\bar{\theta}}_k$. Rewrite the Bellman equation using the target AE representation (3.14) revealing

$$\underset{\tau,\gamma}{E}\left(\bar{\theta}_{k+1}^T \bar{w}_k\right) = \underset{\tau,\gamma}{E}\left[r(z_k,u_k) + \bar{\theta}_{k+1}^T \bar{w}_{k+1}\right]$$

which can be expressed as

$$\underset{\tau,\gamma}{E}[r(z_k,u_k)] = \underset{\tau,\gamma}{E}\left[\bar{\theta}_{k+1}^T \varphi_k - \bar{\theta}_{k+1}^T \varphi_{k+1}\right] = -\underset{\tau,\gamma}{E}\left(\bar{\theta}_{k+1}^T \Delta\varphi_k\right) \tag{3.51}$$

Substituting $r(z_k,u_k)$ into Equation 3.46 and utilizing Equation 3.45 with $\underset{\tau,\gamma}{E}(e_{hk+1}) = \underset{\tau,\gamma}{E}(\alpha_h e_{hk})$ from Equation 3.50 yields

$$\underset{\tau,\gamma}{E}\left(\tilde{\bar{h}}_{k+1}^T \Delta W_k\right) = \underset{\tau,\gamma}{E}\left[-\alpha_h r(z_{k-1},u_{k-1}) - \alpha_h \hat{\bar{h}}_k^T \Delta W_{k-1}\right] \tag{3.52}$$

Similar to $r(z_k,u_k)$, we define $\underset{\tau,\gamma}{E}\left[r(z_{k-1},u_{k-1})\right] = -\underset{\tau,\gamma}{E}\left[\tilde{\varphi}_k^T \Delta\varphi_{k-1}\right]$, and substitute this expression into Equation 3.23, to get

$$\underset{\tau,\gamma}{E}\left(\tilde{\bar{\theta}}_{k+1}^T \Delta\varphi_k\right) = \underset{\tau,\gamma}{E}\left(\alpha_h \tilde{\bar{\theta}}_k^T \Delta\varphi_{k-1}\right) \tag{3.53}$$

Next, the convergence of the stochastic cost function estimation error with adaptive estimation error dynamics $\tilde{\bar{\theta}}$ given by Equation 3.53 is demonstrated for an initial admissible control policy. The linear NCS time-varying system dynamics are shown to be asymptotically stable in the mean if an initial admissible control policy can be applied provided the system matrices are known. However, introducing the estimated value function results in estimation errors for the stochastic value function $V(x_k)$, and the stability of the estimated stochastic cost function needs to be studied. Subsequently, the results of Theorem 3.3 will be used for proving the overall closed-loop system stability in Theorem 3.4.

Theorem 3.3: Asymptotic Stability of the Cost AE Errors

Given the initial conditions for the AE parameter vectors $\hat{\bar{h}}_0$ be bounded in the set **S**, let $u_0(z_k)$ be an initial admissible control policy for the LNCS (3). Let the AE parameter update law be given by Equation 3.20. Then, there exists a positive constant α_h satisfying $0 < \alpha_h < 1$ such that the mean value of the adaptive parameter estimator error vector converges to zero asymptotically.

Proof: Consider the positive definite Lyapunov candidate

$$V_J(\tilde{\bar{\theta}}_k) = \underset{\tau,\gamma}{E}\left(\tilde{\bar{\theta}}_k^T \Delta\varphi_{k-1}\right)^2 \tag{3.54}$$

The first difference is given by $\Delta V_J\left(\tilde{\bar{\theta}}_k\right)=\left(\tilde{\bar{\theta}}_{k+1}\Delta\phi_k\right)^2-\left(\tilde{\bar{\theta}}_k\Delta\varphi_{k-1}\right)^2$, and using Equation 3.53 yields

$$\Delta V_J\left(\tilde{\bar{\theta}}_k\right)=\underset{\tau,\gamma}{E}\left(\alpha_h\tilde{\bar{\theta}}_k\Delta\varphi_{k-1}\right)^2-\underset{\tau,\gamma}{E}\left(\tilde{\bar{\theta}}_k\Delta\varphi_{k-1}\right)^2\le-(1-\alpha_h^2)\underset{\tau,\gamma}{E}\left(\tilde{\bar{\theta}}_k\Delta\varphi_{k-1}\right)^2 \tag{3.55}$$

Since $V_J\left(\tilde{\bar{\theta}}_k\right)$ is positive definite and $\Delta V_J\left(\tilde{\bar{h}}_k\right)$ is negative definite (due to PE condition), tuning parameter α_h is selected properly. Therefore, the mean value of the parameter errors converges to zero asymptotically. This implies that $\hat{V}_k\to V_k$ and $\underset{\tau,\gamma}{E}\left(\hat{\bar{\theta}}_k\right)\to 0$ when $k\to\infty$.

Next, we show that the estimated control input based on this estimated matrix will indeed converge to the certainty-equivalent optimal control input.

$$\underset{\tau,\gamma}{E}(\hat{u}_{1k})=-\underset{\tau,\gamma}{E}(\hat{K}_k z_k)=-\underset{\tau,\gamma}{E}\left[\left(\hat{\bar{H}}_k^{uu}\right)^{-1}\hat{\bar{H}}_k^{uz}z_k\right] \tag{3.56}$$

Moreover, the initial system states are considered to reside in the same set as that of the initial stabilizing control input u_{0k}. Further sufficient condition for the AE tuning gain α_h is derived to ensure that all future states will converge to zero. Then it can be shown that the actual control input approaches the optimal control asymptotically in the mean. Table 3.2 presents the proposed design.

Theorem 3.4: Convergence of the Optimal Control Signal

Given the initial conditions for the system state z_0, cost function and AE parameter vectors $\hat{\bar{h}}_0$ be bounded in the set **S**, let u_{0k} be any initial admissible control policy for the NCS (3.12) with random delays and packet losses satisfying the bounds given by Equation 3.37 for $0<l_o<1/2$. Let the AE parameter be tuned and estimation control policy be provided by Equations 3.49 and 3.56, respectively. Then, there exist positive constants α_h given by Theorem 3.1 such that the system states z_k and stochastic cost function parameter estimator errors $\tilde{\bar{h}}_k$ are all asymptotically stable in the mean (Xu et al. 2012). In other words, as $k\to\infty, \underset{\tau,\gamma}{E}(z_k)\to 0, \underset{\tau,\gamma}{E}\left(\tilde{\bar{\theta}}_k\right)\to 0, \hat{V}(x_k)\to V(x_k)$ and $\underset{\tau,\gamma}{E}(\hat{u}_k)\to\underset{\tau,\gamma}{E}(u_k^*)$.

Proof: Refer to Appendix 3D.

Remark 3.4

It is important to note that when the network-induced delay bound is increased in an NCS, the dimension of the augmented state z_k increases and the computational complexity also goes up. While the recent embedded processors can handle the

TABLE 3.2
Infinite-Horizon Stochastic Optimal Design

Compute Infinite-Horizon Control Input

$$\underset{\tau,\gamma}{E}(\hat{u}_{1k}) = -\underset{\tau,\gamma}{E}(\hat{K}_k z_k) = -\underset{\tau,\gamma}{E}\left[\left(\hat{\bar{H}}_k^{uu}\right)^{-1}\hat{\bar{H}}_k^{uz} z_k\right]$$

where $\hat{\bar{H}}_k = \text{vec}^{-1}\left(\hat{\bar{h}}_k\right)$.

Compute Bellman Equation Residual Error

Bellman Equation Residual Error

$$\underset{\tau,\gamma}{E}(\Xi_{hk+1}) = \underset{\tau,\gamma}{E}\left[\Gamma_k + \hat{\bar{\theta}}_{k+1}^T \Omega_k\right]$$

Parameter Update

$$\underset{\tau,\gamma}{E}\left(\hat{\bar{\theta}}_{k+1}\right) = \underset{\tau,\gamma}{E}\left[\Omega_k(\Omega_k^T\Omega_k)^{-1}\left(\alpha_h \Xi_{hk}^T - \Gamma_k^T\right)\right]$$

where α_h is the tuning parameter.

computational complexity to some extent, the delay bound due to the network phenomenon can be reduced by a suitable design of networking protocols as shown in Chapter 10.

3.3.2 Simulation Results

In this section, stochastic suboptimal and optimal control of an NCS is evaluated. At the same time, the standard suboptimal and optimal control of an NCS with known dynamics and network imperfections is also simulated for comparison.

EXAMPLE 3.2

The continuous-time version of a batch reactor system dynamics are given by (Carnevale et al. 2007)

$$\dot{x} = \begin{bmatrix} 1.38 & -0.2077 & 6.715 & -5.676 \\ -0.5814 & -4.29 & 0 & 0.675 \\ 1.067 & 4.273 & -6.654 & 5.893 \\ 0.048 & 4.273 & 1.343 & -2.104 \end{bmatrix} x + \begin{bmatrix} 0 & 0 \\ 5.679 & 0 \\ 1.136 & -3.146 \\ 1.136 & 0 \end{bmatrix} u \tag{3.57}$$

where $x \in \Re^{4\times1}$ and $u \in \Re^{2\times1}$. It is important to note that this example has developed over the years as a benchmark example for an NCS (see, e.g., Walsh et al. 2002; Carnevale et al. 2007).

The parameters for an NCS are selected as follows:

1. The sampling time: $T_s = 100$ ms.
2. The delay bound is two, that is, $\bar{d} = 2$.

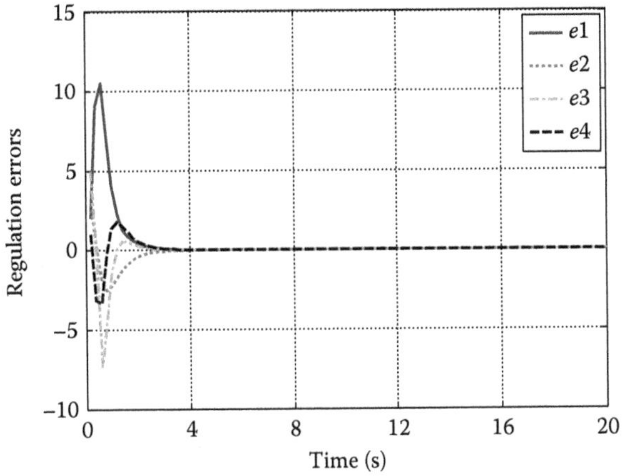

FIGURE 3.11 Performance of conventional stochastic optimal controller.

3. The mean random delay values are $E(\tau_{sc}) = 80$ ms, $E(\tau) = 150$ ms.
4. Packet losses follow Bernoulli distribution with $p = 0.3$.
5. To obtain the ergodic behavior, Monte Carlo simulations were run with 1000 iterations.
6. The initial state vector is taken to be deterministic.

First, Figure 3.11 indicates the stochastic optimal control of an NCS with known dynamics and information on network imperfections (e.g., random delays and packet losses) obtained by solving the SRE backward-in-time. The controllers can force the state regulation errors converge while ensuring the NCS is stable when the delays and packet losses are accurately known.

Next, the ADP VI control (Al-Tamimi et al. 2007) input

$$u_k = -\begin{bmatrix} 0.801 & 0.868 & -0.242 & 1.377 \\ -1.150 & 0.035 & -1.961 & 1.855 \end{bmatrix} x_k$$

is designed by using policy iteration scheme. This ADP VI controller though does not require system dynamics and cannot maintain the batch reactor system stable in the presence of random delays and packet losses as shown in Figure 3.12.

Finally, the adaptive stochastic optimal and stochastic suboptimal controller designs are implemented for an NCS with unknown system dynamics in the presence of random delays and packet losses. The augment state z_k is generated as $z_k = [x_k \quad u_{k-1} \quad u_{k-2}]^T \in \Re^{8\times 1}$ or $w = [z \quad u] \in \Re^{10\times 1}$. The initial stabilizing policy for the algorithm was selected as

$$u_0(z_k) = -\begin{bmatrix} 0.88 & 0.77 & -0.11 & 1.07 & 0.25 & 0.01 & 0.14 & 0.02 \\ -1.65 & -0.08 & -2.93 & 2.61 & -0.02 & 0.68 & -0.03 & 0.51 \end{bmatrix} z_k$$

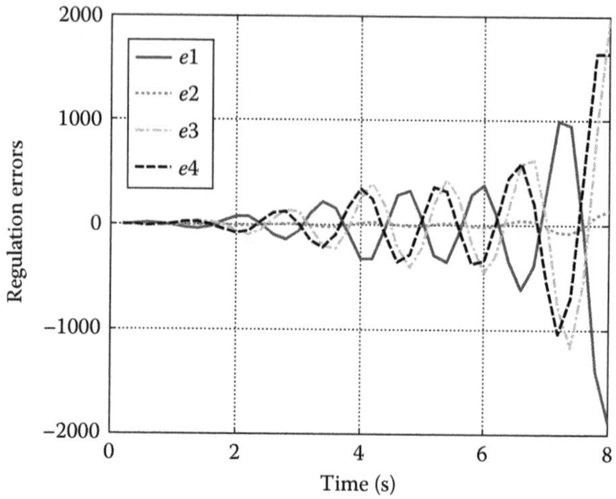

FIGURE 3.12 State regulation errors of ADP VI controller when random delays and packet losses are present.

while the regression function for value function was generated as $\{w_1^2, w_1w_2, w_1w_3,\ldots, w_2^2,\ldots, w_9^2,\ldots, w_{10}^2\}$ as per Equation 3.25.

The design parameter for value function $V(z_k,u_k)$ was selected as $\alpha_h = 10^{-6}$ while initial parameters for the AE were set to zero at the beginning of the simulation. The initial parameters of the action control network were chosen to reflect the initial stabilizing control. The simulation was run for 200 times steps, and for the first 50 times steps, exploration noise with mean zero and variance 0.006 was added to the system at odd time steps and exploration noise with mean zero and variance 0.003 was added to the system at even time steps in order to ensure the PE condition holds.

In Figures 3.13 and 3.14, the performance of the proposed AE-based optimal controller is evaluated. As shown in Figure 3.13a, the AE-based optimal controller can also make the mean value of the NCS state regulation errors converge to zero even when the NCS dynamics are unknown, which implies that the proposed controller can make the NCS closed-loop system stable. The cost-to-go function of the proposed optimal and suboptimal controllers is compared in Figure 3.13b where the proposed AE-based optimal controller can minimize the cost-to-go $(V(x_k) = \underset{\tau,\gamma}{E}[\sum_{m=k}^{\infty}(z_m^TQ_zz_m + u_m^TR_zu_m)])$ function more than the proposed suboptimal controller based on the certainty-equivalence deterministic NCS model.

This clearly shows that the AE-based optimal controller is more effective than the suboptimal control based on certainty-equivalence deterministic NCS representation. Now in Figure 3.14a and b, the control inputs of the proposed AE-based optimal and suboptimal controllers are compared. The proposed AE-based optimal controller can force the NCS states converge quicker than the suboptimal control based on the certainty-equivalence deterministic NCS model. The proposed suboptimal controller has a smaller overshoot initially when compared to the stochastic optimal controller.

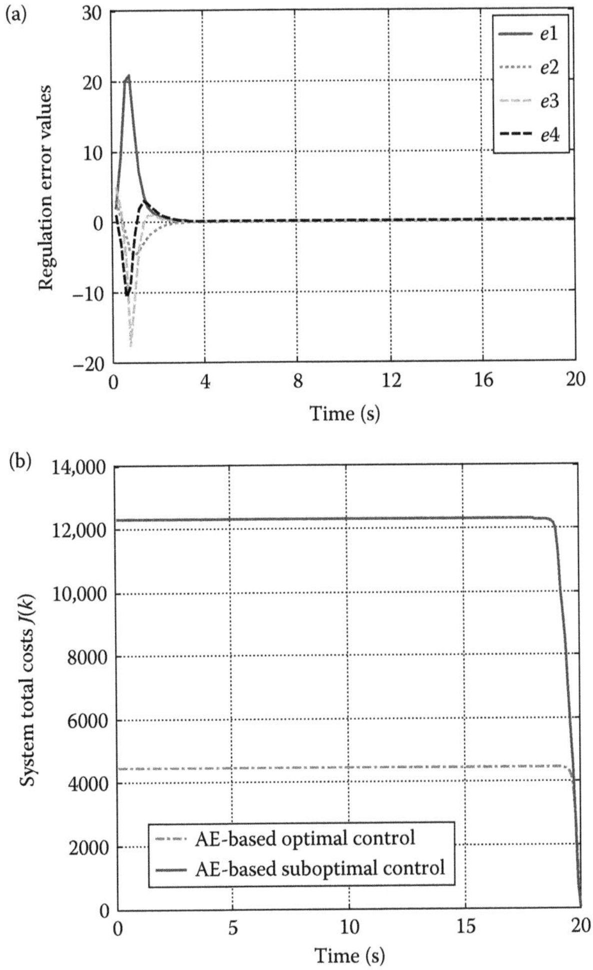

FIGURE 3.13 Performance of suboptimal and optimal controller for an NCS with unknown dynamics: (a) State regulation errors with AE-based optimal control. (b) Comparison of system costs with AE-based optimal and suboptimal controllers.

According to the above results that are depicted in Figures 3.11 through 3.14, the performance of AE-based stochastic optimal and suboptimal controllers for an NCS with uncertain dynamics has nearly the same performance of stochastic optimal/suboptimal control with known system dynamics. The slightly higher overshoot observed at the beginning for the proposed optimal/suboptimal controller is due to an initial online learning phase needed to tune the optimal/suboptimal controller. After a short time, the proposed AE-based stochastic optimal and suboptimal controllers will have a similar performance even when NCS system dynamics due to network imperfections are unknown.

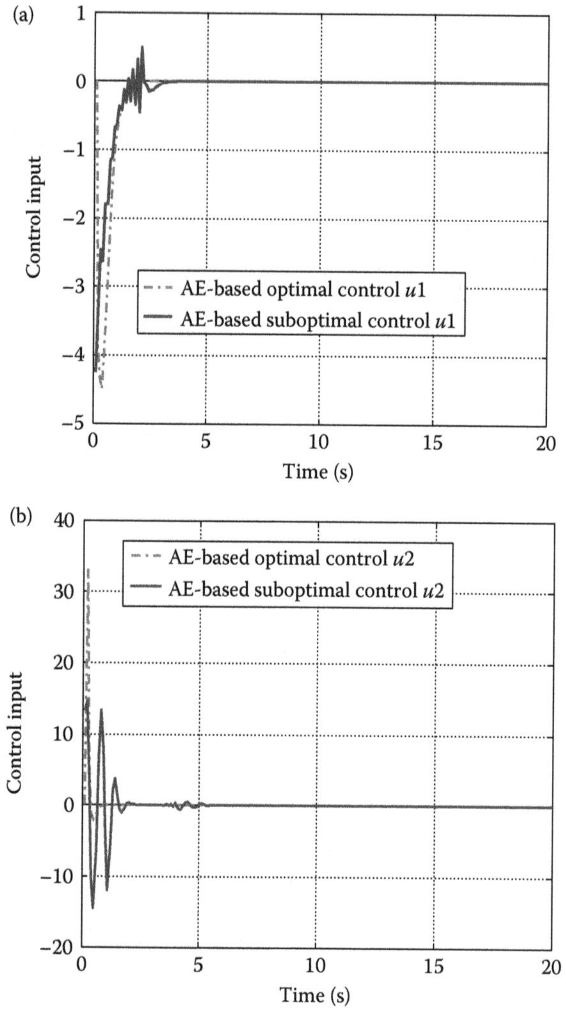

FIGURE 3.14 Comparison of control inputs with stochastic optimal and suboptimal controllers $u = (u_1 u_2)^T \in \mathbb{R}^{2\times 1}$.

3.4 CONCLUSIONS

This chapter presented both finite- and infinite-horizon optimal adaptive control of linear systems in the presence of network imperfections, mainly random delays and packet losses. The incorporation of delays and packet losses results in an augmented stochastic linear time-varying system with uncertain dynamics. The traditional RE-based solution is not suitable since the system dynamics are needed while the proposed optimal adaptive control solution based on the Bellman equation and terminal constraint errors result in satisfactory performance as demonstrated by simulation studies.

PROBLEMS

Section 3.2

Problem 3.2.1: Repeat Example 3.1 with delay bound greater than 2.

Problem 3.2.2: Consider the linearized version of the power system dynamics given by

$$\dot{x} = \begin{bmatrix} -0.665 & 11.5 & 0 & 0 \\ 0 & -2.5 & 2.5 & 0 \\ -9.5 & 0 & -13.736 & -13.736 \\ 0.6 & 0 & 0 & 0 \end{bmatrix} x + \begin{bmatrix} 0 \\ 0 \\ 13.136 \\ 0 \end{bmatrix} u$$

Using the new dynamics and information from Example 3.1, repeat the simulation results.

Problem 3.2.3: The networked control of inverted pendulum system with random delays and packet losses is given

$$\begin{bmatrix} \dot{x} \\ \ddot{x} \\ \dot{\phi} \\ \ddot{\phi} \end{bmatrix} = \begin{bmatrix} 0 & 1 & 0 & 0 \\ 0 & -0.1818 & 2.6727 & 0 \\ 0 & 0 & 0 & 1 \\ 0 & -0.4545 & 31.1818 & 0 \end{bmatrix} \begin{bmatrix} x \\ \dot{x} \\ \phi \\ \dot{\phi} \end{bmatrix} + \begin{bmatrix} 0 \\ 1.8182 \\ 0 \\ 4.5455 \end{bmatrix} u \tag{2.75}$$

The parameters of this system are given as

1. The sampling time: $T_s = 35$ ms.
2. The bound of delays is two, that is, $\bar{d} = 2$.
3. The random delay: $(\tau_1) = 30$ ms, $E(\tau_2) = 60$ ms.
4. Packet losses follow Bernoulli distribution with $p = 0.3$.

Apply the proposed stochastic Q-learning and plot the system response under perfect measurement assumption and no process noise.

Section 3.3

Problem 3.3.1: Repeat Example 3.2 with delay bound greater than 2.

Problem 3.3.2: Repeat Problem 3.2.2 with infinite time horizon.

Problem 3.3.3: Repeat Problem 3.2.3 with infinite time horizon.

Problem 3.3.4: Repeat Problem 3.2.3 with infinite time horizon and different initial weights.

APPENDIX 3A

Proof of Lemma 3.1

Without loss of generality, we consider the Lyapunov function $L(x_k)$ at time instant k. Then, the summation term $\sum_{i=0}^{k-1} \Delta L(x_i)$ can be expressed as

$$\sum_{i=0}^{k-1} \Delta L(x_i) = \underbrace{L(x_k) - L(x_{k-1})}_{\Delta L(x_{k-1})} + \cdots + \underbrace{L(x_1) - L(x_0)}_{\Delta L(x_0)} = L(x_k) - L(x_0) \quad (3A.1)$$

Using Equations 3.14 and 3A.1, the Lyapunov function at time k can be represented as

$$L(x_k) = L(x_0) + \sum_{i=0}^{k-1} \Delta L(x_i) \leq L(x_0) - \sum_{i=0}^{k-1} \alpha L(x_i)$$

$$\leq L(x_0) - \left[\sum_{i=0}^{k-1} \alpha(1-\alpha)^i\right] L(x_0) \quad (3A.2)$$

Then, according to Brockett et al. (1983), the summation of geometric sequence $\sum_{i=0}^{k-1} \alpha(1-\alpha)^i$ can be calculated as $\sum_{i=0}^{k-1} \alpha(1-\alpha)^i = 1-(1-\alpha)^k$. Further, Equation 3A.2 can be derived as

$$L(x_k) \leq L(x_0) - [1-(1-\alpha)^k] L(x_0) \leq (1-\alpha)^k L(x_0) \quad (3A.3)$$

Since the initial Lyapunov function $L(x_0)$ is bounded and $0 < \alpha < 1$, the term $B_{L,k}°(1 - \alpha)^k L(x_0)$ will also be bounded and decreases if time instant k is increased. Therefore, according to Equation 3A.3, for any time kT_s, the Lyapunov function $L(x_k)$, $\forall k = 1,2,\ldots$ will be bounded with ultimate bound $B_{L,k}$. When time goes to infinity, that is, $k \to \infty$, $L(x_k)$ and $B_{L,k}$ will also converge to zero *exponentially* since$0< 1 - \alpha<$ 1 and $(1 - \alpha)^k \to 0$ as $k \to \infty$.

APPENDIX 3B

Proof of Theorem 3.1

Consider the positive definite Lyapunov function candidate as

$$L_V\left(\tilde{W}_k, k\right) = \underset{\tau,\gamma}{E} \left\|\tilde{W}_k^T \Theta(z, k-1)\right\|^2 \quad (3B.1)$$

where $\Theta(z, k-1)$ is given by Equation 3.30. Since the Lyapunov function candidate (3B.1) is a nonautonomous function and $\Theta(z, k-1)$ is bounded, the Lyapunov function candidate satisfies the inequality $L_{V0}(\tilde{W}_k) \le L_V(\tilde{W}_k, k) \le L_{V1}(\tilde{W}_k)$ where $L_{V1}\left(\tilde{W}_k\right) = \underset{\tau,\gamma}{E}\left\|\tilde{W}_k^T\Theta_{\max}(z)\right\|^2 > 0$ and $L_{V0}\left(\tilde{W}_k\right) = \underset{\tau,\gamma}{E}\left\|\tilde{W}_k^T\Theta_{\min}(z)\right\|^2 > 0$, with positive definite time-invariant functions $L_{V0}(\tilde{W}_k)$ and $L_{V1}(\tilde{W}_k)$. According to Sarangapani (2006), time-varying nonautonomous $L_V(\tilde{W}_k, k)$ is positive definite (i.e., $L_{V0}(\tilde{W}_k) \le L_V(\tilde{W}_k, k)$) and decrescent (i.e., $L_V(\tilde{W}_k, k) \le L_{V1}(\tilde{W}_k)$).

Using Equation 3.36, the first difference of Equation 3B.1 can be expressed as

$$
\begin{aligned}
\Delta L_V(\tilde{W}_k, k) &= \underset{\tau,\gamma}{E}\left\|\tilde{W}_{k+1}^T\Theta(z,k)\right\|^2 - \underset{\tau,\gamma}{E}\left\|\tilde{W}_k^T\Theta(z,k-1)\right\|^2 \\
&\le \underset{\tau,\gamma}{E}\left\|\tilde{W}_{k+1}^T[\Delta\varphi_k - \sigma(0)]\right\|^2 - \underset{\tau,\gamma}{E}\left\|\tilde{W}_k^T[\Delta\varphi_{k-1} - \sigma(0)]\right\|^2 \\
&\le \alpha_W^2 \underset{\tau,\gamma}{E}\left\|\tilde{W}_k^T[\Delta\varphi_{k-1} - \sigma(0)]\right\|^2 - \underset{\tau,\gamma}{E}\left\|\tilde{W}_k^T[\Delta\varphi_{k-1} - \sigma(0)]\right\|^2 \\
&\le -(1-\alpha_W^2)\underset{\tau,\gamma}{E}\left\|\tilde{W}_k^T[\Delta\varphi_{k-1} - \sigma(0)]\right\|^2 \\
&\le -(1-\alpha_W^2)\underset{\tau,\gamma}{E}\left\|\tilde{W}_k^T\Theta(z,k-1)\right\|^2 \\
&\le -(1-\alpha_W^2)\underset{\tau,\gamma}{E}\left\|\tilde{W}_k^T\Theta_{\min}(z)\right\|^2 \equiv -(1-\alpha_W^2)L_{V0}(\tilde{W}_k) < 0
\end{aligned}
\tag{3B.2}
$$

Since $0 < \alpha_W < 1$, $\Delta L_V\left(\tilde{W}_k, k\right)$ is negative definite and $L_V\left(\tilde{W}_k, k\right)$ is positive definite (Sarangapani 2006).

APPENDIX 3C

Proof of Theorem 3.2

Consider the positive definite Lyapunov function candidate as

$$L_k = L_z(z_k) + L_V(\tilde{W}_k, k) \tag{3C.1}$$

where $L_z(z_k)$ is given as $L_z(z_k) = \underset{\tau,\gamma}{E}(z_k^T\Pi z_k)$ with $\Pi = \left(\left(R_z(1-\alpha_W^2)\right)/\left(2B_{zM}^2\right)\right)I$ a positive definite matrix and **I** an identity matrix, $\|B_{zk}\| \le B_{zM}$ with B_{Zm} being positive constant, and time-varying nonautonomous function $L_V\left(\tilde{W}_k, k\right)$ is defined as Equation 3B.1. Similarly to Theorem 3.1, $L_V\left(\tilde{W}_k, k\right)$ can be bounded as $L_{V0}\left(\tilde{W}_k\right) \le L_V\left(\tilde{W}_k, k\right) \le L_{V1}\left(\tilde{W}_k\right)$ with positive definite time-invariant functions $L_{V0}\left(\tilde{W}_k\right)$ and $L_{V1}\left(\tilde{W}_k\right)$. According to Sarangapani (2006), time-varying nonautonomous function $L_V(\tilde{W}_k, k)$ is positive definite (i.e., $L_{V0}(\tilde{W}_k) \le L_V(\tilde{W}_k, k)$) and decrescent (i.e., $L_V\left(\tilde{W}_k\right) \le L_{V1}\left(\tilde{W}_k, k\right)$). Then, the first difference of Equation 3C.1 can be represented as $\Delta L_k = \Delta L_z(z_k) + \Delta L_V\left(\tilde{W}_k, k\right)$, and considering the first part $\Delta L_z(z_k) = \underset{\tau,\gamma}{E}\left(z_{k+1}^T \Pi z_{k+1}\right) - \underset{\tau,\gamma}{E}\left(z_k^T \Pi z_k\right)$ by applying the Cauchy–Schwartz inequality and Lemma 3.2, we have

$$
\begin{aligned}
\Delta L_z(z_k) &= \underset{\tau,\gamma}{E}(z_{k+1}^T \Pi z_{k+1}) - \underset{\tau,\gamma}{E}(z_k^T \Pi z_k) \\
&\le \|\Pi\| \underset{\tau,\gamma}{E} \left\| A_{zk} z_k + B_{zk} u_k^* - B_{zk} u_k^* + B_{zk} \hat{u}_k \right\|^2 - \|\Pi\| \underset{\tau,\gamma}{E} \|z_k\|^2 \\
&\le 2\|\Pi\| \underset{\tau,\gamma}{E} \left\| A_{zk} z_k + B_{zk} u_k^* \right\|^2 + 2\|\Pi\| \underset{\tau,\gamma}{E} \left\| B_{zk} \tilde{u}_k \right\|^2 - \|\Pi\| \underset{\tau,\gamma}{E} \|z_k\|^2 \qquad (3C.2) \\
&\le -(1-2l_o)\|\Pi\| \underset{\tau,\gamma}{E} \|z_k\|^2 + 2\|\Pi\| \underset{\tau,\gamma}{E} \left\| B_{zk} \tilde{u}_k \right\|^2
\end{aligned}
$$

Next, using Theorem 3.1 and Equation 3C.2, ΔL can be expressed as

$$
\begin{aligned}
\Delta L_k &= \Delta L_z(z_k) + \Delta L_V(\tilde{W}_k, k) \\
&\le \underset{\tau,\gamma}{E}(z_{k+1}^T \Pi z_{k+1}) - \underset{\tau,\gamma}{E}(z_k^T \Pi z_k) + \underset{\tau,\gamma}{E} \left\| \tilde{W}_{k+1} \Theta(z,k) \right\|^2 - \underset{\tau,\gamma}{E} \left\| \tilde{W}_k \Theta(z,k-1) \right\|^2 \\
&\le -(1-2l_o)\|\Pi\| \underset{\tau,\gamma}{E} \|z_k\|^2 + 2\|\Pi\| \underset{\tau,\gamma}{E} \left\| B_{zk} \tilde{u}_k \right\|^2 \\
&\quad + \alpha_W^2 \underset{\tau,\gamma}{E} \left\| \tilde{W}_k^T [\Delta\varphi(z_{k-1}, k-1) - \sigma(0)] \right\|^2 \\
&\quad - \underset{\tau,\gamma}{E} \left\| \tilde{W}_k^T [\Delta\varphi(z_{k-1}, k-1) - \sigma(0)] \right\|^2 \qquad (3C.3) \\
&\le -(1-2l_o)\|\Pi\| \underset{\tau,\gamma}{E} \|z_k\|^2 + \frac{1}{2}(1-\alpha_W^2) \underset{\tau,\gamma}{E} \left\| \tilde{W}_k \Theta(z,k-1) \right\|^2 \\
&\quad + \alpha_W^2 \underset{\tau,\gamma}{E} \left\| \tilde{W}_k^T \Theta(z,k-1) \right\|^2 - \underset{\tau,\gamma}{E} \left\| \tilde{W}_k^T \Theta(z,k-1) \right\|^2 \\
&\le -(1-2l_o)\|\Pi\| \underset{\tau,\gamma}{E} \|z_k\|^2 - \frac{1}{2}(1-\alpha_W^2) \underset{\tau,\gamma}{E} \left\| \tilde{W}_k \Theta(z,k-1) \right\|^2 \\
&\le -(1-2l_o)\|\Pi\| \underset{\tau,\gamma}{E} \|z_k\|^2 - \frac{1}{2}(1-\alpha_W^2) \underset{\tau,\gamma}{E} \left\| \tilde{W}_k \Theta_{\min}(z) \right\|^2 \\
&\equiv -(1-2l_o)\|\Pi\| \underset{\tau,\gamma}{E} \|z_k\|^2 - L_{V0}(\tilde{W}_k)
\end{aligned}
$$

Since $0 < l_o < 1/2$ provided by Lemma 3.2 and $0 < \alpha_W < 1$ given by Theorem 3.1, ΔL_k is negative definite and L_k is positive definite (Sarangapani 2006).

APPENDIX 3D

Proof of Theorem 3.4

Consider the positive definite Lyapunov function candidate

$$V = V_D(z_k) + V_J\left(\tilde{\bar{h}}_k\right) \tag{3D.1}$$

where $V_D(z_k)$ is defined as $V_D(z_k) = z_k^T z_k$ and $V_J\left(\tilde{\bar{\theta}}_k\right)$ is defined as

$$V_J\left(\tilde{\bar{\theta}}_k\right) = \mathop{E}_{\tau,\gamma}\left(\tilde{\bar{\theta}}_k \bar{\varphi}_k - \tilde{\bar{\theta}}_k \bar{\varphi}_{k-1}\right)^2 = \mathop{E}_{\tau,\gamma}\left(\tilde{\bar{\theta}}_k \Delta\varphi_{k-1}\right)^2 \tag{3D.2}$$

The first difference of Equation 3D.2 can be expressed as $\Delta V = \Delta V_D(z_k) + \Delta V_J\left(\tilde{\bar{\theta}}_k\right)$, and considering that $\Delta V_J\left(\tilde{\bar{\theta}}_k\right) = \mathop{E}_{\tau,\gamma}\left(\tilde{\bar{\theta}}_{k+1}^T \Delta\varphi_k\right)^2 - \mathop{E}_{\tau,\gamma}\left(\tilde{\bar{\theta}}_{k+1}^T \Delta\varphi_{k-1}\right)^2$ with the AE, we have

$$\begin{aligned}
\Delta V_J\left(\tilde{\bar{\theta}}_k\right) &= \mathop{E}_{\tau,\gamma}\left(\tilde{\bar{\theta}}_{k+1}^T \Delta\varphi_k\right)^2 - \mathop{E}_{\tau,\gamma}\left(\tilde{\bar{\theta}}_k^T \Delta\varphi_{k-1}\right)^2 \\
&= \mathop{E}_{\tau,\gamma}\left(\alpha_h \tilde{\bar{\theta}}_k^T \Delta\varphi_{k-1}\right)^2 - \mathop{E}_{\tau,\gamma}\left(\tilde{\bar{\theta}}_k^T \Delta\varphi_{k-1}\right)^2 \\
&= -(1-\alpha_h^2) \mathop{E}_{\tau,\gamma}\left(\tilde{\bar{\theta}}_k^T \Delta\varphi_{k-1}\right)^2 \\
&\le -\left(1-\alpha_h^2\right) \mathop{E}_{\tau,\gamma}\left[\left\|\Delta\varphi_{k-1}\right\|^2 \left\|\tilde{\bar{\theta}}_k\right\|^2\right]
\end{aligned} \tag{3D.3}$$

Next, considering the first part $\Delta V_D(z_k) = z_{k+1}^T z_{k+1} - z_k^T z_k$ and applying the NCS and Cauchy–Schwartz inequality reveals

$$\begin{aligned}
\Delta V_D(z_k) &\le \mathop{E}_{\tau,\gamma}\left\|A_{zk} z_k + B_{zk} u_k - B_{zk}\tilde{u}_k\right\|^2 - \mathop{E}_{\tau,\gamma}\left(z_k^T z_k\right) \\
&\le 2\mathop{E}_{\tau,\gamma}\left\|A_{zk} z_k + B_{zk} u_k\right\|^2 + 2\mathop{E}_{\tau,\gamma}\left\|B_{zk}\tilde{u}_k\right\|^2 - \mathop{E}_{\tau,\gamma}\left(z_k^T z_k\right)
\end{aligned} \tag{3D.4}$$

Applying the Lemma 3.2 (bounds on the optimal closed loop system in Equation 3.37), we have

$$\begin{aligned}
\Delta V_D(z_k) &\le -(1-2l_o)\mathop{E}_{\tau,\gamma}\left\|z_k\right\|^2 + 2\mathop{E}_{\tau,\gamma}\left\|B_{zk}\tilde{u}_k\right\|^2 \le -(1-2k^*)\mathop{E}_{\tau,\gamma}\left\|z_k\right\|^2 \\
&\quad + 2B_M^2 \mathop{E}_{\tau,\gamma}\left\|-\left(\hat{\bar{\Psi}}_k^{uu}\right)^{-1}\hat{\bar{\Psi}}_k^{uz} z_k + \frac{1}{2}R^{-1}B_{zk}^T \frac{\partial \hat{V}(x_{k+1})}{\partial z_{k+1}}\right\|^2 \\
&\le -(1-2k^*)\mathop{E}_{\tau,\gamma}\left\|z_k\right\|^2
\end{aligned} \tag{3D.5}$$

At final step, combining the Equations 3D.5 and 3D.3, we have

$$\Delta V \le -(1-2l_o)\underset{\tau,\gamma}{E}\|z_k\|^2 - \left(1-\alpha_h^2\right)\underset{\tau,\gamma}{E}\left[\|\Delta\varphi_{k-1}\|^2\left\|\tilde{\bar{\theta}}_k\right\|^2\right] \quad (3D.6)$$

Since $0 < l_o < 1/2$ and $0 < \alpha_h < 1$, ΔV is negative definite and V is positive definite. Note that $\left|\sum_{k=k_0}^{\infty}\Delta V_k\right| = \left|V_\infty - V_0\right| < \infty$ since $\Delta V < 0$ as long as Equation 3D.6 holds. Therefore, system states z_k and $\tilde{\bar{h}}_k$ are all asymptotically stable in the mean. In other words, as $k \to \infty, \underset{\tau,\gamma}{E}(z_k) \to 0, \underset{\tau,\gamma}{E}\left(\tilde{\bar{\theta}}_k\right) \to 0, \hat{V}(x_k) \to V(x_k)$. Moreover, $\underset{\tau,\gamma}{E}(\hat{u}_k) \to \underset{\tau,\gamma}{E}(u_k^*)$ when $\hat{V}(x_k) \to V(x_k)$.

REFERENCES

Al-Tamimi, A., Lewis, F.L., and Abu-Khalaf, M. 2007. Model-free Q-learning designs for linear discrete-time zero-sum games with application to H-infinite control. *Automatica*, 43, 473–481.

Åstrom, K.J. 1970. *Introduction to Stochastic Control Theory*. Academic Press, New York.

Barto, A.G., Sutton, R.S., and Anderson, C.W. 1983. Neuron-like adaptive elements that can solve difficult learning control problems. *IEEE Transactions on Systems, Man, and Cybernetics, Part B: Cybernetics*, 13, 834–846.

Brockett, R.W., Millman, R.S., and Sussmann, H.J. 1983. *Differential Geometric Control Theory*. Birkhauser, USA.

Carnevale, D., Teel, A.R., and Nesic, D. 2007. A Lyapunov proof of improved maximum allowable transfer interval for networked control systems. *IEEE Transactions on Automatic Control*, 52, 892–897.

Chen, C.T. 1999. *Linear System Theory and Design*. 3rd edition, Oxford University Press, Oxford.

Dierks, T. and Jagannathan, S. 2012. Online optimal control of affine nonlinear discrete-time systems with unknown internal dynamics by using time-based policy update. *IEEE Transactions on Neural Networks and Learning Systems*, 23, 1118–1129.

Goldsmith, A. 2003. *Wireless Communication*. Cambridge University Press, Cambridge, UK.

Halevi, Y. and Ray, A. 1988. Integrated communication and control systems: Part I—Analysis. *Journal of Dynamic Systems, Measurement, and Control*, 110, 367–373.

Heydari, A. and Balakrishnan, S.N. 2011. Finite-horizon input-constrained nonlinear optimal control using single network adaptive critics. *Proceedings of the American Control Conference*, San Francisco, USA, 3047–3052.

Hu, S.S. and Zhu, Q.X. 2003. Stochastic optimal control and analysis of stability of networked control systems with long delay. *Automatica*, 39, 1877–1884.

Lewis, F.L. and Syrmos, V.L. 1995. *Optimal Control*. 2nd ed., Wiley, New York.

Luck, R. and Ray, A. 1990. An observer-based compensator for distributed delays. *Automatica*, 26, 903–908.

Mahmoud, M.S. and Ismail, A. 2005. New results on delay-dependent control of time-delay systems. *IEEE Transactions on Automatic Control*, 50, 95–100.

Nilsson, J., Bernhardsson, B., and Wittenmark, B. 1998. Stochastic analysis and control of real-time systems with random time delays. *Automatica*, 1, 57–64.

Sarangapani, J. 2006. *Neural Network Control of Nonlinear Discrete-Time Systems*. CRC Press, Boca Raton, Florida.

Stengel, R.F. 1986, *Stochastic Optimal Control: Theory and Application*. Wiley-Interscience, New York.

Walsh, G.C., Ye, H., and Bushnell, L.G., 2002. Stability analysis of networked control systems. *IEEE Transactions on Control Systems Technology*, 10(3), 438–446.

Wang, F.Y., Jin, N., Liu, D., and Wei, Q.L. 2011. Adaptive dynamic programing for finite horizon optimal control of discrete-time nonlinear system of ε-error bound. *IEEE Transactions on Neural Networks and Learning Systems*, 22, 24–36.

Werbos, P.J. 1983. A menu of designs for reinforcement learning over time. *Journal of Neural Networks Control*, 3, 835–846.

Xu, H. and Jagannathan, S., 2012. Stochastic optimal controller design for uncertain nonlinear networked control system via neuro dynamic programming. *IEEE Transactions on Neural Networks and Learning Systems*, 24, 471–484.

Xu, H. and Jagannathan, S. 2013. Finite-horizon stochastic optimal control of uncertain linear networked control systems. *Proceedings of the IEEE Symposium on Adaptive Dynamic Programming and Reinforcement Learning*, Singapore, pp. 24–30, April 2013.

Xu, H., Jagannathan, S., and Lewis, F.L. 2012. Stochastic optimal control of unknown linear networked control system in the presence of random delays and packet losses. *Automatica*, 48, 1017–1030.

Xu, H., Jagannathan, S., and Lewis, F.L. 2014. Stochastic optimal output feedback designs for unknown linear discrete-time system zero-sum games under communication constraints. *Asian Journal of Control*, 16(5), 1263–1276.

Zhang, W., Branicky, M.S., and Phillips, S. 2001. Stability of networked control systems. *IEEE Control Systems Magazine*, 21, 84–99.

4 Optimal Control of Unknown Quantized Network Control Systems

In traditional feedback control systems, it is quite common to assume that the measured signals are transmitted to the controller and the control inputs are delivered back to the plant with arbitrarily high precision. However, in practice, the interface between the plant and the controller is often connected via analog-to-digital (A/D) and digital-to-analog (D/A) devices, which quantize the signals. In an NCS, the signals are quantized prior to transmission onto the communication network. As a result, the quantized networked control system (QNCS) has attracted a great deal of attention from control researchers since the quantization process always existed in computer-based control systems but now it exists in a communication network within the feedback loop.

In the earlier literature, the study of the effect of quantization in feedback control systems was normally categorized based on whether or not the quantizer is static or dynamic. The static quantizer, for which the quantization region does not change with time, was first analyzed for unstable linear systems by Delchamps (1990) by means of quantized state feedback. Later, Elia and Mitter (2001) pointed out that logarithmic quantizers are preferred.

In the case of the dynamic quantizer, for which the quantization region can be adjusted over time based on the idea of scaling quantization levels, Brockett and Liberzon (2000) addressed a hybrid quantized control methodology for feedback stabilization for both continuous and linear discrete-time systems while demonstrating globally asymptotic stability. Liberzon (2003 introduced a "zoom" parameter to extend the idea of changing the sensitivity of the quantizer to both linear and nonlinear systems. For these systems, however, stabilization of the closed-loop system in the presence of quantization is the major issue that was addressed when the system dynamics are known, whereas the quantization effects in the presence of uncertain system dynamics and optimal control of such systems are not yet considered.

As mentioned in Chapter 1, traditional optimal control theory (Stengel 1986) addresses the linear quadratic regulation (LQR) of linear discrete-time systems for both finite and infinite time horizon in an offline and backward-in-time manner, provided the linear system dynamics are known beforehand. In the past couple of decades, significant effort has been made to obtain optimal control in the absence of system dynamics in a forward-in-time manner by using adaptive dynamic programming (ADP) schemes (Werbos 1983; Watkins 1989; Si et al. 2004). Normally, to relax the system dynamics and attain optimality, the ADP schemes use policy and/or value iterations to solve the Bellman or Hamilton–Jacobi–Bellman (HJB) equation to generate infinite-horizon-based adaptive optimal control (Bradtke and Ydsie 1994; Chen and Jagannathan 2008). To overcome the significant number of iterations within each sampling interval in the iterative-based schemes for convergence, Dierks and Jagannathan (2012) introduced a time-driven ADP to generate infinite-horizon optimal control for a class of nonlinear discrete-time systems in affine form.

Finite-horizon optimal control, in contrast, is quite difficult to solve since a terminal constraint has to be satisfied while the control is generally time-varying in contrast with the infinite-horizon scenario wherein the terminal constraint is ignored and the control input becomes time invariant. For finite-horizon optimal regulation, Beard (1995), Heydari and Balakrishnan (2013), and Wang et al. (2011) provided a good insight using either backward-in-time, or iterative and offline ADP techniques. However, no known technique exists for the optimal adaptive control of uncertain quantized linear discrete-time systems until recently by Zhao et al. (2012a, 2015b).

Similar to network imperfections, actuator saturation, on the other hand, is very common in practical industrial applications due to physical limitations (Abu-Khalaf and Lewis 2005). Control of systems with saturating actuators has been one of the focuses of many researchers, including Sussman et al. (1994) and Saberi et al. (1996) for several years. However, most of these approaches considered only stabilization whereas optimality is not considered. To address the optimal control problem with actuator constraint, a general framework for the design of optimal control laws based on dynamic programing is introduced by Lyshevski (1998). It has been shown by Lyshevski (1998) that the use of a nonquadratic functional can effectively tackle the input constraint while achieving optimality.

Motivated by quantization and actuator saturation issues, in this chapter, the time-driven ADP technique via the Bellman equation from Zhao et al. (2012a, 2013, 2015a) is used to solve the optimal regulation of uncertain linear QNCS over a finite time horizon in an online and forward-in-time manner without performing value and/or policy iterations. First, to handle the quantization effect within the control loop, a dynamic quantizer with finite number of bits is proposed. The quantization error will be addressed as part of the optimal controller design. Subsequently, the Bellman equation, utilized for optimal control, is investigated with an approximated action-dependent value function (Watkins 1989) by using the quantized state and input vector such that the requirement for system dynamics is not needed. Finally, a terminal constraint error is defined and incorporated in the novel parameter update law of the value function such that this term will be minimized at each time step in order to solve the optimal control. The Lyapunov approach is utilized to show the stability of our closed-loop system with the value function parameter update law.

The addition of state and input quantization makes optimal control design and its analysis more involved. Now, different from Zhao et al. (2012a, 2013, 2015a), a network is included within the feedback loop. Though a lossless network is considered, deterministic analysis from Zhao et al. (2012a, 2013, 2015a) is extended to the stochastic domain by assuming the certainty equivalence principle and perfect state measurements.

The proposed approach can be viewed as a variant of a rollout scheme (Bertsekas 2005). In this approach, an initial admissible control policy is selected as the base policy and the control policy is enhanced by using a one-time policy improvement at each sampling interval. For cost improvement, as shown by Bertsekas (2005), it is sufficient that the base policy satisfies a certain mild assumption that bears a relation to the concepts of Lyapunov stability. This in turn connects the proposed approach to the general idea of policy iteration in dynamic programming. In this chapter, closed-loop stability is demonstrated unlike convergence of the ADP scheme since the cost function is implicitly taken as a linear in the unknown parameter function with its parameters being adjusted online through a novel update law in real time.

To summarize, the main contributions of the first part of this chapter, the stochastic extension of the deterministic case from Zhao et al. (2012b, 2013, 2015a) includes

- The design of the dynamic quantizer (Liberzon 2003) to NCS with a newly developed update law, which is independent of NCS system dynamics, both to overcome the saturation effect and shrink the quantization error over time by appropriately updating the zoom parameter of the input and state quantizers
- The investigation of the optimal control design of such a quantized linear NCS with uncertain dynamics by using time-based ADP to overcome the knowledge of system dynamics by selecting a novel parameter update law for the value function
- The demonstration of closed-loop stability by using both Lyapunov and geometric sequence analysis when the time horizon of the optimal control design is considered finite whereas for infinite horizon, asymptotic stability in the mean can be achieved with proposed dynamic quantizer and optimal adaptive controller design

Next, the optimal regulation of nonlinear systems with quantization and actuator saturation effects will be addressed for infinite and finite fixed time scenario. The finite-horizon optimal regulation scheme for uncertain nonlinear quantized NCS with an actuator constraint, which can be implemented in an online and forward-in-time manner with output measurements and value and policy iterations, is presented in this chapter. Similar to the linear case, a nonlinear system is converted first into a nonlinear NCS after the incorporation of network imperfections.

Subsequently, an extended neural network (NN)-based Luenberger observer is proposed by Zhao et al. (2015b) to estimate the system state vector as well as the control coefficient matrix. Note that the proposed observer can maintain the system stability while the NN learns the system dynamics. The actor–critic architecture is then utilized to generate the near-optimal control policy wherein the certainty-equivalent value function is approximated by using the critic NN and the certainty-equivalent optimal policy is generated by using the approximated value function and the control coefficient matrix generated from the observer given an initial admissible control. Finally, a novel dynamic quantizer is proposed to mitigate the effect of quantization error for the control inputs. Because of the presence of observer errors, the control policy will be near optimal.

To handle the time-varying nature of the solution to the HJB equation or value function, NNs with constant weights and time-varying activation functions are utilized. In addition, the control policy is updated once a sampling instant and hence value/policy iterations are not performed. An error term corresponding to the terminal constraint is defined and minimized over time so as to satisfy the terminal constraint. A novel update law for tuning the NN is developed such that the critic NN weights will be tuned by using not only the Bellman error but also the terminal constraint errors. Finally, the stability of our proposed design scheme is demonstrated by using Lyapunov stability analysis.

Therefore, the main contribution of the second part of this chapter includes a stochastic extension of the work of Zhao et al. (2015b). It includes the development of a novel approach to solve the finite-horizon output feedback-based near-optimal control of uncertain quantized nonlinear networked discrete-time systems in affine form in an online and forward-in-time manner without utilizing value and/or policy iterations. A novel dynamic quantizer as well as an online NN observer is introduced for eliminating the quantization errors and generating both the state vector and control coefficient matrix, respectively, so that an explicit need for an identifier is relaxed. Tuning laws for all the NNs are also derived. Lyapunov stability is also demonstrated.

The net result is the design of the optimal controller for uncertain quantized linear and nonlinear NCS in a forward-in-time manner thus yielding an online and forward-in-time scheme without using an iterative approach. The remainder of this chapter is organized as follows. In Section 4.1, the background is briefly introduced for nonlinear and linear quantized control system (QCS). In Section 4.2, the finite-horizon optimal control of linear QNCS is presented. In Section 4.3, the finite-horizon optimal control of nonlinear QNCS is introduced. Concluding remarks are provided in Section 4.4.

4.1 BACKGROUND

Before proceeding, the following notation needs to be defined. The superscript q for the system state and control input vectors represent quantized signals, which are denoted as $\boldsymbol{x}_k^q$ and $\boldsymbol{u}_{qk}^q$, respectively. The control input, $\boldsymbol{u}_k$, represents the input signal without quantization, whereas if it is computed based on the quantized system state vector, it is then denoted as $\boldsymbol{u}_{qk}$ with a subscript q.

4.1.1 Quantized Linear Networked Control Systems

Now, under this closed-loop configuration, consider the time-invariant linear discrete-time system in the state-space form under the influence of both state and input quantization described by

$$\boldsymbol{x}_{k+1} = \boldsymbol{A}\boldsymbol{x}_k + \boldsymbol{B}\boldsymbol{u}_k^a \tag{4.1}$$

where $\boldsymbol{x}_k \in \Omega_x \subset \Re^n$ is the system state vector and $\boldsymbol{u}_k^a \in \Omega_u \subset \Re^m$ is the control input vector received at the actuator at time step k. Because of the quantization, we have $\boldsymbol{u}_k^a = \boldsymbol{u}_{qk}^q$, where $\boldsymbol{u}_{qk}^q$ represents the quantized control input. Here, the communication network is lossless and perfect state measurements are considered available.

As a consequence, the dynamics (4.1) will be considered uncertain and stochastic but not time-varying.

Remark 4.1

It should be noted that in the QCS, only the quantized system state vector, $\boldsymbol{x}_k^q$, instead of the true state vector $\boldsymbol{x}_k$, is available to the controller. In contrast, the controller has the information of both $\boldsymbol{u}_{qk}$ and $\boldsymbol{u}_{qk}^q$, and hence $\boldsymbol{u}_{qk}$ will be used in the problem formulation. On the other hand, the quantized control inputs, $\boldsymbol{u}_{qk}^q$, will be considered in the error analysis in Section 4.2 and the comprehensive closed-loop stability analysis in Section 4.2.

Assumption 4.1

The system dynamics given by $(\boldsymbol{A}, \boldsymbol{B})$ are controllable but are considered unknown while the system state vector $\boldsymbol{x}_k \in \Omega_x$ is considered measurable. The order of the system is considered known.

Assumption 4.2

The input matrix, $\boldsymbol{B}$, satisfies $\|\boldsymbol{B}\|_F \leq B_M$, with $\|\bullet\|_F$ denoting the Frobenius norm (Xu and Jagannathan 2013).

It should be noted that the open-loop system can be unstable whereas the goal of the optimal controller proposed here is to stabilize the closed-loop system by optimizing the value function, with details given in the next section. For LQR problems, the ADP technique is proven to be an effective approach to relax the requirement on the system dynamics (Bradtke and Ydstie 1994) and is adopted in this chapter, which is described in detail in Section 4.2.1. In a practical situation, the system state and control input vector is quantized before transmission.

Now, in the presence of state and input quantizers, the general structure of the QNCS considered is shown in Figure 4.1 under the assumption that the network is

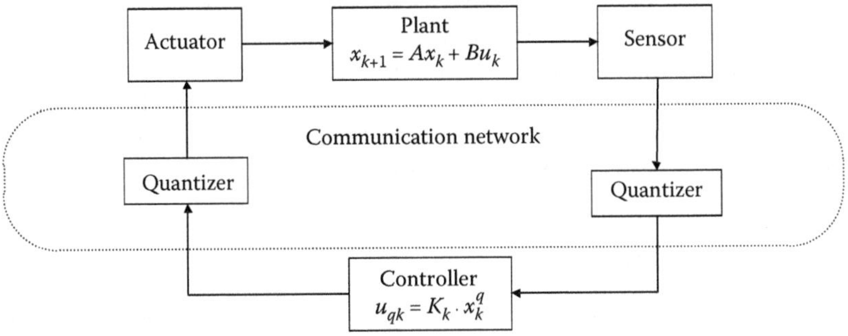

FIGURE 4.1 Block diagram of the QCS with a lossless network.

lossless. The state measurements are first quantized by a dynamic quantizer before being transmitted to the controller. Similarly, the control inputs are also quantized before being applied on the actuator. Next, a brief background on dynamic quantizer is introduced before introducing the controller design with the quantized state and control input.

Separating the quantization error from the actual control inputs $\boldsymbol{u}_k^a$, the system dynamics (4.1) can be represented as

$$\boldsymbol{x}_{k+1} = \boldsymbol{A}\boldsymbol{x}_k + \boldsymbol{B}(\boldsymbol{u}_{qk} + \boldsymbol{e}_{u,k}) = \boldsymbol{A}\boldsymbol{x}_k + \boldsymbol{B}\boldsymbol{u}_{qk} + \boldsymbol{B}\boldsymbol{e}_{u,k} \tag{4.2}$$

where $\boldsymbol{e}_{u,k} = \boldsymbol{u}_{qk}^q - \boldsymbol{u}_{qk}$ is the quantization error for the control inputs.

Remark 4.2

From Equation 4.2, the system dynamics can be viewed as the system with state quantization (incorporated in $\boldsymbol{u}_{qk}$) plus an additional error term ($\boldsymbol{e}_{u,k}$) caused by the control input quantization. Note $\boldsymbol{e}_{u,k}$ is always bounded as long as the control inputs are within the quantization range. The boundedness of quantization error can be ensured by the novel dynamic quantizer design proposed in the next section so that the control inputs do not saturate. Moreover, the proposed approach can force the quantization error bounds converge close to zero thereby the quantization error also converges close to zero.

The objective of the controller design is to determine a state feedback control policy that minimizes the following certainty-equivalent time-varying stochastic cost function:

$$J_k = \underset{\tau,\gamma}{E}\left[\boldsymbol{x}_N^T \boldsymbol{S}_N \boldsymbol{x}_N + \sum_{i=k}^{N-1} r(\boldsymbol{x}_i, \boldsymbol{u}_{qi}, i)\right] \tag{4.3}$$

where $[k, N]$ is the time interval of interest, $r(\boldsymbol{x}_k, \boldsymbol{u}_{qk}, k)$ is a positive definite cost-to-go function that penalizes the system states $\boldsymbol{x}_k$ and the control inputs $\boldsymbol{u}_{qk}$ at each intermediate time k in $[k, N]$. In this chapter, the cost-to-go function is taken in the form $r(\boldsymbol{x}_k, \boldsymbol{u}_{qk}, k) = \boldsymbol{x}_k^T \boldsymbol{Q}_k \boldsymbol{x}_k + \boldsymbol{u}_{qk}^T \boldsymbol{R}_k \boldsymbol{u}_{qk}$, where the weighting matrix $\boldsymbol{Q}_k \in \Re^{n\times n}$ is positive semidefinite, $\boldsymbol{R}_k \in \Re^{m\times m}$ is positive definite and symmetric, while $\boldsymbol{S}_N \in \Re^{n\times n}$ is a positive semidefinite symmetric penalty matrix for the terminal state $\boldsymbol{x}_N$. A nonquadratic cost functional can also be selected, whereas for simplicity, a standard quadratic cost function is selected.

4.1.2 Quantizer Representation

In order to incorporate the quantization effect on the control inputs, the uniform quantizer with finite number of bits that is shown in Figure 4.2 has been utilized. Let z be the signal to be quantized and M be the quantization range for the quantizer.

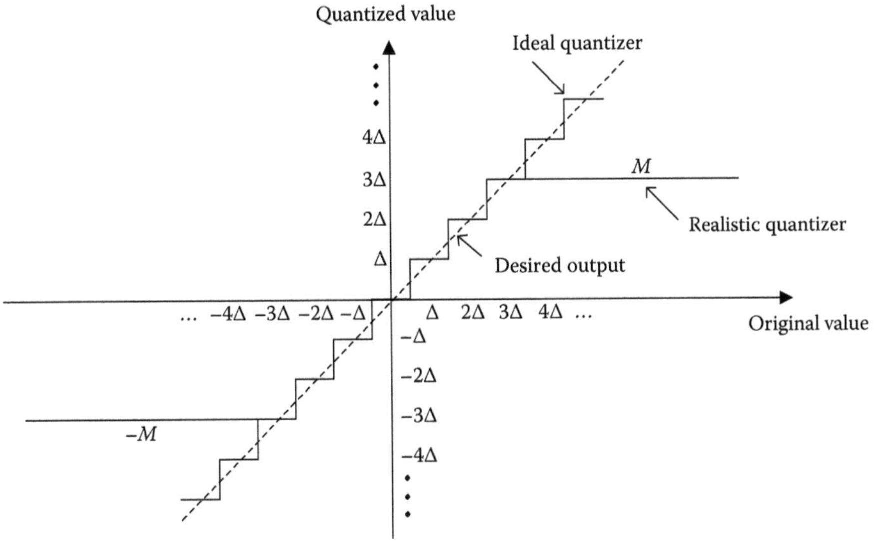

FIGURE 4.2 Ideal and realistic quantizer.

If z does not belong to the quantization range, the quantizer saturates. Let e be the quantization error; it is assumed that the following two conditions hold:

$$\begin{aligned} &1.\ \text{If} \quad |z| \le M, \quad \text{then} \quad 0 \le e = |q(z) - z| \le \frac{\Delta}{2} \\ &2.\ \text{If} \quad |z| > M, \quad \text{then} \quad 0 \le e = |q(z) - z| < \infty \end{aligned} \tag{4.4}$$

where the quantization value

$$q(z) = \begin{cases} M, & z > M \\ \Delta \cdot \left(\left\lfloor \dfrac{z}{\Delta} \right\rfloor + \dfrac{1}{2} \right), & -M \le z \le M \\ -M, & z < -M \end{cases}$$

is a nonlinear mapping that represents a general uniform quantizer representation with the step-size Δ defined as $\Delta = M/2^R$ with R being the number of bits of the quantizer.

In addition, theoretically, when the number of bits of the quantizer approaches to infinity, the quantization error will reduce to zero and hence infinite precision of the quantizer can be achieved. In a realistic scenario, however, both the quantization range and the number of bits cannot be arbitrarily large. To circumvent these drawbacks, a dynamic quantizer scheme is proposed in the form similar to that of Liberzon (2003) as

$$z_q = q_d(z) = \mu q\left(\frac{z}{\mu}\right) \tag{4.5}$$

where μ is a scaling factor.

4.1.3 Quantized Nonlinear Networked Control System

Consider the nonlinear discrete-time system of the form

$$\boldsymbol{x}_{k+1} = f(\boldsymbol{x}_k) + g(\boldsymbol{x}_k)\boldsymbol{u}_{qk}, \boldsymbol{y}_k = \boldsymbol{C}\boldsymbol{x}_k \tag{4.6}$$

where $\boldsymbol{x}_k \in \Omega_x \subset \Re^n$ and $\boldsymbol{y}_k \in \Omega_y \subset \Re^p$ are the system state and output vectors, respectively; $\boldsymbol{u}_{qk} = q_d(\boldsymbol{u}_k) \in \Omega_u \subset \Re^m$ is the quantized control input vector, where $q_d(\cdot)$ is the dynamic quantizer defined later; $\boldsymbol{u}_k \in \boldsymbol{U} \subset \Re^m$, where $\boldsymbol{U} = \{\boldsymbol{u} = (u_1, u_2,\ldots, u_m) \in \Re^m{:}a_i \le u_i \le b_i, i = 1, 2,\ldots, m\}$ with a_i, b_i being the constant bounds (Lyshevski 1998); $f(\boldsymbol{x}_k){:}\Re^n \to \Re^n$, $g(\boldsymbol{x}_k){:}\Re^n \to \Re^{n\times m}$ represents the unknown nonlinear control coefficient matrix; and $\boldsymbol{C} \in \Re^{p\times n}$ is the known output matrix. In addition, the input matrix $g(\boldsymbol{x}_k)$ is considered to be bounded such that $0 < \|g(\boldsymbol{x}_k)\| < g_M$, where g_M is a positive constant.

The general structure of the quantized nonlinear discrete-time system considered in this chapter is illustrated in Figure 4.3. Here, similar to the linear case, the communication network between the nonlinear system and the controller is lossless while the state measurements are perfect. It is important to note that a digital communication network is usually used to connect sensor, controller and actuator in practical (Xu and Jagannathan 2013) scenarios. Because of limited communication bandwidth, system states and control inputs should be quantized before transmission (Tse and Viswanath 2005). In Zhao et al. (2012b), state quantization has been considered with a communication network. Therefore, control input quantization is considered here.

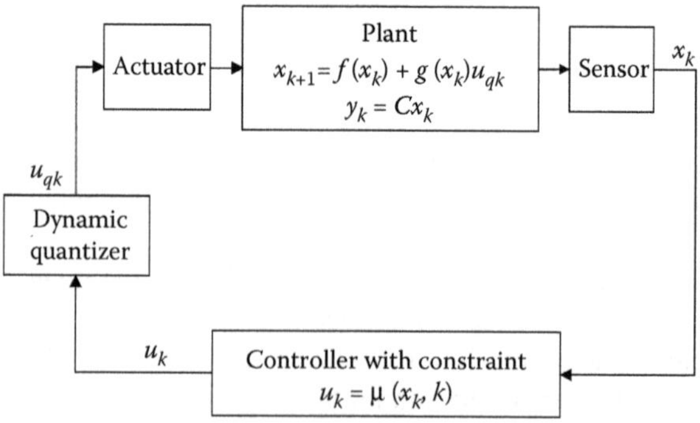

FIGURE 4.3 Block diagram of the quantized system with input saturation.

Assumption 4.3

The nonlinear system given in Equation 4.6 is controllable and observable (Khalil and Praly 2014). Here, the system output, $\boldsymbol{y}_k \in \Omega_y$, is considered measurable.

The controllability ensures that a stabilizing controller can be designed while the observability guarantees that the state vector can be estimated from the output measurements. The objective of the control design is to determine a feedback control policy that minimizes the following certainty-equivalent time-varying stochastic cost function:

$$V(\boldsymbol{x}_k,k)=\underset{\tau,\gamma}{E}\left[\psi(\boldsymbol{x}_N)+\sum_{i=k}^{N-1}(\boldsymbol{Q}(\boldsymbol{x}_i,i)+W(\boldsymbol{u}_i))\right] \tag{4.7}$$

which is subjected to the system dynamics (4.6), $[k,N]$ is the time interval of interest, $\psi(\boldsymbol{x}_N)$ is the terminal constraint that penalizes the terminal state $\boldsymbol{x}_N \in \Omega_x$, $\boldsymbol{Q}(\boldsymbol{x}_k,k) \in \Re$ is a positive semidefinite function, and $W(\boldsymbol{u}_k) \in \Re$ is positive definite. It should be noted that in the finite-horizon scenario, the control inputs can be time-varying, that is, $\boldsymbol{u}_k = \mu(\boldsymbol{x}_k, k) \in \Omega_u$.

Setting $k = N$, the terminal constraint for the value function is given as

$$V(\boldsymbol{x}_N,N)=\underset{\tau,\gamma}{E}\left[\psi(\boldsymbol{x}_N)\right] \tag{4.8}$$

For unconstrained control inputs, $W(\boldsymbol{u}_k)$ generally takes the form $W(\boldsymbol{u}_k)=\boldsymbol{u}_k^T\boldsymbol{R}\boldsymbol{u}_k$, with $\boldsymbol{R} \in \Re^{m\times m}$ being a positive definite and symmetric weighting matrix. However, in this chapter, to confront the actuator saturation, we employ a nonquadratic functional (Lyshevski 1998) as

$$W(\boldsymbol{u}_k)=2\int_0^{u_k}(\varphi^{-1}(\boldsymbol{v}))^T\boldsymbol{R}d\boldsymbol{v} \tag{4.9}$$

with

$$\varphi(\boldsymbol{v})=[\varphi(v_1)\ldots\varphi(v_m)]^T;\varphi^{-1}(\boldsymbol{u}_k)=[\varphi^{-1}(u_{1,k})\ldots\varphi^{-1}(u_{m,k})] \tag{4.10}$$

where $\boldsymbol{v} \in \Omega_v \subset R^m$, $\varphi \in \Omega_\varphi \subset \Re^m$, and $\varphi(\cdot)$ is a bounded one-to-one function that belongs to $\boldsymbol{C}^p$ $(p \geq 1)$. Define $w(\boldsymbol{v}) = \varphi^{-1}(\boldsymbol{v})\boldsymbol{R}$, and

$$\int_0^{u_k}\boldsymbol{w}^T(\boldsymbol{v})d\boldsymbol{v}\equiv\int_0^{u_1(k)}w_1(v_1)dv_1+\cdots+\int_0^{u_m(k)}w_m(v_m)dv_m \tag{4.11}$$

is a *scalar*, for $\boldsymbol{u}_k \in \Omega_u \subset \Re^m$, $\boldsymbol{v} \in \Omega_v \subset \Re^m$, and $\boldsymbol{w}(\boldsymbol{v}) = [w_1\ldots w_m] \in \Omega_w \subset \Re^m$.

Moreover, it is a monotonic odd function with its first derivative bounded by a constant U. An example is the hyperbolic tangent function $\varphi(\cdot) = \tanh(\cdot)$. Note that $W(\boldsymbol{u}_k)$ is positive definite since $\varphi^{-1}(\boldsymbol{u}_k)$ is monotonic odd and $\boldsymbol{R}$ is positive definite. By Bellman's principle of optimality (Stengel 1986), the optimal value function should satisfy the HJB equation

$$V^*(\boldsymbol{x}_k,k) = \min_{u_k} \underset{\tau,\gamma}{E}\left\{\boldsymbol{Q}(\boldsymbol{x}_k,k) + 2\int_0^{u_k} (\varphi^{-1}(v))^T \boldsymbol{R} dv + V^*(\boldsymbol{x}_{k+1},k+1)\right\} \tag{4.12}$$

The certainty-equivalent optimal control policy $\boldsymbol{u}_k^* \in \Omega_u$ that minimizes the value function $V^*(\boldsymbol{x}_k,k)$ is revealed to be

$$\underset{\tau,\gamma}{E}(\boldsymbol{u}_k^*) = \arg\min_{u_k} \underset{\tau,\gamma}{E}\left\{\boldsymbol{Q}(\boldsymbol{x}_k,k) + 2\int_0^{u_k} (\varphi^{-1}(v))^T \boldsymbol{R} dv + V^*(\boldsymbol{x}_{k+1},k+1)\right\}$$

According to Lagrange theory, optimal control can be attained by solving $(\partial\{\boldsymbol{Q}(\boldsymbol{x}_k,k) + 2\int_0^{u_k} (\varphi^{-1}(v))^T \boldsymbol{R} dv + V^*(\boldsymbol{x}_{k+1},k+1)\})/\partial u_k = 0$, that is,

$$\underset{\tau,\gamma}{E}(\boldsymbol{u}_k^*) = -\underset{\tau,\gamma}{E}\left[\varphi\left(\frac{1}{2}\boldsymbol{R}^{-1} g^T(\boldsymbol{x}_k)\frac{\partial V^*(\boldsymbol{x}_{k+1},k+1)}{\partial \boldsymbol{x}_{k+1}}\right)\right] \tag{4.13}$$

It is clear from Equation 4.13 that the optimal control policy cannot be obtained for the nonlinear discrete-time system even with the available system state vector due to its dependency on the future state vector $\boldsymbol{x}_{k+1} \in \Omega_x$. To avoid this drawback and relax the requirement for system dynamics, iteration-based schemes are normally utilized by using NNs with offline training (Chen and Jagannathan 2008). However, iteration-based schemes are not preferable for hardware implementation since the number of iterations to ensure stability cannot be easily determined (Xu and Jagannathan 2013). Moreover, iterative methods require the control coefficient matrix $g(\boldsymbol{x}_k)$ to generate the control policy (Dierks and Jagannathan 2012). Therefore, in this book, a solution is found with system outputs and completely unknown system dynamics without utilizing the iterative approach and in the presence of quantization effect. The treatment is done in the stochastic domain due to NCS while it is presented in a deterministic manner in Zhao et al. (2015b).

4.2 FINITE-HORIZON OPTIMAL CONTROL OF LINEAR QNCS

In this section, the finite-horizon optimal regulation of linear QNCS with uncertain system dynamics is addressed from Zhao et al. (2012a, 2013, 2015a). Under the ideal case when no saturation occurs, the traditional uniform quantizer only yields a bounded response, which is not preferable. The process of reducing the quantization error over time poses a great obstacle for the optimal control design. Therefore, the dynamic quantizer design is first proposed to overcome this difficulty.

Next, to relax the requirement on system dynamics, an action-dependent value function (Werbos 1983; Watkins 1989), which is defined and estimated adaptively via the temporal difference error (TDE) of the reinforcement learning scheme, will be in turn utilized to design the optimal controller. The Bellman equation error, which is essential to achieve optimality, is considered under quantization effect and parameter estimation. In addition, to satisfy the terminal constraint, an additional error term corresponding to the terminal constraint is defined and minimized as time evolves. Therefore, the objective of the controller design is to minimize both the errors so that the finite-horizon optimal regulation problem is properly investigated.

4.2.1 Action-Dependent Value-Function Setup

Before proceeding, it is important to note that in the case of finite horizon, the value function becomes time-varying (Stengel 1986) and is a function of both system states and time-to-go, and it is denoted as $V(\boldsymbol{x}_k, N-k)$. Since the certainty-equivalent stochastic value function is equal to the stochastic cost function J_k (Stengel 1986), the certainty-equivalent stochastic value function $V(\boldsymbol{x}_k, N-k)$ for LQR can also be expressed in the quadratic form of the system states as

$$V(\boldsymbol{x}_k, N-k) = \underset{\tau,\gamma}{E}\left[\boldsymbol{x}_k^T \boldsymbol{S}_k \boldsymbol{x}_k\right] \tag{4.14}$$

with $\boldsymbol{S}_k$ being the solution sequence to the certainty-equivalent SRE obtained backward-in-time from the terminal value $\boldsymbol{S}_N$ as

$$\underset{\tau,\gamma}{E}[\boldsymbol{S}_k] = \underset{\tau,\gamma}{E}[\boldsymbol{A}^T[\boldsymbol{S}_{k+1} - \boldsymbol{S}_{k+1}\boldsymbol{B}(\boldsymbol{B}^T\boldsymbol{S}_{k+1}\boldsymbol{B} + \boldsymbol{R}_k)^{-1}\boldsymbol{B}^T\boldsymbol{S}_{k+1}]\boldsymbol{A} + \boldsymbol{Q}_k] \tag{4.15}$$

Next, define the Hamiltonian for the QNCS as

$$\begin{aligned} H(\boldsymbol{x}_k, \boldsymbol{u}_{qk}, N-k) = \underset{\tau,\gamma}{E}[r(\boldsymbol{x}_k, \boldsymbol{u}_{qk}, k) \\ + V(\boldsymbol{x}_{k+1}, N-k-1) - V(\boldsymbol{x}_k, N-k)] \end{aligned} \tag{4.16}$$

By using Stengel (1986), the certainty-equivalent optimal control inputs are obtained via stationarity condition, that is, $H(\boldsymbol{x}_k, \boldsymbol{u}_{qk}, N-k)/\boldsymbol{u}_{qk} = 0$, which yields

$$\begin{aligned} \underset{\tau,\gamma}{E}(\boldsymbol{u}_{qk}^*) = -\underset{\tau,\gamma}{E}[(\boldsymbol{R}_k + \boldsymbol{B}^T\boldsymbol{S}_k\boldsymbol{B})^{-1}\boldsymbol{B}^T\boldsymbol{S}_k\boldsymbol{A}\cdot\boldsymbol{x}_k + (\boldsymbol{R}_k + \boldsymbol{B}^T\boldsymbol{S}_k\boldsymbol{B})^{-1} \\ \times(\boldsymbol{B}^T\boldsymbol{S}_k\boldsymbol{A}\boldsymbol{e}_{\boldsymbol{x},k} - \boldsymbol{R}_k\boldsymbol{e}_{\boldsymbol{u},k} - \boldsymbol{B}^T\boldsymbol{S}_k\boldsymbol{B}\boldsymbol{e}_{\boldsymbol{u},k} - \boldsymbol{B}^T\boldsymbol{S}_k\boldsymbol{e}_{\boldsymbol{x},k+1})] \end{aligned} \tag{4.17}$$

It can be seen clearly from Equation 4.17 that the optimal control input calculated based on quantized system states enjoy the same optimal control gain, $\boldsymbol{K}_k = \underset{\tau,\gamma}{E}[(\boldsymbol{R}_k + \boldsymbol{B}^T\boldsymbol{S}_k\boldsymbol{B})^{-1}\boldsymbol{B}^T\boldsymbol{S}_k\boldsymbol{A}]$, as that of the case when quantization is not taken into account, plus an additional term corresponding to the quantization errors that would vanish with the proposed design as shown later. Since the only available signal to the controller is the quantized measurement $\boldsymbol{x}_k^q$, then using the "certainty equivalence" principle (Stengel 1986), the control inputs applied to the system is calculated as

$$\underset{\tau,\gamma}{E}(\boldsymbol{u}_{qk}) = -\underset{\tau,\gamma}{E}[(\boldsymbol{R}_k + \boldsymbol{B}^T \boldsymbol{S}_k \boldsymbol{B})^{-1} \boldsymbol{B}^T \boldsymbol{S}_k \boldsymbol{A} \cdot \boldsymbol{x}_k^q] \tag{4.18}$$

Remark 4.3

From Equation 4.15, it is clear that the conventional approach of finding the optimal solution is essentially an offline scheme given the system matrices $\boldsymbol{A}$ and $\boldsymbol{B}$ as given in Equation 4.17. To relax the system dynamics, under the infinite-horizon case, policy iterations are utilized to estimate the value function and derive the control inputs in a forward-in-time manner (Bradtke and Ydstie 1994). However, an inadequate number of iterations will lead to the instability of the system (Dierks and Jagannathan 2012). The iterative approach is not utilized (Xu and Jagannathan 2013) and the proposed online estimator parameters are updated once a sampling interval.

Next, we will show that by applying ADP methodology, the need for the system dynamics can be avoided by estimating the value function $V(\boldsymbol{x}_k, N-k)$. Since the Kalman gain in Equation 4.18 is the same as the standard Kalman gain without quantization, assume that there is no quantization effect in the system. Recalling the time-varying nature of finite-horizon control, define a certainty-equivalent time-varying stochastic optimal *action-dependent* value function given by

$$\underset{\tau,\gamma}{E}\{V_{AD}(\boldsymbol{x}_k, \boldsymbol{u}_k, N-k)\} = \underset{\tau,\gamma}{E}\{r(\boldsymbol{x}_k, \boldsymbol{u}_k, k) + J_{k+1}\} = \underset{\tau,\gamma}{E}\left\{\begin{bmatrix} \boldsymbol{x}_k \\ \boldsymbol{u}_k \end{bmatrix}^T \boldsymbol{G}_k \begin{bmatrix} \boldsymbol{x}_k \\ \boldsymbol{u}_k \end{bmatrix}\right\} \tag{4.19}$$

The standard Bellman equation is given by

$$\begin{aligned}
\underset{\tau,\gamma}{E}\left\{\begin{bmatrix} \boldsymbol{x}_k \\ \boldsymbol{u}_k \end{bmatrix}^T \boldsymbol{G}_k \begin{bmatrix} \boldsymbol{x}_k \\ \boldsymbol{u}_k \end{bmatrix}\right\} &= \underset{\tau,\gamma}{E}\left[r(\boldsymbol{x}_k, \boldsymbol{u}_k, k) + J_{k+1}\right] \\
&= \underset{\tau,\gamma}{E}\left[\boldsymbol{x}_k^T \boldsymbol{Q}_k \boldsymbol{x}_k + \boldsymbol{u}_k^T \boldsymbol{R}_k \boldsymbol{u}_k + \boldsymbol{x}_{k+1}^T \boldsymbol{S}_{k+1} \boldsymbol{x}_{k+1}\right] \\
&= \underset{\tau,\gamma}{E}\left\{\begin{bmatrix} \boldsymbol{x}_k \\ \boldsymbol{u}_k \end{bmatrix}^T \begin{bmatrix} \boldsymbol{Q}_k + \boldsymbol{A}^T \boldsymbol{S}_{k+1} \boldsymbol{A} & \boldsymbol{A}^T \boldsymbol{S}_{k+1} \boldsymbol{B} \\ \boldsymbol{B}^T \boldsymbol{S}_{k+1} \boldsymbol{A} & \boldsymbol{R}_k + \boldsymbol{B}^T \boldsymbol{S}_{k+1} \boldsymbol{B} \end{bmatrix} \begin{bmatrix} \boldsymbol{x}_k \\ \boldsymbol{u}_k \end{bmatrix}\right\}
\end{aligned} \tag{4.20}$$

Therefore, define the *time-varying* matrix $\boldsymbol{G}_k$ as

$$\begin{aligned} \underset{\tau,\gamma}{E}\{G_k\} &= \underset{\tau,\gamma}{E}\left\{\begin{bmatrix} Q_k + A^T S_{k+1} A & A^T S_{k+1} B \\ B^T S_{k+1} A & R_k + B^T S_{k+1} B \end{bmatrix}\right\} \\ &= \underset{\tau,\gamma}{E}\left\{\begin{bmatrix} G_k^{xx} & G_k^{xu} \\ G_k^{ux} & G_k^{uu} \end{bmatrix}\right\} \end{aligned} \tag{4.21}$$

Compared to Equation 4.49, the control gain can be expressed in terms of G_k as

$$\underset{\tau,\gamma}{E}(K_k) = \underset{\tau,\gamma}{E}[(G_k^{uu})^{-1} G_k^{ux}] \tag{4.22}$$

From the above analysis, the estimation of the certainty-equivalent time-varying action-dependent stochastic value function $V_{AD}(x_k, u_k, N-k)$ includes the information of G_k matrix, which can be solved online. Therefore, the control inputs can be obtained from Equation 4.21 instead of using A and B as given in Equation 4.18.

4.2.2 Model-Free Online Tuning of Action-Dependent Value Function with Quantized Signals

In this section, finite-horizon stochastic optimal control design is proposed without using an iteration-based scheme. Recalling the definition of the action-dependent value function $V_{AD}(x_k, u_k, N-k)$, the following assumption and lemma are introduced before proceeding.

Assumption 4.4

The action-dependent value function $V_{AD}(x_k, u_k, N-k)$ is slowly varying and can be expressed as the linear in the unknown parameters (LIP).

By adaptive control theory and the definition of value function, using Assumption 4.4, $V_{AD}(x_k, u_k, N-k)$ can be written in the vector form as

$$V_{AD}(x_k, u_k, N-k) = \underset{\tau,\gamma}{E}(z_k^T G_k z_k) = \underset{\tau,\gamma}{E}(g_k^T \bar{z}_k) \tag{4.23}$$

where $z_k = [x_k^T \; u_k^T]^T \in \Re^{n+m=l}$ is the regression function, $\bar{z}_k = (z_{k1}^2, \ldots, z_{k1}z_{kl}, z_{k2}^2, \ldots, z_{kl-1}z_{kl}, z_{kl}^2)$ is the Kronecker product quadratic polynomial basis vector, and $g_k = \text{vec}(G_k)$, with vec(·) a vector function that acts on a $l \times l$ matrix and gives a $l \times (l+1)/2 = L$ column vector. The output of vec(G_k) is constructed by stacking the columns of the square matrix into a one column vector with the off-diagonal elements summed as $G_{mn}^k + G_{nm}^k$.

Lemma 4.1

Let $g(k)$ be a smooth and uniformly piecewise-continuous function in a compact set $\Omega \subset \Re$. Then, for any $\varepsilon > 0$, there exist constant elements $\theta_1, \ldots, \theta_m \in \Re$ with $m \in N$ as well as the elements $\varphi_1(k), \ldots, \varphi_m(k) \in \Re$ of basis function, such that

$$\underset{\tau,\gamma}{E}\left|g(k)-\sum_{i=1}^{m}\theta_i\varphi_i(k)\right|<\varepsilon, \qquad k\in[0,N] \tag{4.24}$$

Based on Assumption 4.3 and Lemma 4.1, the smooth and uniformly piecewise-continuous function, the smooth and uniformly piecewise-continuous function $\boldsymbol{g}_k$ can be represented as

$$\underset{\tau,\gamma}{E}(\boldsymbol{g}_k^T)=\underset{\tau,\gamma}{E}\left[\theta^T\varphi(N-k)\right] \tag{4.25}$$

where $\theta \in R^L$ is the target parameter vector and $\varphi(N-k) \in R^{L\times L}$ is the time-varying basis or regression function matrix with entries as functions of time-to-go, that is,

$$\varphi(N-k)=\begin{bmatrix}\varphi_{11}(N-k) & \varphi_{12}(N-k) & \dots & \varphi_{1L}(N-k)\\ \varphi_{21}(N-k) & \varphi_{22}(N-k) & \dots & \varphi_{2L}(N-k)\\ \vdots & \vdots & \ddots & \vdots\\ \varphi_{L1}(N-k) & \varphi_{L2}(N-k) & \dots & \varphi_{LL}(N-k)\end{bmatrix}$$

where $\varphi_{ij}(N-k)=\exp(-\tanh(N-k)^{L+1-j})$, for $i, j = 1,2,\ldots,L$. This time-based function reflects the time-dependent nature of finite horizon. Further, based on universal approximation theory, $\varphi(N-k)$ is piecewise-continuous (Sandberg 1998; Cybenko 1989).

Therefore, the action-dependent value function can be written in terms of θ as

$$\underset{\tau,\gamma}{E}\{V_{AD}(\boldsymbol{x}_k,\boldsymbol{u}_k,N-k)\}=\underset{\tau,\gamma}{E}\left[\theta^T\varphi(N-k)\overline{z}_k\right] \tag{4.26}$$

From Stengel (1986), the standard Bellman equation can be written in terms of $V_{AD}(\boldsymbol{x}_k, \boldsymbol{u}_k, N-k)$ as

$$\underset{\tau,\gamma}{E}\left[V_{AD}(\boldsymbol{x}_{k+1},\boldsymbol{u}_{k+1},N-k-1)-V_{AD}(\boldsymbol{x}_k,\boldsymbol{u}_k,N-k)+r(\boldsymbol{x}_k,\boldsymbol{u}_k,k)\right]=0 \tag{4.27}$$

Remark 4.4

In the infinite-horizon case, Equation 4.26 does not have the time-varying term $\varphi(N-k)$, since the desired value of vector $\boldsymbol{g}$ is a constant, or it becomes time invariant (Xu and Jagannathan 2013). By contrast, in the finite-horizon case, the desired value of $\boldsymbol{g}_k$ is considered to be slowly time-varying. Hence the basis function should be a function of time and can take the form of a product of the time-dependent basis function and the system states (Lewis et al. 1999).

To approximate the time-varying matrix $\boldsymbol{G}_k$, or alternatively $\boldsymbol{g}_k$, define

$$\underset{\tau,\gamma}{E}(\hat{\boldsymbol{g}}_k^T) = \underset{\tau,\gamma}{E}[\hat{\theta}_k^T \varphi(N-k)] \tag{4.28}$$

where $\hat{\theta}_k$ is the estimation of the time-invariant part of the target parameter vector $\boldsymbol{g}_k$. Next, when taking both the quantization effect and the estimated value of $\boldsymbol{g}_k$ into consideration, the Bellman Equation 4.27 becomes

$$\underset{\tau,\gamma}{E}(e_{BQ,k}) = \underset{\tau,\gamma}{E}[(\boldsymbol{z}_{k+1}^q)^T \hat{\boldsymbol{G}}_{k+1} \boldsymbol{z}_{k+1}^q - (\boldsymbol{z}_k^q)^T \hat{\boldsymbol{G}}_k \boldsymbol{z}_k^q + (\boldsymbol{x}_k^q)^T \boldsymbol{Q}_k \boldsymbol{x}_k^q + (\boldsymbol{u}_{qk}^q)^T \boldsymbol{R}_k \boldsymbol{u}_{qk}^q] \tag{4.29}$$

where $\boldsymbol{z}_k^q = [(\boldsymbol{x}_k^q)^T \quad (\boldsymbol{u}_{qk}^q)^T]^T \in \Re^{n+m=l}$ is the regression function with quantized information, $e_{BQ,k}$ is the error in the Bellman equation, which can be regarded as TDE in reinforcement learning.

Furthermore, $e_{BQ,k}$ can be represented as

$$\begin{aligned}
\underset{\tau,\gamma}{E}\{e_{BQ,k}\} &= \underset{\tau,\gamma}{E}\{\psi(z_k, z_k^q) + [(z_k^q)^T \boldsymbol{G}_k z_k^q - (z_{k+1}^q)^T \boldsymbol{G}_{k+1} z_{k+1}^q \\
&\quad - (\boldsymbol{x}_k^q)^T \boldsymbol{Q} \boldsymbol{x}_k^q - (\boldsymbol{u}_{qk}^q)^T \boldsymbol{R} \boldsymbol{u}_{qk}^q] - [(z_k^q)^T \hat{\boldsymbol{G}}_k z_k^q \\
&\quad - (z_{k+1}^q)^T \hat{\boldsymbol{G}}_{k+1} z_{k+1}^q - (\boldsymbol{x}_k^q)^T \boldsymbol{Q} \boldsymbol{x}_k^q - (\boldsymbol{u}_{qk}^q)^T \boldsymbol{R} \boldsymbol{u}_{qk}^q]\} \\
&= \underset{\tau,\gamma}{E}\{\psi(z_k, z_k^q) + (z_k^q)^T \tilde{\boldsymbol{G}}_k z_k^q - (z_{k+1}^q)^T \tilde{\boldsymbol{G}}_{k+1} z_{k+1}^q\} \\
&= \underset{\tau,\gamma}{E}\{\psi(z_k, z_k^q) + \tilde{\theta}_k^T [\varphi(N-k)\bar{z}_k^q - \varphi(N-k-1)\bar{z}_{k+1}^q]\} \\
&= \underset{\tau,\gamma}{E}\{\psi(z_k, z_k^q) + \tilde{\theta}_k^T \Delta\xi(z_k^q, k)\}
\end{aligned} \tag{4.30}$$

where $\psi(z_k, z_k^q) = z_k^T \boldsymbol{G}_k z_k - z_{k+1}^T \boldsymbol{G}_{k+1} z_{k+1} - \boldsymbol{x}_k^T \boldsymbol{Q}_k \boldsymbol{x}_k - \boldsymbol{u}_k^T \boldsymbol{R}_k \boldsymbol{u}_k - [(z_k^q)^T \boldsymbol{G}_k z_k^q - (z_{k+1}^q)^T \boldsymbol{G}_{k+1} z_{k+1}^q - (\boldsymbol{x}_k^q)^T \boldsymbol{Q}_k \boldsymbol{x}_k^q - (\boldsymbol{u}_{qk}^q)^T \boldsymbol{R}_k \boldsymbol{u}_{qk}^q]$ and $\Delta\xi(z_k^q, k) = \varphi(N-k)\bar{z}_k^q - \varphi(N-k-1)\bar{z}_{k+1}^q$

Since the action-dependent value function and the cost-to-go are in quadratic form, by Lipchitz continuity, we have

$$\left\| \psi(z_k, z_k^q) \right\| \le L_\psi \left\| \begin{bmatrix} e_{Mx,k} \\ e_{Mu,k} \end{bmatrix} \right\|^2 \le L_\psi e_{Mx,k}^2 + L_\psi e_{Mu,k}^2$$

with $L_\psi > 0$ being the Lipchitz constant. Therefore, we have

$$\underset{\tau,\gamma}{E}(e_{BQ,k}) \le \underset{\tau,\gamma}{E}\{L_\psi e_{Mx,k}^2 + L_\psi e_{Mu,k}^2 + \tilde{\theta}_k^T \Delta\xi(z_k^q, k)\} \tag{4.31}$$

Recall that for the optimal control with finite horizon, the terminal constraint of cost/value function should be taken into account properly. Therefore, define the estimated value function at the terminal stage as

$$\underset{\tau,\gamma}{E}\{\hat{V}_{AD}(\boldsymbol{x}_N,0)\} = \underset{\tau,\gamma}{E}\left[\hat{\theta}_k^T\varphi(0)\bar{z}_N\right] \tag{4.32}$$

In Equation 4.32, note that the time-dependent basis function $\varphi(N-k)$ is taken as $\varphi(0)$ at the terminal stage, since from definition, $\varphi(\cdot)$ is a function of time-to-go and the time index is taken in the reverse order. Next, define the terminal constraint error vector as

$$\underset{\tau,\gamma}{E}(\boldsymbol{e}_{N,k}) = \underset{\tau,\gamma}{E}[\boldsymbol{g}_N - \hat{\boldsymbol{g}}_{N,k}] = \underset{\tau,\gamma}{E}[\boldsymbol{g}_N - \hat{\theta}_k^T\varphi(0)] = \underset{\tau,\gamma}{E}[\tilde{\theta}_k^T\varphi(0)] \tag{4.33}$$

where $\boldsymbol{g}_N$ is upper bounded by $\|\mathbf{g}_N\| \le g_M$.

Remark 4.5

For both infinite- and finite-horizon cases, the TDE $e_{BQ,k}$ is always required for tuning the parameter (see Dierks and Jagannathan 2012; Xu and Jagannathan 2013) for the infinite-horizon case without quantization. In the finite-horizon case, the terminal error $\boldsymbol{e}_{N,k}$, which indicates the difference between the estimated value and true value of the terminal constraint, or "target" (in our case, $\boldsymbol{g}_N$), is critical for the controller design. The terminal constraint is satisfied by minimizing $\boldsymbol{e}_{N,k}$ along the system evolution.

Remark 4.6

The Bellman equation with and without quantization effects are not same. The former requires $[\boldsymbol{x}_k^q, \boldsymbol{u}_{qk}^q]$ whereas the latter uses $[\boldsymbol{x}_k, \boldsymbol{u}_k]$. In order to design the optimal adaptive controller, the estimated Bellman equation with quantization effects need to eventually converge to the standard Bellman equation. Next, define the update law for the adaptive estimator as

$$\begin{aligned}\underset{\tau,\gamma}{E}\{\hat{\theta}_{k+1}\} = \underset{\tau,\gamma}{E}\Bigg\{&\hat{\theta}_k + \alpha_\theta\frac{\Delta\xi(z_k^q,k)e_{BQ,k}^T}{\Delta\xi^T(z_k^q k)\Delta\xi(z_k^q,k)+1}\\ &+\alpha_\theta\frac{\left[\Delta\xi(z_k^q,k)(L_\psi e_{Mx,k}^2 + L_\psi e_{Mu,k}^2)\times \operatorname{sgn}(\Delta\xi(z_k^q,k))\right]}{\Delta\xi^T(z_k^q,k)\Delta\xi(z_k^q,k)+1}\\ &+\alpha_\theta\frac{\varphi(0)\boldsymbol{e}_{N,k}^T}{\|\varphi(0)\|^2+1}\Bigg\}\end{aligned} \tag{4.34}$$

Define the estimation error as $\tilde{\theta}_k = \theta - \hat{\theta}_k$. Then we have

$$\underset{\tau,\gamma}{E}\{\tilde{\theta}_{k+1}\} = \underset{\tau,\gamma}{E}\left[\tilde{\theta}_k - \alpha_\theta \frac{\Delta\xi(z_k^q,k)e_{BQ,k}^T}{\Delta\xi^T(z_k^q k)\Delta\xi(z_k^q,k)+1}\right.$$

$$\left.-\alpha_\theta \frac{\left[\Delta\xi(z_k^q,k)(L_\psi e_{Mx,k}^2 + L_\psi e_{Mu,k}^2)\times \mathrm{sgn}(\Delta\xi(z_k^q,k))\right]}{\Delta\xi^T(z_k^q,k)\Delta\xi(z_k^q,k)+1} - \alpha_\theta \frac{\varphi(0)e_{N,k}^T}{\|\varphi(0)\|^2+1}\right]$$

$$= \underset{\tau,\gamma}{E}\left[\tilde{\theta}_k - \alpha_\theta \frac{\Delta\xi(z_k^q,k)\psi^T(z_k,z_k^q)}{\Delta\xi^T(z_k^q k)\Delta\xi(z_k^q,k)+1} - \alpha_\theta \frac{\Delta\xi(z_k^q,k)\tilde{\theta}_k\Delta^T\xi(z_k^q,k)}{\Delta\xi^T(z_k^q k)\Delta\xi(z_k^q,k)+1}\right.$$

$$\left.-\alpha_\theta \frac{\Delta\xi(z_k^q,k)(L_\psi e_{Mx,k}^2 + L_\psi e_{Mu,k}^2)\mathrm{sgn}(\Delta\xi(z_k^q,k))}{\Delta\xi^T(z_k^q,k)\Delta\xi(z_k^q,k)+1} - \alpha_\theta \frac{\varphi(0)\varphi^T(0)\tilde{\theta}_k}{\|\varphi(0)\|^2+1}\right]$$

$$= \underset{\tau,\gamma}{E}\left[\tilde{\theta}_k - \alpha_\theta \frac{\Delta\xi(z_k^q,k)\Delta^T\xi(z_k^q,k)\tilde{\theta}_k}{\Delta\xi^T(z_k^q k)\Delta\xi(z_k^q,k)+1} - \alpha_\theta \frac{\varphi(0)\varphi^T(0)\tilde{\theta}_k}{\|\varphi(0)\|^2+1}\right.$$

$$\left.-\alpha_\theta \frac{\left[\Delta\xi(z_k^q,k)\mathrm{sgn}(\Delta\xi(z_k^q,k))\times((L_\psi e_{Mx,k}^2 + L_\psi e_{Mu,k}^2) - \psi^T(z_k,z_k^q))\right]}{\Delta\xi^T(z_k^q k)\Delta\xi(z_k^q,k)+1}\right]$$

Hence, we have

$$\underset{\tau,\gamma}{E}(\tilde{\theta}_{k+1}) \le \underset{\tau,\gamma}{E}\left\{\tilde{\theta}_k - \alpha_\theta \frac{\Delta\xi(z_k^q,k)\Delta^T\xi(z_k^q,k)\tilde{\theta}_k}{\Delta\xi^T(z_k^q k)\Delta\xi(z_k^q,k)+1} - \alpha_\theta \frac{\varphi(0)\varphi^T(0)\tilde{\theta}_k}{\|\varphi(0)\|^2+1}\right\} \tag{4.35}$$

Remark 4.7

It is observed from the definition (4.33) that the value function becomes zero when $\|z_k^q\| = 0$. Hence, when the quantized system states have converged to zero, the online adaptive estimator is no longer updated. This can be viewed as a persistency of excitation (PE) requirement for the inputs to the value function estimator wherein the system states must be persistently exciting long enough for the estimator to learn the value function (Narendra and Annaswamy 1989). The PE condition can be satisfied by adding exploration noise (Green and Moore 1986) to the augmented state vector. In this chapter, exploration noise is added to satisfy the PE condition and it is removed once the parameters converge.

Remark 4.8

It is important to observe that the parameter update law (4.34) is novel in the sense that it uses several terms, one for reducing the Bellman error, the second for the quantization error, while a last term addresses the terminal constraint.

4.2.3 Estimation of the Optimal Feedback Control

The optimal control can be obtained by minimizing the value function. Recall from Equation 4.54, the estimated certainty-equivalent stochastic optimal control policy can be obtained as

$$\underset{\tau,\gamma}{E}(\boldsymbol{u}_{qk}) = -\underset{\tau,\gamma}{E}\left[\hat{\boldsymbol{K}}_k \cdot \boldsymbol{x}_k^q\right] = -\underset{\tau,\gamma}{E}[(\hat{\boldsymbol{G}}_k^{uu})^{-1}\hat{\boldsymbol{G}}_k^{ux} \cdot \boldsymbol{x}_k^q] \tag{4.36}$$

From Equation 4.36, the optimal control gain can be calculated based on the information of $\hat{\boldsymbol{G}}_k$ matrix, which is obtained by estimating the action-dependent value function. This relaxes the requirement of the system dynamics while the parameter estimate is updated by Equation 4.34 once a sampling interval, which relaxes the iterative approach. The flowchart of the proposed scheme is shown in Figure 4.4.

4.2.4 Convergence Analysis

In this section, convergence of quantization error, parameter estimation error, and closed-loop stability will be analyzed. It will be shown that for the finite-horizon case, all the errors, that is, $\boldsymbol{e}_{x,k}$, $\boldsymbol{e}_{u,k}$, and $\tilde{\theta}_k$ are stochastically bounded in the mean and furthermore, the bounds converge to zero as time span approaches infinity. Since the design scheme is similar to policy iteration, we need to solve a fixed-point equation rather than a recursive equation. The initial admissible control guarantees that the solution of the fixed-potion equation exists; thus the approximation process can be effectively done by our proposed scheme. Now, we are ready to show our main mathematical claims.

Theorem 4.1: Convergence of the Adaptive Estimation Error

Let the initial conditions for $\hat{\boldsymbol{g}}_0$ be bounded in a set Ω_g, which contains the ideal parameter vector $\boldsymbol{g}_k$. Let $\boldsymbol{u}_0(k) \in \Omega_u$ be an initial admissible control policy for the linear system (4.39). Let the assumptions stated in the chapter hold including the controllability of the system and system state vector $\boldsymbol{x}_k \in \Omega_x$ being measurable. Let the update law for tuning $\hat{\theta}_k$ be Equation 4.66. Then, with a positive constant α_θ satisfying $0 < \alpha_\theta < 1/4$, there exists a $\varepsilon > 0$ depending on the initial value $B_{\tilde{\theta},0}$ and terminal stage N, such that for a fixed final time instant N, we have $\|\tilde{\theta}_k\| \le \varepsilon(\tilde{\theta}_k, N)$. Further, the term $\varepsilon(\tilde{\theta}_k, N)$ will converge to zero asymptotically in the mean when $N \to \infty$ or the parameter estimation error converges to zero.

After establishing the convergence of the parameter estimation, we are ready to show the convergence of the quantization error for both system states and control inputs. Before proceeding, the following lemma is needed.

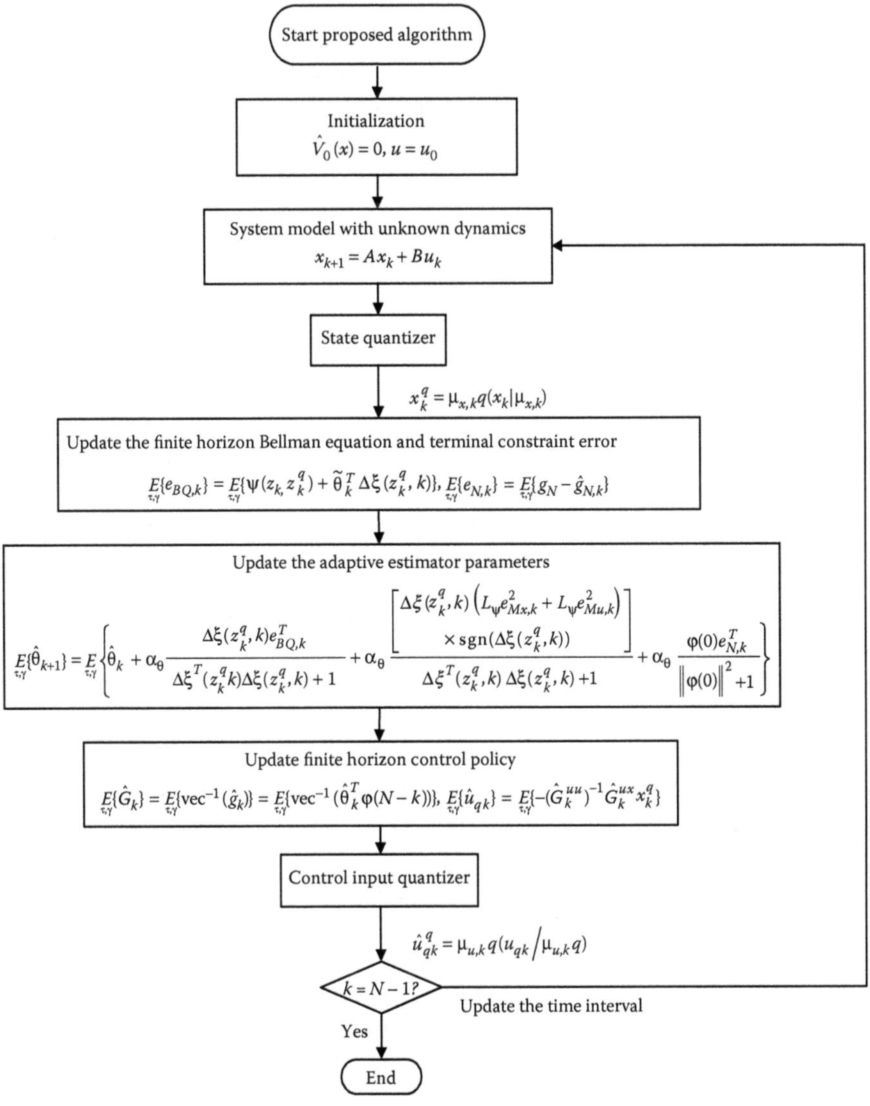

FIGURE 4.4 Flowchart of the optimal adaptive linear quadratic regulator for QNCS with a lossless network.

Lemma 4.2

Consider the linear discrete-time system, and then with the optimal control policy $\boldsymbol{u}_k^*$ for the system such that the closed-loop system dynamics $\boldsymbol{A}\boldsymbol{x}_k + \boldsymbol{B}\boldsymbol{u}_k^*$ can be written as

$$\underset{\tau,\gamma}{E}\{\|\boldsymbol{A}\boldsymbol{x}_k + \boldsymbol{B}\boldsymbol{u}_k^*\|^2\} \leq \underset{\tau,\gamma}{E}\{\rho\|\boldsymbol{x}_k\|^2\} \tag{4.37}$$

where $0 < \rho < 1/3$ is a constant.

Lemma 4.3

Consider the dynamic quantizer for the system states given in Equation 4.2. Let the zoom parameter for the state quantizer be updated by Equation 4.66. Let the adaptive estimator be updated according to Equation 4.34. Then, there exists a $\varepsilon > 0$ depending on the initial value $e_{M\mathbf{x}}$,0 and the terminal stage N, such that for a fixed final time instant N, we have $\|\boldsymbol{e}_{x,k}\| \leq \varepsilon(\boldsymbol{e}_{x,k}, N)$. Furthermore, $\varepsilon(\boldsymbol{e}_{x,k}, N)$ will converge to zero asymptotically in the mean with $N \to \infty$ and the quantization error converges to zero asymptotically.

Proof: See Appendix 4B.

Lemma 4.4

Consider the dynamic quantizer for the control inputs given in Equation 4.2. Let the zoom parameter for the input quantizer be updated by Equation 4.66. Let the adaptive estimator be updated according to Equation 4.34. Then, there exists a $\varepsilon > 0$ depending on the initial value $e_{M\mathbf{u},0}$ and the terminal stage N, such that for a fixed final time instant N, we have $\|\boldsymbol{e}_{u,k}\| \leq \varepsilon(\boldsymbol{e}_{u,k}, N)$. Further, the term $\varepsilon(\boldsymbol{e}_{u,k}, N)$ will converge to zero asymptotically with $N \to \infty$ or input quantization error converges to zero asymptotically.

Proof: See Appendix 4C.

Theorem 4.2: Boundedness of the Closed-Loop System

Let the linear discrete-time system be controllable and the system state be measurable. Let the initial conditions for $\hat{\boldsymbol{g}}_0$ be bounded in a set Ω_g, which contains the ideal parameter vector $\boldsymbol{g}_k$. Let $\boldsymbol{u}_0(k) \in \Omega_{\boldsymbol{u}}$ be an initial admissible control policy for the system (4.2) such that Equation 4.37 holds for some ρ. Let the scaling parameter $\mu_{x,k}$ and $\mu_{u,k}$ be updated by Equation 4.66 with both input and state quantizers present. Further, let the parameter vector of the action-dependent value function estimator be tuned based on Equation 4.34. Then, with the positive constants α_θ, β, and γ satisfying $0 < \alpha_\theta < 1/4$, $0 < \beta < 1$, and $0 < \gamma < 1$, there exists an $\varepsilon > 0$ depending on the initial value of $B_{\mathbf{x},0}, B_{\tilde{\theta},0}, e_{Mx,0}, e_{Mu,0}$ and the terminal stage N, such that for a fixed final time instant N, we have $\|\boldsymbol{x}_k\| \leq \varepsilon(\boldsymbol{x}_k, N)$, $\left\|\tilde{\theta}_k\right\| \leq \varepsilon(\tilde{\theta}_k, N)$, $\|\boldsymbol{e}_{x,k}\| \leq \varepsilon(\boldsymbol{e}_{x,k}, N)$, and $\|\boldsymbol{e}_{u,k}\| \leq \varepsilon(\boldsymbol{e}_{u,k}, N)$. Furthermore, by using geometric analysis, when $N \to \infty$, $\varepsilon(\boldsymbol{x}_k, N)$, $\varepsilon(\tilde{\theta}_k, N)$, $\varepsilon(\boldsymbol{e}_{x,k}, N)$, and $\varepsilon(\boldsymbol{e}_{u,k}, N)$ will converge to zero, that is, the closed-loop system is asymptotically stable in the mean. Moreover, the estimated control input with quantization will converge to optimal control input, $\hat{\boldsymbol{u}}_{qk}^q \to \boldsymbol{u}_k^*$, when $N \to \infty$.

Proof: See Appendix 4D.

4.2.5 Simulation Results

In this section, an example is given to illustrate the feasibility of our proposed dynamic quantizer scheme and the finite-horizon optimal control scheme.

EXAMPLE 4.1

Consider the continuous-time version of the batch reactor dynamics described in Heemels et al. (2010). By using a sampling interval of $T_s = 0.1$ s, we obtain the following discrete-time version of the batch reactor dynamics:

$$x_{k+1} = \begin{bmatrix} 1.1782 & 0.0015 & 0.5116 & -0.4033 \\ -0.0515 & 0.6619 & -0.011 & 0.0613 \\ 0.0762 & 0.3351 & 0.5606 & 0.3824 \\ -0.0006 & 0.3353 & 0.0893 & 0.8494 \end{bmatrix} x_k + \begin{bmatrix} 0.0045 & -0.0876 \\ 0.4672 & 0.0012 \\ 0.2132 & -0.2353 \\ 0.2131 & -0.0161 \end{bmatrix} u_k \tag{4.38}$$

Note that the batch reactor is open-loop unstable with poles located at $s_1 = 1.2203$, $s_2 = 1.0064$, $s_3 = 0.6031$, and $s_4 = 0.4204$, whereas the objective of the optimal controller is to regulate the batch reactor system. The weighting matrices and terminal constraint matrix given in Equation 4.42 are selected as $Q = 0.001I$, $R = 0.1I$, and $S_N = I$, where I is the identity matrix with appropriate dimension. The augmented states z_k^q are generated as $z_k^q = [(x_k^q)^T \; (u_{qk}^q)^T]^T \in \Re^6$ and the regression function was generated per Equation 4.55. Note that the regression function (4.55) is given by using the Kronecker product quadratic polynomial that guarantees that the value function is positive. The initial system state and initial admissible control gain vectors are chosen as deterministic $x_0 = [0.6, 0.9, 0.9, 1.5]^T$ and $K_0 = [-0.2, -0.5, 0.5, 0.4; -2, -0.2, -2, -0.8]$, respectively. The sampling interval is taken as $T_s = 0.1$ s. Moreover, to obtain the ergodic behavior, Monte Carlo simulation runs with 1000 iterations were utilized.

For the dynamic quantizer design, the parameters are selected as $\beta = 0.9$ and $\gamma = 0.9$. For the action-dependent value function estimator, the designing parameter is chosen as $\alpha_\theta = 0.01$. The time-dependent basis function $\varphi(N-k)$ is selected in Equation 4.25 with saturation. Note that for finite time period, $\varphi(N-k)$ is always bounded. Saturation for $\varphi(N-k)$ is to ensure the magnitude of $\varphi(N-k)$ is within a reasonable range such that the parameter estimation is computable. The initial values for $\hat{\theta}_k$ are randomly selected. A lossless network is considered. The simulation results are given as below.

First, the system response and control input are plotted in Figure 4.5. It is clearly shown in the figure that both system states and control signal converges close to zero within a finite time span, which illustrates the stability of our proposed algorithm. Next, to show the feasibility of the quantizer design, the quantization errors with a 4-bit quantizer and 8-bit quantizer are plotted in Figures 4.6 and 4.7, respectively, by using our proposed quantizer and the traditional static quantizer. From Figure 4.6b, it can be seen that with a small number of bits, the traditional static quantizer cannot even guarantee the stability of the system due to the relatively large quantization errors, while the proposed dynamic quantizer in Figure 4.6a can keep the system stable. This aspect will be advantageous in the NCS since a fewer number of bits for the quantizer indicates lower network traffic preventing congestion.

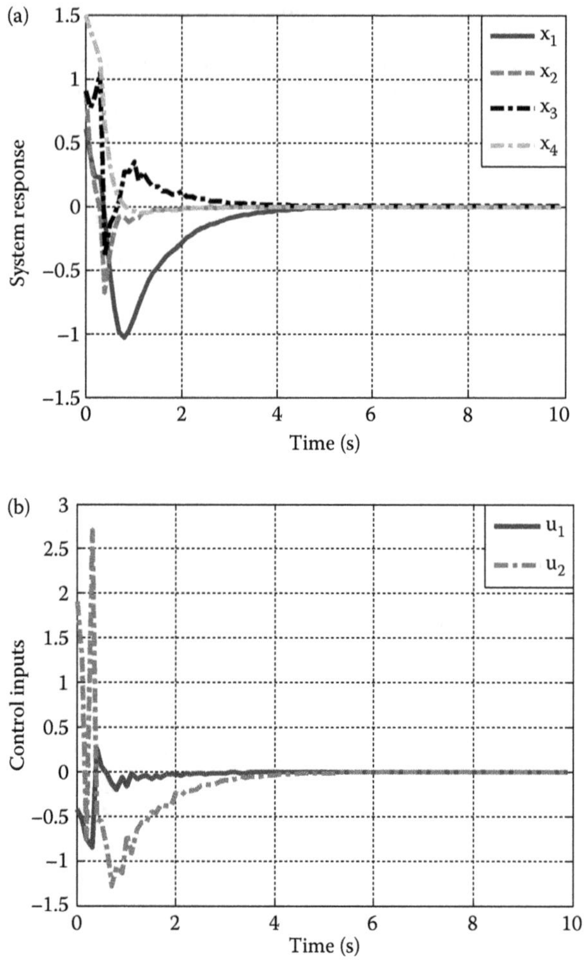

FIGURE 4.5 (a) System response. (b) Control inputs.

On the other hand, when the number of bits for the quantizer is increased to eight, it is clearly shown from Figure 4.7b that with the proposed dynamic quantizer, the quantization error shrinks over time, whereas in the case of traditional static quantizer, the quantization error remains bounded as time evolves. This illustrates the fact that the effect of the quantization error can be properly handled by our proposed quantizer design.

Next, to show the optimality of our proposed scheme, the error history is given in Figure 4.8. It can be seen from this figure that the Bellman error converges to zero, which shows that the optimality is indeed achieved. More importantly, the terminal constraint error shown in this figure also converges close to zero as time evolves, which illustrates that the terminal constraint is also properly satisfied with our finite-horizon optimal control design algorithm. It should be noted that the terminal constraint error does not converge exactly to zero due to the choice of the time-dependent regression

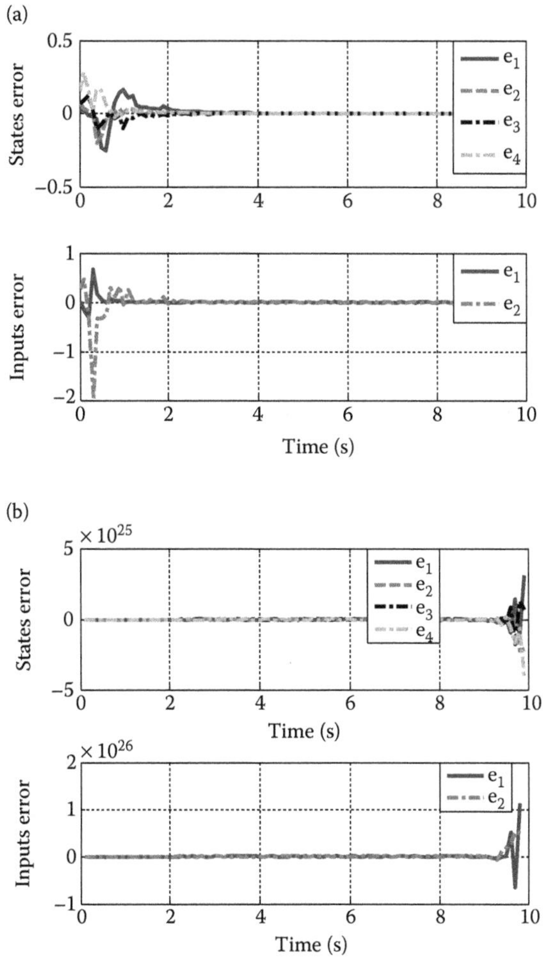

FIGURE 4.6 (a) Quantization error from dynamic quantizer with $R = 4$. (b) Quantization error from traditional static quantizer with $R = 4$.

function. A more appropriate regression function would yield a better convergence of the terminal constraint error, which will be considered as our future work.

Finally, for comparison purposes, the cost function difference between the backward-in-time RE-based approach with known system dynamics and our proposed forward-in-time scheme with unknown system dynamics is shown in Figure 4.9. The simulation result clearly shows that the difference in the cost also converges to zero similar to the system response validating the proposed scheme.

4.3 FINITE-HORIZON OPTIMAL CONTROL OF NONLINEAR QNCS

In this section, the output feedback-based near-optimal regulation scheme over finite horizon for uncertain quantized nonlinear NCS systems with an input constraint

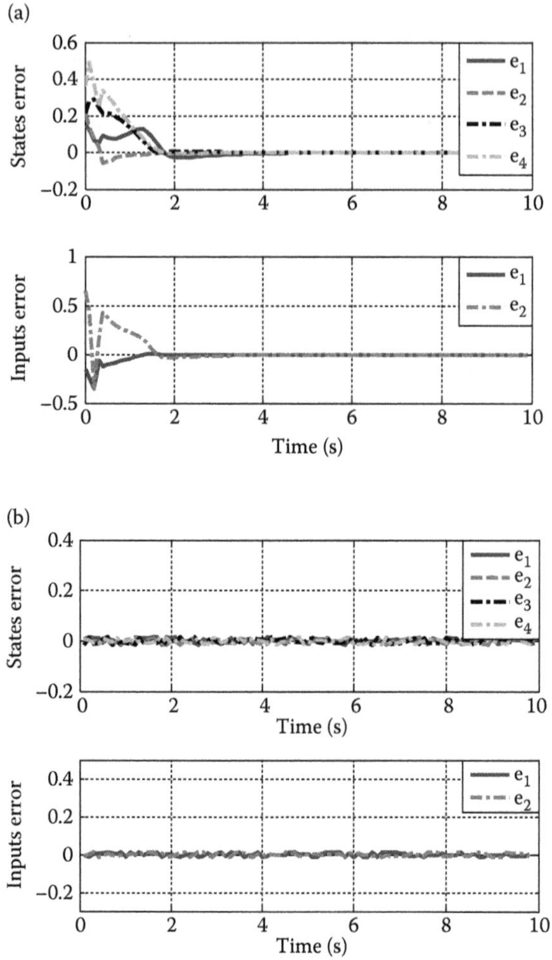

FIGURE 4.7 (a) Quantization error from traditional static quantizer with $R = 8$. (b) Quantization error from dynamic quantizer with $R = 8$.

is addressed from Zhao et al. (2015b). Similar to the previous section, a lossless communication network is considered with perfect output measurements. First, owing to the unavailability of the system state vector and uncertain system dynamics, an extended version of the Luenberger observer by using an NN is proposed to reconstruct both the system state vector and control coefficient matrix in an online manner. Thus the proposed observer design relaxes the need for an explicit identifier and naturally lends itself to stability.

Next, the ADP framework is utilized to approximate the time-varying value function with actor–critic structure, while both NNs are represented by constant weights and time-varying activation functions. An error term corresponding to the terminal constraint is defined and minimized over time. Finally, a novel dynamic quantizer is proposed to reduce the quantization error over time. The stability of

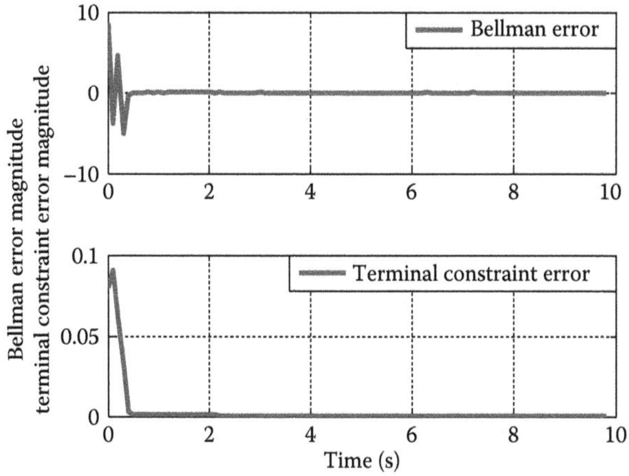

FIGURE 4.8 Cost difference between the proposed and traditional approach.

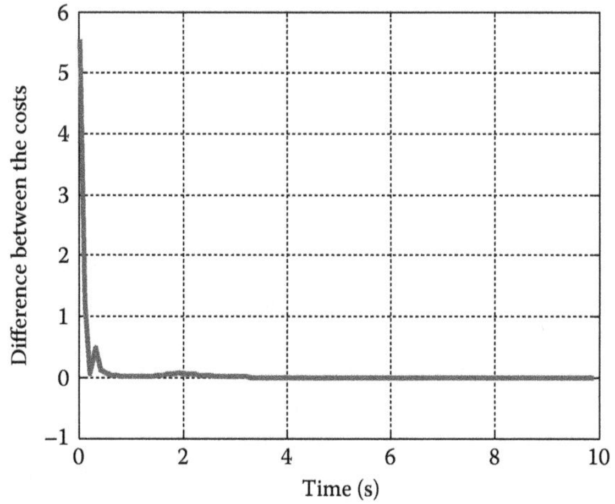

FIGURE 4.9 Error history.

the closed-loop system is demonstrated by using Lyapunov theory to show that the parameter estimation remains bounded as the system evolves provided an initial admissible control input is chosen.

4.3.1 Observer Design

The system dynamics (4.6) can be reformulated as

$$\underset{\tau,\gamma}{E}[\boldsymbol{x}_{k+1}] = \underset{\tau,\gamma}{E}[\boldsymbol{A}\boldsymbol{x}_k + F(\boldsymbol{x}_k) + g(\boldsymbol{x}_k)\boldsymbol{u}_{qk}], \quad \underset{\tau,\gamma}{E}(\boldsymbol{y}_k) = \underset{\tau,\gamma}{E}(\boldsymbol{C}\boldsymbol{x}_k) \tag{4.39}$$

where $\boldsymbol{A}$ is a Hurwitz matrix such that $(\boldsymbol{A}, \boldsymbol{C})$ is observable and $F(\boldsymbol{x}_k) = f(\boldsymbol{x}_k) - \boldsymbol{A}\boldsymbol{x}_k$. An NN has been proven to be an effective method in the estimation and control of nonlinear systems due to its online learning capability. According to the universal approximation property (Narendra and Parthasarathy 1990; Sarangapani 2006), the system state vector can be represented by using NN on a compact set Ω as

$$
\begin{aligned}
\underset{\tau,\gamma}{E}(x_{k+1}) &= \underset{\tau,\gamma}{E}[\boldsymbol{A}\boldsymbol{x}_k + F(\boldsymbol{x}_k) + g(\boldsymbol{x}_k)\boldsymbol{u}_{qk}] \\
&= \underset{\tau,\gamma}{E}[\boldsymbol{A}\boldsymbol{x}_k + \boldsymbol{W}_F^T\sigma_F(\boldsymbol{x}_k) + \boldsymbol{W}_g^T\sigma_g(\boldsymbol{x}_k)\boldsymbol{u}_{qk} + \varepsilon_{Fk} + \varepsilon_{gk}\boldsymbol{u}_{qk}] \\
&= \underset{\tau,\gamma}{E}\left(\boldsymbol{A}\boldsymbol{x}_k + \begin{bmatrix} \boldsymbol{W}_F \\ \boldsymbol{W}_g \end{bmatrix}^T \begin{bmatrix} \sigma_F(\boldsymbol{x}_k) & 0 \\ 0 & \sigma_g(\boldsymbol{x}_k) \end{bmatrix} \begin{bmatrix} 1 \\ \boldsymbol{u}_{qk} \end{bmatrix} + [\varepsilon_{Fk} \quad \varepsilon_{gk}] \begin{bmatrix} 1 \\ \boldsymbol{u}_{qk} \end{bmatrix}\right) \\
&= \underset{\tau,\gamma}{E}(\boldsymbol{A}\boldsymbol{x}_k + W^T\sigma(\boldsymbol{x}_k)\bar{\boldsymbol{u}}_{qk} + \bar{\varepsilon}_k)
\end{aligned}
\tag{4.40}
$$

where

$$
\boldsymbol{W} = \begin{bmatrix} \boldsymbol{W}_F \\ \boldsymbol{W}_g \end{bmatrix} \in \Re^{L\times n}, \quad \sigma(\boldsymbol{x}_k) = \begin{bmatrix} \sigma_F(\boldsymbol{x}_k) & 0 \\ 0 & \sigma_g(\boldsymbol{x}_k) \end{bmatrix} \in \Re^{L\times(1+m)},
$$

$$
\bar{\boldsymbol{u}}_{qk} = \begin{bmatrix} 1 \\ \boldsymbol{u}_{qk} \end{bmatrix} \in \Re^{(1+m)}
$$

and $\bar{\varepsilon}_k = [\varepsilon_{Fk} \ \ \varepsilon_{gk}]\bar{\boldsymbol{u}}_{qk} \in \Re^n$, with L being the number of hidden neurons. In addition, the target NN weights, activation function, and reconstruction errors are assumed to be upper bounded by $\|\boldsymbol{W}\| \le W_M$, $\|\sigma(x_k)\| \le \sigma_M$ and $\|\bar{\varepsilon}_k\| \le \bar{\varepsilon}_M$, where W_M, σ_M, and $\bar{\varepsilon}_M$ are positive constants. Then, the system states $\boldsymbol{x}_{k+1} = \boldsymbol{A}\boldsymbol{x}_k + F(\boldsymbol{x}_k) + g(\boldsymbol{x}_k)\boldsymbol{u}_{qk}$ can be identified by updating the target NN weight matrix $\boldsymbol{W}$.

Since the true system state vector is unavailable for the controller, we propose the following extended Luenberger observer by using an NN described by

$$
\underset{\tau,\gamma}{E}(\hat{\boldsymbol{x}}_{k+1}) = \underset{\tau,\gamma}{E}[\boldsymbol{A}\hat{\boldsymbol{x}}_k + \hat{\boldsymbol{W}}_k^T\sigma(\hat{\boldsymbol{x}}_k)\bar{\boldsymbol{u}}_{qk} + \boldsymbol{L}(\boldsymbol{y}_k - \boldsymbol{C}\hat{\boldsymbol{x}}_k)], \ \underset{\tau,\gamma}{E}(\hat{\boldsymbol{y}}_k) = \underset{\tau,\gamma}{E}(\boldsymbol{C}\hat{\boldsymbol{x}}_k) \tag{4.41}
$$

where $\hat{\boldsymbol{W}}_k$ is the estimated value of the target NN weights $\boldsymbol{W}$, $\hat{\boldsymbol{x}}_k$ is the reconstructed system state vector, $\hat{\boldsymbol{y}}_k$ is the estimated output vector, and $\boldsymbol{L} \in \Re^{n\times p}$ is the observer gain selected by the designer. Now define the state estimation error as

$$
\begin{aligned}
\underset{\tau,\gamma}{E}(\tilde{\boldsymbol{x}}_{k+1}) &= \underset{\tau,\gamma}{E}(\boldsymbol{x}_{k+1} - \hat{\boldsymbol{x}}_{k+1}) \\
&= \underset{\tau,\gamma}{E}[\boldsymbol{A}\boldsymbol{x}_k + \boldsymbol{W}^T\sigma(\boldsymbol{x}_k)\bar{\boldsymbol{u}}_{qk} + \bar{\varepsilon}_k - (\boldsymbol{A}\hat{\boldsymbol{x}}_k + \hat{\boldsymbol{W}}_{k+1}^T\sigma(\hat{\boldsymbol{x}}_k)\bar{\boldsymbol{u}}_{qk} + \boldsymbol{L}(\boldsymbol{y}_k - \boldsymbol{C}\hat{\boldsymbol{x}}_k))] \\
&= \underset{\tau,\gamma}{E}\left[\boldsymbol{A}_c\tilde{\boldsymbol{x}}_k + \tilde{\boldsymbol{W}}_k^T\sigma(\hat{\boldsymbol{x}}_k)\bar{\boldsymbol{u}}_{qk} + \bar{\varepsilon}_{Ok}\right]
\end{aligned}
\tag{4.42}
$$

where $A_c = A - LC$ is the closed-loop matrix, $\tilde{W}_k = W - \hat{W}_k$ is the NN weights estimation error, and $\tilde{\sigma}(x_k, \hat{x}_k) = \sigma(x_k) - \sigma(\hat{x}_k)$ and $\bar{\varepsilon}_{Ok} = W^T \tilde{\sigma}(x_k, \hat{x}_k)\bar{u}_{qk} + \bar{\varepsilon}_k$ are bounded terms due to the bounded values of ideal NN weights, activation functions, and reconstruction errors.

Remark 4.9

The observer presented in Equation 4.41 is novel since it generates both the reconstructed system state vector and the control coefficient matrix $g(x_k)$ for the near-optimal controller design, which can be viewed as an NN-based identifier.

Now select the tuning law for the NN weights as

$$\underset{\tau,\gamma}{E}(\hat{W}_{k+1}) = \underset{\tau,\gamma}{E}[(1-\alpha_I)\hat{W}_k + \beta_I \sigma(\hat{x}_k)\bar{u}_{qk}\tilde{y}_{k+1}^T l^T] \tag{4.43}$$

where α_I, β_I are the tuning parameters, $\tilde{y}_{k+1} = y_{k+1} - \hat{y}_{k+1}$ is the output error, and $l \in \Re^{n\times p}$ is selected as column vectors with all ones to match the dimension. Hence, the NN weight estimation error dynamics, by recalling from Equation 4.42, are revealed to be

$$\begin{aligned}\underset{\tau,\gamma}{E}(\tilde{W}_{k+1}) &= \underset{\tau,\gamma}{E}(W - \hat{W}_{k+1})\\ &= \underset{\tau,\gamma}{E}[(1-\alpha_I)\tilde{W}_k + \alpha_I W - \beta_I \sigma(\hat{x}_k)\bar{u}_{qk}\tilde{y}_{k+1}^T l^T]\\ &= \underset{\tau,\gamma}{E}[(1-\alpha_I)\tilde{W}_k + \alpha_I W - \beta_I \sigma(\hat{x}_k)\bar{u}_{qk}\tilde{x}_k^T A_c^T C^T l^T\\ &\quad - \beta_I \sigma(\hat{x}_k)\bar{u}_{qk}\bar{u}_{qk}^T \sigma^T(\hat{x}_k)\tilde{W}_k C^T l^T - \beta_I \sigma(\hat{x}_k)\bar{u}_{qk}\bar{\varepsilon}_{Ok}^T C^T l^T]\end{aligned} \tag{4.44}$$

Next, the boundedness of the NN weights estimation error $\tilde{W}_k$ will be demonstrated in Theorem 4.3. Before proceeding, the following definition is required.

Definition 4.1: Persistence of Excitation

The function $x_k \in \Re^n$ is said to be PE if and only if there exists a positive constant δ_1, l such that for all $k_0 \geq 0$,

$$\sum_{i=k_0}^{k_0+l-1} \underset{\tau,\gamma}{E}[x_k x_k^T] > \delta I_n$$

where $\mathbf{I}_n \in \Re^{n\times n}$ is the identity matrix and l is termed as the excitation period of x_k.

Theorem 4.3: Boundedness of the Observer Error

Let the initial NN observer weights $\hat{\boldsymbol{W}}_k$ be selected within the compact set Ω_{ID}. Given an admissible control input $\boldsymbol{u}_0 \in \Omega_u$ and Assumption 4.3, let the proposed observer be given by Equation 4.41 while its NN weight tuning law be given by Equation 4.43. By using both the PE condition on the control input signals and pole placement (Chen 2012) method to identify $\boldsymbol{A}$, $\boldsymbol{L}$ satisfying $\|A_c\| = \|A - LC\| \le \sqrt{1/(1+4C_M^2(1+\sigma_{\min}^2))}$, there exist positive constants α_I and β_I satisfying $0 < \beta_I < 2(1-\alpha_I)C_M/(\sigma_{\min}^2+1)$ with $0 < \|\boldsymbol{C}\| \le C_M$, and $(2-\sqrt{2})/2 < \alpha_I < 1$ such that the observer error $\underset{\tau,\gamma}{E}(\tilde{\boldsymbol{x}}_k)$ and the NN weights estimation errors $\underset{\tau,\gamma}{E}(\tilde{\boldsymbol{W}}_k)$ are all bounded in the mean.

Proof: Follow steps similar to Zhao et al. (2015b).

4.3.2 Near-Optimal Regulator Design

According to the universal approximation property of NNs (Sarangapani 2006) and actor–critic methodology, the certainty-equivalent value function and control inputs can be represented by a "critic" NN and an "actor" NN, respectively, as

$$
\begin{aligned}
V(\boldsymbol{x}_k,k) &= \underset{\tau,\gamma}{E}[\boldsymbol{W}_V^T\sigma_V(\boldsymbol{x}_k,k)+\varepsilon_V(\boldsymbol{x}_k,k)] \\
\boldsymbol{u}(\boldsymbol{x}_k,k) &= \underset{\tau,\gamma}{E}[\boldsymbol{W}_u^T\boldsymbol{\sigma}_u(\boldsymbol{x}_k,k)+\boldsymbol{\varepsilon}_u(\boldsymbol{x}_k,k)]
\end{aligned}
\tag{4.45}
$$

where $\boldsymbol{W}_V \in \Re^{L_V}$ and $\boldsymbol{W}_u \in \Re^{L_u\times m}$ are the constant target NN weights, with L_V and L_u the number of hidden neurons, $\sigma_V(\boldsymbol{x}_k,k) \in \Re^{L_V}$ and $\boldsymbol{\sigma}_u(\boldsymbol{x}_k,k) \in \Re^{L_u}$ are the *time-varying* stochastic activation functions, and $\varepsilon_V(\boldsymbol{x}_k, k)$ and $\boldsymbol{\varepsilon}_u(\boldsymbol{x}_k, k)$ are the NN reconstruction errors for the critic and action network, respectively. Under standard assumption, the target NN weights are considered bounded above such that $\underset{\tau,\gamma}{E}\|\boldsymbol{W}_V\| \le W_{VM}$ and $\underset{\tau,\gamma}{E}\|\boldsymbol{W}_u\| \le W_{uM}$, respectively, where both W_{VM} and W_{uM} are positive constants (Sarangapani 2006).

The NN activation functions and the reconstruction errors are also assumed to be bounded above such that $\underset{\tau,\gamma}{E}\|\sigma_V(\boldsymbol{x}_k,k)\| \le \sigma_{VM}$, $\underset{\tau,\gamma}{E}\|\boldsymbol{\sigma}_u(\boldsymbol{x}_k,k)\| \le \sigma_{uM}$, $|\varepsilon_V(\boldsymbol{x}_k, k)| \le \varepsilon_{VM}$ and $|\boldsymbol{\varepsilon}_u(\boldsymbol{x}_k, k)| \le \varepsilon_{uM}$, with σ_{VM}, σ_{uM}, ε_{VM}, and ε_{uM} all positive constants (Sarangapani 2006). In addition, in this book, the gradient of the reconstruction error is also assumed to be bounded above such as $\|\partial\varepsilon_{V,k}/\partial\boldsymbol{x}_{k+1}\| \le \varepsilon'_{VM}$, with ε'_{VM} a positive constant (Dierks and Jagannathan 2012). The terminal constraint of the value function is defined, similar to Equation 4.20, as

$$
V(\boldsymbol{x}_N,N) = \underset{\tau,\gamma}{E}[\boldsymbol{W}_V^T\sigma_V(\boldsymbol{x}_N,N)+\varepsilon_V(\boldsymbol{x}_N,N)] \tag{4.46}
$$

with $\sigma_V(\boldsymbol{x}_N, N)$ and $\varepsilon_V(\boldsymbol{x}_N, N)$ representing the activation and construction error corresponding to the terminal state $\boldsymbol{x}_N$.

4.3.2.1 Value Function Approximation

According to Equation 4.45, the certainty-equivalent time-varying stochastic value function $V(\mathbf{x}_k, k)$ can be approximated by using an NN as

$$\hat{V}(\hat{\boldsymbol{x}}_k, k) = \underset{\tau,\gamma}{E}[\hat{\boldsymbol{W}}_{Vk}^T \sigma_V(\hat{\boldsymbol{x}}_k, k)] \tag{4.47}$$

where $\hat{V}(\hat{\boldsymbol{x}}_k, k)$ represents the approximated value function at time step k. $\hat{\boldsymbol{W}}_{Vk}$ and $\sigma_V(\hat{\boldsymbol{x}}_k, k)$ are the estimated critic NN weights and "reconstructed" activation function with the estimated states vector $\hat{\boldsymbol{x}}_k$ as the inputs. The value function at the terminal stage can be represented by

$$\hat{V}(\boldsymbol{x}_N, N) = \underset{\tau,\gamma}{E}(\hat{\boldsymbol{W}}_{Vk}^T \sigma_V(\hat{\boldsymbol{x}}_N, N)) \tag{4.48}$$

where $\hat{\boldsymbol{x}}_N$ is an estimation of the terminal state. It should be noted that since the true value of $\boldsymbol{x}_N$ is not known, $\hat{\boldsymbol{x}}_N$ can be considered to be an "estimate" of $\boldsymbol{x}_N$ and can be chosen at random as long as $\hat{\boldsymbol{x}}_N$ lies within a region for a stabilizing control policy (Chen and Jagannathan 2008).

To ensure optimality, the Bellman equation should hold along the system trajectory. According to the principle of optimality, the true Bellman equation is given by

$$\underset{\tau,\gamma}{E}[\boldsymbol{Q}(\boldsymbol{x}_k, k) + W(\boldsymbol{u}_k^*) + V^*(\boldsymbol{x}_{k+1}, k+1) - V^*(\boldsymbol{x}_k, k)] = 0 \tag{4.49}$$

However, Equation 4.49 no longer holds when the reconstructed system state vector $\hat{\boldsymbol{x}}_k$ and NN approximation are considered. Therefore, with estimated values, the Bellman Equation 4.24 becomes

$$\begin{aligned}\underset{\tau,\gamma}{E}(e_{B,k}) &= \underset{\tau,\gamma}{E}[\boldsymbol{Q}(\hat{\boldsymbol{x}}_k, k) + W(\boldsymbol{u}_k) + \hat{V}(\hat{\boldsymbol{x}}_{k+1}, k+1) - \hat{V}(\hat{\boldsymbol{x}}_k, k)] \\ &= \underset{\tau,\gamma}{E}[\boldsymbol{Q}(\hat{\boldsymbol{x}}_k, k) + W(\boldsymbol{u}_k) + \hat{\boldsymbol{W}}_{Vk}^T \sigma_V(\hat{\boldsymbol{x}}_{k+1}, k+1) - \hat{\boldsymbol{W}}_{Vk}^T \sigma_V(\hat{\boldsymbol{x}}_k, k)] \\ &= \underset{\tau,\gamma}{E}[\boldsymbol{Q}(\hat{\boldsymbol{x}}_k, k) + W(\boldsymbol{u}_k) + \hat{\boldsymbol{W}}_{Vk}^T \Delta\sigma_V(\hat{\boldsymbol{x}}_k, k)]\end{aligned} \tag{4.50}$$

where $e_{B,k}$ is the residual error of the Bellman equation *along the system trajectory*, and $\Delta\sigma_V(\hat{\boldsymbol{x}}_k, k) = \sigma_V(\hat{\boldsymbol{x}}_k, k) - \sigma_V(\hat{\boldsymbol{x}}_{k+1}, k+1)$.

Next, using Equation 4.48, define an additional error term corresponding to the terminal constraint as

$$\underset{\tau,\gamma}{E}(e_{N,k}) = \underset{\tau,\gamma}{E}[\psi(\boldsymbol{x}_N) - \hat{\boldsymbol{W}}_{Vk}^T \sigma_V(\hat{\boldsymbol{x}}_N, N)] \tag{4.51}$$

The objective of the optimal control design is thus to minimize both the Bellman equation residual error $e_{B,k}$ and the terminal constraint error $e_{N,k}$. Next, based on the gradient descent approach, the update law for critic NN can be defined as

$$\underset{\tau,\gamma}{E}(\hat{\boldsymbol{W}}_{Vk+1}) = \underset{\tau,\gamma}{E}\left(\hat{\boldsymbol{W}}_{Vk} + \alpha_V \frac{(\Delta\sigma_V(\hat{\boldsymbol{x}}_k,k) + \sigma_V'(\hat{\boldsymbol{x}}_{k+1},k+1) + 2\sigma_{VM}\boldsymbol{B}_l)e_{B,k}}{1 + \left\|\Delta\sigma_V(\hat{\boldsymbol{x}}_k,k) + \sigma_V'(\hat{\boldsymbol{x}}_{k+1},k+1) + 2\sigma_{VM}\boldsymbol{B}_l\right\|^2}\right.$$

$$\left. - \alpha_V \frac{\sigma_V(\hat{\boldsymbol{x}}_N,N)e_{N,k}}{1 + \sigma_V^T(\hat{\boldsymbol{x}}_N,N)\sigma_V(\hat{\boldsymbol{x}}_N,N)}\right) \tag{4.52}$$

where α_V is a design parameter, $\sigma_V'(\hat{\boldsymbol{x}}_{k+1},k+1)$ is the gradient of $\sigma_V(\hat{\boldsymbol{x}}_{k+1},k+1)$, and $\boldsymbol{B}_l \in \Re^{L_V}$ is a constant vector.

Now define $\underset{\tau,\gamma}{E}(\tilde{\boldsymbol{W}}_{Vk}) = \underset{\tau,\gamma}{E}[\boldsymbol{W}_V - \hat{\boldsymbol{W}}_{Vk}]$. The standard Bellman Equation 4.49 can be expressed by NN representation as

$$0 = \underset{\tau,\gamma}{E}[\boldsymbol{Q}(\boldsymbol{x}_k,k) + W(\boldsymbol{u}_k^*) - \boldsymbol{W}_V^T\Delta\sigma_V(\boldsymbol{x}_k,k) - \Delta\varepsilon_V(\boldsymbol{x}_k,k)] \tag{4.53}$$

where $\Delta\sigma_V(\boldsymbol{x}_k,k) = \sigma_V(\boldsymbol{x}_k,k) - \sigma_V(\boldsymbol{x}_{k+1},k+1)$ and $\Delta\varepsilon_V(\boldsymbol{x}_k,k) = \varepsilon_V(\boldsymbol{x}_k,k) - \varepsilon_V(\boldsymbol{x}_{k+1},k+1)$. Subtracting Equations 4.50 from Equation 4.53,

$$\begin{aligned}
\underset{\tau,\gamma}{E}\{e_{B,k}\} &= \underset{\tau,\gamma}{E}\Big[\boldsymbol{Q}(\hat{\boldsymbol{x}}_k,k) + W(\boldsymbol{u}_k) + \hat{\boldsymbol{W}}_{Vk}^T\Delta\sigma_V(\hat{\boldsymbol{x}}_k,k) + \hat{\boldsymbol{W}}_{Vk}^T\Delta\sigma_V(f(\hat{\boldsymbol{x}}_k) + g(\hat{\boldsymbol{x}}_k)\boldsymbol{u}_k,\ k+1) \\
&\quad - \hat{\boldsymbol{W}}_{Vk}^T\Delta\sigma_V(f(\hat{\boldsymbol{x}}_k) + g(\hat{\boldsymbol{x}}_k)\boldsymbol{u}_k, k+1) \\
&\quad - \boldsymbol{Q}(\boldsymbol{x}_k,\ k) - W(\boldsymbol{u}_k^*) + \boldsymbol{W}_V^T\Delta\sigma_V(\boldsymbol{x}_k,\ k) + \Delta\varepsilon_V(\boldsymbol{x}_k,k)\Big] \\
&\le L_Q \underset{\tau,\gamma}{E}\left\|\tilde{\boldsymbol{x}}_k\right\|^2 + 2\underset{\tau,\gamma}{E}\left[\int_{\boldsymbol{u}_k^*}^{\boldsymbol{u}_k}(\varphi^{-1}(v))^T\boldsymbol{R}dv + \tilde{\boldsymbol{W}}_{Vk}^T\Delta\sigma_V(\hat{\boldsymbol{x}}_k,k)\right. \\
&\quad + W_{VM}L_{\sigma_V}\underset{\tau,\gamma}{E}\left\|\boldsymbol{u}(\hat{\boldsymbol{x}}_k,k) + \varphi\left(\frac{1}{2}\boldsymbol{R}^{-1}g^T(\hat{\boldsymbol{x}}_k)\nabla\sigma_V^T(\hat{\boldsymbol{x}}_{k+1},k+1)\hat{\boldsymbol{W}}_{Vk}\right)\right\| \\
&\quad + \tilde{\boldsymbol{W}}_{Vk}^T\sigma_{VM}\boldsymbol{B}_l\operatorname{sgn}(\tilde{\boldsymbol{W}}_{Vk}) + \boldsymbol{W}_V^T\Delta\tilde{\sigma}_V(\boldsymbol{x}_k,\hat{\boldsymbol{x}}_k,\ k) + \Delta\varepsilon_V(\boldsymbol{x}_k,\ k) \\
&\le (L_Q + W_{VM}L_{\sigma_V})\underset{\tau,\gamma}{E}\left\|\tilde{\boldsymbol{x}}_k\right\|^2 + 2Ru_M\underset{\tau,\gamma}{E}\left\|\tilde{\boldsymbol{u}}_k\right\| + \tilde{\boldsymbol{W}}_{Vk}^T\Delta\sigma_V(\hat{\boldsymbol{x}}_k,k) \\
&\quad + 2W_{VM}\sigma_{VM} + \tilde{\boldsymbol{W}}_{Vk}^T\sigma_{VM}\boldsymbol{B}_l\operatorname{sgn}(\tilde{\boldsymbol{W}}_{Vk}) + \Delta\varepsilon_{VB}(\boldsymbol{x}_k,\ k)
\end{aligned} \tag{4.54}$$

where L_Q is a positive Lipschitz constant for $\boldsymbol{Q}(\bullet,k)$ due to selected quadratic form in system states and L_{σ_V} is a positive Lipschitz constant for $\sigma_V(\bullet,k)$. In addition, $\Delta\tilde{\sigma}_V(\boldsymbol{x}_k,\hat{\boldsymbol{x}}_k,k) = \Delta\sigma_V(\boldsymbol{x}_k,k) - \Delta\sigma_V(\hat{\boldsymbol{x}}_k,k)$ and $\Delta\varepsilon_{VB}(\boldsymbol{x}_k,k) = \boldsymbol{W}_V^T\Delta\tilde{\sigma}_V(\boldsymbol{x}_k,\hat{\boldsymbol{x}}_k,k) + \Delta\varepsilon_V(\boldsymbol{x}_k,k)$

are all bounded terms due to boundedness of ideal NN weights, activation functions, and reconstruction errors.

Recalling from Equation 4.51, the terminal constraint error $e_{N,k}$ can be further expressed as

$$\begin{aligned}
\underset{\tau,\gamma}{E}\{e_{N,k}\} &= \underset{\tau,\gamma}{E}\left[\psi(\boldsymbol{x}_N) - \hat{\boldsymbol{W}}_{Vk}^T\sigma_V(\hat{\boldsymbol{x}}_N, N)\right] \\
&= \underset{\tau,\gamma}{E}\left[\boldsymbol{W}_V^T\sigma_V(\boldsymbol{x}_N, N) - \boldsymbol{W}_V^T\sigma_V(\hat{\boldsymbol{x}}_N, N) + \varepsilon_V(\boldsymbol{x}_N, N) + \boldsymbol{W}_V^T\sigma_V(\hat{\boldsymbol{x}}_N, N)\right] \\
&\quad - \underset{\tau,\gamma}{E}\left[\hat{\boldsymbol{W}}_{Vk}^T\sigma_V(\hat{\boldsymbol{x}}_N, N)\right] \\
&= \underset{\tau,\gamma}{E}\left[\tilde{\boldsymbol{W}}_{Vk}^T\sigma_V(\hat{\boldsymbol{x}}_N, N) + \boldsymbol{W}_V^T\tilde{\sigma}_V(\boldsymbol{x}_N, \hat{\boldsymbol{x}}_N, N) + \varepsilon_V(\boldsymbol{x}_N, N)\right] \\
&= \underset{\tau,\gamma}{E}\left[\tilde{\boldsymbol{W}}_{Vk}^T\sigma_V(\hat{\boldsymbol{x}}_N, N)\right] + \varepsilon_{VN}
\end{aligned} \tag{4.55}$$

where $\tilde{\sigma}_V(\boldsymbol{x}_N, \hat{\boldsymbol{x}}_N, N) = \sigma_V(\boldsymbol{x}_N, N) - \sigma_V(\hat{\boldsymbol{x}}_N, N)$ and $\varepsilon_{VN} = \boldsymbol{W}_V^T\tilde{\sigma}_V(\boldsymbol{x}_N, \hat{\boldsymbol{x}}_N, N) + \varepsilon_V(\boldsymbol{x}_N, N)$ are bounded due to bounded ideal NN weights, activation function, and reconstruction errors. Finally, the error dynamics for critic NN weights are revealed to be

$$\begin{aligned}
\underset{\tau,\gamma}{E}\tilde{\boldsymbol{W}}_{Vk+1} &= \underset{\tau,\gamma}{E}\left\| \tilde{\boldsymbol{W}}_{Vk} - \alpha_V \frac{(\Delta\sigma_V(\hat{\boldsymbol{x}}_k, k) + \sigma_V'(\hat{\boldsymbol{x}}_{k+1}, k+1) + 2\sigma_{VM}\boldsymbol{B}_l)e_{B,k}}{1 + \left\|\Delta\sigma_V(\hat{\boldsymbol{x}}_k, k) + \sigma_V'(\hat{\boldsymbol{x}}_{k+1}, k+1) + 2\sigma_{VM}\boldsymbol{B}_l\right\|^2} \right\| \\
&\quad - \alpha_V \underset{\tau,\gamma}{E} \frac{\sigma_V(\hat{\boldsymbol{x}}_N, N)e_{N,k}}{1 + \sigma_V^T(\hat{\boldsymbol{x}}_N, N)\sigma_V(\hat{\boldsymbol{x}}_N, N)}
\end{aligned} \tag{4.56}$$

4.3.2.2 Control Input Approximation

In this section, the near-optimal control policy is obtained such that the estimated value function (4.47) is minimized. Recalling Equation 4.45, the approximation of the control inputs by using NN can be represented as

$$\underset{\tau,\gamma}{E}[\boldsymbol{u}(\hat{\boldsymbol{x}}_k, k)] = \underset{\tau,\gamma}{E}(\hat{\boldsymbol{W}}_{uk}^T\sigma_u(\hat{\boldsymbol{x}}_k, k)) \tag{4.57}$$

where $\boldsymbol{u}(\hat{\boldsymbol{x}}_k, k)$ represents the approximated control input vector at time step k, $\hat{\boldsymbol{W}}_{uk}$ and $\sigma_u(\hat{\boldsymbol{x}}_k, k)$ are the estimated values of the actor NN weights and "reconstructed" activation function with the estimated state vector $\hat{\boldsymbol{x}}_k$ as the input. Define the control input error as

$$\underset{\tau,\gamma}{E}(\boldsymbol{e}_{uk}) = \underset{\tau,\gamma}{E}[\boldsymbol{u}(\hat{\boldsymbol{x}}_k, k) - \boldsymbol{u}_1(\hat{\boldsymbol{x}}_k, k)] \tag{4.58}$$

where $u_1(\hat{x}_k,k) = -\varphi(1/2R^{-1}\hat{g}^T(\hat{x}_k)\nabla\sigma_V^T(\hat{x}_{k+1},k+1)\hat{W}_{Vk}) \in \Omega_u$ is the control policy that minimizes the approximated value function $\hat{V}(\hat{x}_k,k)$, ∇ denotes the gradient of the estimated value function with respect to the system states, $\hat{g}(\hat{x}_k)$ is the approximated control coefficient matrix generated by the NN-based observer, and $\hat{V}(\hat{x}_{k+1},k+1)$ is the approximated value function from the critic network.

Therefore, the control error (4.58) becomes

$$\begin{aligned} \underset{\tau,\gamma}{E}\{e_{uk}\} &= \underset{\tau,\gamma}{E}\left[u(\hat{x}_k,k) - u_1(\hat{x}_k,k)\right] \\ &= \underset{\tau,\gamma}{E}\,\hat{W}_{uk}^T\sigma_u(\hat{x}_k,k) + \varphi\,\underset{\tau,\gamma}{E}\left(\frac{1}{2}R^{-1}\hat{g}^T(\hat{x}_k)\nabla\sigma_V^T(\hat{x}_{k+1},\,k+1)\hat{W}_{Vk}\right) \end{aligned} \tag{4.59}$$

The actor NN weights tuning law is then defined as

$$\underset{\tau,\gamma}{E}(\hat{W}_{uk+1}) = \underset{\tau,\gamma}{E}(\hat{W}_{uk}) - \alpha_u\,\underset{\tau,\gamma}{E}\left(\frac{\sigma_u(\hat{x}_k,k)e_{uk}^T}{1+\sigma_u^T(\hat{x}_k,k)\sigma_u(\hat{x}_k,k)}\right) \tag{4.60}$$

where $\alpha_u > 0$ is a design parameter. To find the error dynamics for the actor NN weights, first observe that

$$\begin{aligned} 0 = \underset{\tau,\gamma}{E}\Big(&W_u^T\sigma_u(x_k,k) + \varepsilon_u(x_k,k) \\ &+\varphi\left(\frac{1}{2}R^{-1}g^T(x_k)(\nabla\sigma_V^T(x_{k+1},k+1)W_V + \nabla\varepsilon_V(x_{k+1},k+1))\right)\Big) \end{aligned} \tag{4.61}$$

Subtracting Equation 4.61 from Equation 4.59, we have

$$\begin{aligned} \underset{\tau,\gamma}{E}\{e_{uk}\} = &-\underset{\tau,\gamma}{E}\,\tilde{W}_{uk}^T\sigma_u(\hat{x}_k,k) - \frac{1}{2}L_\varphi R^{-1}g^T(x_k)\nabla\sigma_V^T(\hat{x}_{k+1},k+1)\,\underset{\tau,\gamma}{E}\,\tilde{W}_{Vk} \\ &-\frac{1}{2}L_\varphi R^{-1}\tilde{g}^T(\hat{x}_k)\nabla\sigma_V^T(\hat{x}_{k+1},k+1)W_V \\ &-\frac{1}{2}L_\varphi R^{-1}(g^T(\hat{x}_k) - g^T(x_k))\nabla\sigma_V^T(\hat{x}_{k+1},k+1)\tilde{W}_{Vk} \\ &+\frac{1}{2}L_\varphi R^{-1}\tilde{g}^T(\hat{x}_k)\nabla\sigma_V^T(\hat{x}_{k+1},k+1)\tilde{W}_{Vk} + \bar{\varepsilon}_u(x_k,k) \end{aligned} \tag{4.62}$$

where $\tilde{W}_{uk} = W_u - \hat{W}_{uk}$, L_φ is the positive Lipschitz constant for the saturation function $\varphi(\bullet)$,

$$\tilde{\sigma}_u(\boldsymbol{x}_k, \hat{\boldsymbol{x}}_k, k) = \sigma_u(\boldsymbol{x}_k, k) - \sigma_u(\hat{\boldsymbol{x}}_k, k),$$

$$\tilde{\varphi}_k = \varphi\left(\frac{1}{2}\boldsymbol{R}^{-1}\hat{g}^T(\hat{\boldsymbol{x}}_k)\nabla\sigma_V^T(\hat{\boldsymbol{x}}_{k+1}, k+1)\hat{\boldsymbol{W}}_{V,k}\right)$$

$$-\varphi\left(\frac{1}{2}\boldsymbol{R}^{-1}g^T(\boldsymbol{x}_k)(\nabla\sigma_V^T(\boldsymbol{x}_{k+1}, k+1)\boldsymbol{W}_V + \nabla\varepsilon_V(\boldsymbol{x}_{k+1}, k+1))\right)$$

and

$$\bar{\varepsilon}_u(\boldsymbol{x}_k, k) = -\varepsilon_u(\boldsymbol{x}_k, k) + \frac{1}{2}L_\varphi \boldsymbol{R}^{-1}g^T(\boldsymbol{x}_k)\nabla\tilde{\sigma}_V^T(\boldsymbol{x}_{k+1}, \hat{\boldsymbol{x}}_{k+1}, k+1)\boldsymbol{W}_V$$

$$+\frac{1}{2}L_\varphi \boldsymbol{R}^{-1}(g^T(\hat{\boldsymbol{x}}_k) - g^T(\boldsymbol{x}_k))\nabla\sigma_V^T(\hat{\boldsymbol{x}}_{k+1}, k+1)\boldsymbol{W}_V - \frac{1}{2}L_j\boldsymbol{R}^{-1}g^T(\boldsymbol{x}_k)$$

$$\times\nabla\varepsilon_V^T(\hat{\boldsymbol{x}}_{k+1}, k+1) - \boldsymbol{W}_u^T\tilde{\sigma}_u(\boldsymbol{x}_k, \hat{\boldsymbol{x}}_k, k)$$

Note that $\tilde{\sigma}_u(\boldsymbol{x}_k, \hat{\boldsymbol{x}}_k, k)$ and $\bar{\varepsilon}_u(\boldsymbol{x}_k, k)$ are all bounded due to the boundedness of NN activation function and reconstruction error. Then the error dynamics for the actor NN weights are revealed to be

$$\underset{\tau,\gamma}{E}(\tilde{\boldsymbol{W}}_{uk+1}) = \underset{\tau,\gamma}{E}\left(\tilde{\boldsymbol{W}}_{uk} + \alpha_u \frac{\sigma_u(\hat{\boldsymbol{x}}_k, k)\boldsymbol{e}_{uk}^T}{1 + \sigma_u^T(\hat{\boldsymbol{x}}_k, k)\sigma_u(\hat{\boldsymbol{x}}_k, k)}\right) \tag{4.63}$$

Remark 4.10

The actor NN weight tuning based on gradient descent approach is similar to that of Dierks and Jagannathan (2012) with the difference being that the estimated state vector $\hat{\boldsymbol{x}}_k$ is utilized as the input to the actor NN activation function instead of the measured state vector $\boldsymbol{x}_k$. In addition, the total error comprising of the Bellman and terminal constraint errors are utilized to tune the weights whereas in Dierks and Jagannathan (2012), the terminal constraint is ignored. Further, the optimal control scheme in this book utilizes the identified control coefficient matrix $\hat{g}(\hat{\boldsymbol{x}}_k)$, whereas in Dierks and Jagannathan (2012), the control coefficient matrix $g(\boldsymbol{x}_k)$ is considered known. Owing to these differences, the stability analysis differs significantly from that of Dierks and Jagannathan (2012).

4.3.2.3 Dynamic Quantizer Design

To handle the saturation caused by limited quantization range in a realistic quantizer, a new parameter μ_k is introduced. The proposed dynamic quantizer for the control input is defined as

$$\boldsymbol{u}_{qk} = q_d(\boldsymbol{u}_k) = \mu_k q\left(\frac{\boldsymbol{u}_k}{\mu_k}\right) \tag{4.64}$$

where μ_k is a time-varying scaling parameter to be defined later for the control input quantizers. Normally, the dynamics of the quantization error cannot be established since it is mainly a round-off error. Instead, we will consider the quantization error bound as presented next, which will aid in the stability analysis. Given the dynamic quantizer in the form (4.64), the quantization error for the control inputs (4.59) is bounded, as long as saturation does not occur with the bound

$$\underset{\tau,\gamma}{E}\left\|\boldsymbol{e}_{uk}\right\| \leq \frac{1}{2}\mu_k\Delta_k = e_{M,k} \tag{4.65}$$

where $e_{M,k}$ is the upper bound for the control input quantization error.

Next, define the scaling parameter μ_k as

$$\mu_k = \underset{\tau,\gamma}{E}\frac{\left\|\boldsymbol{u}_k\right\|}{(\lambda^k \boldsymbol{M})} \tag{4.66}$$

where $0 < \lambda < 1$. Recall from representation (4.64) that the signals to be quantized can be "scaled" back into the quantization range with the decaying rate of λ^k, thus eliminating the saturation effect.

To complete this section, the flowchart of our proposed finite-horizon near-optimal regulation scheme is shown in Figure 4.10. We initialize the system with an admissible control input and for proper parameter selection, the NN weights are initialized. The control input is then quantized by using the proposed dynamic quantizer. The NNs for observer, critic, and actor are updated based on our proposed weight tuning laws at each sampling interval, beginning with an initial time and until the final fixed time instant in an online and forward-in-time fashion.

4.3.2.4 Stability Analysis

In this section, system stability will be investigated. It will be shown that the overall closed-loop system remains bounded under the proposed near-optimal regulator design. Before proceeding, the following lemma is needed.

Lemma 4.5: Bounds on the Optimal Closed-Loop Dynamics

Consider the nonlinear discrete-time system (4.1) with Assumption 4.3. There exists a certainty-equivalent optimal control policy $\boldsymbol{u}_k^*$ such that the closed-loop system dynamics $f(\boldsymbol{x}_k)+g(\boldsymbol{x}_k)\boldsymbol{u}_k^*$ can be expressed as

$$\underset{\tau,\gamma}{E}\left\|f(\boldsymbol{x}_k)+g(\boldsymbol{x}_k)\boldsymbol{u}_k^*\right\|^2 \leq \rho\underset{\tau,\gamma}{E}\left\|\boldsymbol{x}_k\right\|^2 \tag{4.67}$$

where $0 < \rho < 1$ is a constant.

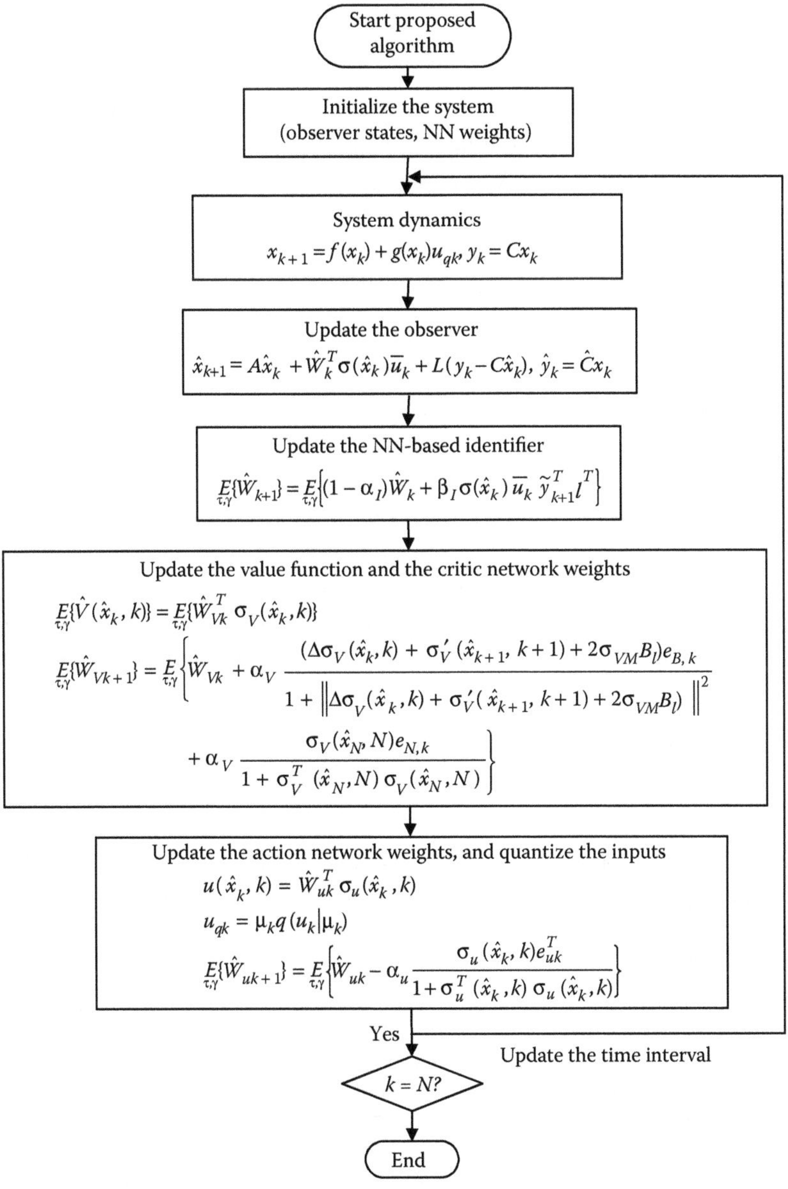

FIGURE 4.10 Flowchart of finite-horizon near-optimal regulator.

Theorem 4.4: Boundedness of the Closed-Loop System

Let Assumption 4.3 holds and an initial control input be admissible. Let the NN weights be selected within a compact set with the observer be provided by Equation 4.41 and the NN weight tuning for the observer, critic network, and action network be given by Equations 4.43, 4.52, and 4.56, respectively. Then, there exists positive constant $(2-\sqrt{2})/2 < \alpha_I < 1$, $0 < \alpha_V < 1/6$, and $0 < \alpha_u < 1$, such that the system state $\boldsymbol{x}_k$, observer error $\underset{\tau,\gamma}{E}(\tilde{\boldsymbol{x}}_k)$, NN observer weight estimation error $\tilde{\boldsymbol{W}}_k$, critic and action network weights estimation errors $\underset{\tau,\gamma}{E}(\tilde{\boldsymbol{W}}_{Vk})$, and $\underset{\tau,\gamma}{E}(\tilde{\boldsymbol{W}}_{uk})$ are all bounded in the mean. In addition, the estimated control input is bounded in the mean close to the optimal value such that $\underset{\tau,\gamma}{E}\|u^*(x_k,k)-\hat{u}(\hat{x}_k,k)\| \le \varepsilon_{uo}$ for a small positive constant ε_{uo}.

4.3.2.5 Simulation Results

In this section, a practical example is considered to illustrate our proposed near-optimal regulation design scheme.

EXAMPLE 4.2

Consider the two-link planar robot arm (Chen and Jagannathan, 2008; Slotine and Li, 1991) given by

$$\dot{\boldsymbol{x}} = f(\boldsymbol{x}) + g(\boldsymbol{x})\boldsymbol{u}, \quad y = \boldsymbol{C}\boldsymbol{x} \tag{4.68}$$

where $f(\boldsymbol{x})$ and $g(\boldsymbol{x})$ can be found from Chen and Jagannathan (2008). The system is discretized with sampling time of $h = 5$ ms and control constraint is set to be $U = 1.5$, that is, $-1.5 \le u_1 \le 1.5$ and $-1.5 \le u_2 \le 1.5$. Define the performance index

$$V(\boldsymbol{x}_k,k) = \underset{\tau,\gamma}{E}\left[\psi(\boldsymbol{x}_N) + \sum_{i=k}^{N-1}(\boldsymbol{Q}(\boldsymbol{x}_i,i) + 2\int_0^{\boldsymbol{u}_i} U\tanh^{-T}\left(\frac{\boldsymbol{v}}{U}\right)\boldsymbol{R}d\boldsymbol{v})\right] \tag{4.69}$$

where $\boldsymbol{Q}(\boldsymbol{x}_k, k)$, for simplicity, is selected as a standard quadratic form of the system states as $\boldsymbol{Q}(\boldsymbol{x}_k,k) = \boldsymbol{x}_k^T\bar{\boldsymbol{Q}}\boldsymbol{x}_k$ with $\boldsymbol{Q} = 0.1\boldsymbol{I}_4$ and the weighting matrix $\boldsymbol{R}$ is selected as $\boldsymbol{R} = 0.001\boldsymbol{I}_2$, where $\boldsymbol{I}$ denotes the identity matrix with appropriate dimension. The Hurwitz matrix $\boldsymbol{A}$ is selected as a 4×4 block diagonal matrix whose blocks $\boldsymbol{A}_{ii}$ are chosen to be $\boldsymbol{A}_{ii} = \begin{bmatrix} 0.9 & 0.1 \\ 0 & 0.9 \end{bmatrix}$. The terminal constraint is chosen as $\psi(\boldsymbol{x}_N) = 3$. The horizon length is 5 s. For the NN setup, the inputs for the NN observer are selected as $\boldsymbol{z}_k = [\hat{\boldsymbol{x}}_k, \boldsymbol{u}_k]$. Inspired by Heydari and Balakrishnan (2013), the time-varying activation functions for the critic and actor network are chosen as sigmoid function with input to be $[\hat{x}_1,\ldots,\hat{x}_4,\tau,\hat{x}_1\hat{x}_2,\ldots,\hat{x}_3\hat{x}_4,\tau^2,\hat{x}_1\tau,\ldots,\hat{x}_4\tau,\hat{x}_1^2,\ldots,\hat{x}_4^2]$ and $[\hat{x}_1,\ldots,\hat{x}_4,\hat{x}_1\tau,\ldots,\hat{x}_4\tau]$, which result in 24 and 8 neurons, respectively, and $\tau = (N-k)/N$ is the normalized time-to-go. A lossless network is selected.

The number of bits for the quantizer is chosen to be 4 while the design parameters are selected as $\alpha_I = 0.7$, $\beta_I = 0.01$, $\alpha_V = 0.1$, $\alpha_u = 0.03$, and $\lambda = 0.9$. The initial system states and the observer states are selected as deterministic $\boldsymbol{x}_0 = [\pi/3, \pi/6, 0, 0]^T$ and $\hat{\boldsymbol{x}}_0 = [0,0,0,0]^T$, respectively. By using the pole placement method, the initial admissible control input is chosen as $\boldsymbol{u}(0) = [0.2; -1]$ and the observer gain is chosen as $\mathbf{L} = [-0.3, 0.1, 0.7,1]^T$ and the matching matrices B_I and $\boldsymbol{I}$ are selected as column vectors with all ones. All the NN weights are initialized at random. To obtain ergodic behavior, Monte Carlo simulation runs with 1000 iterations were utilized.

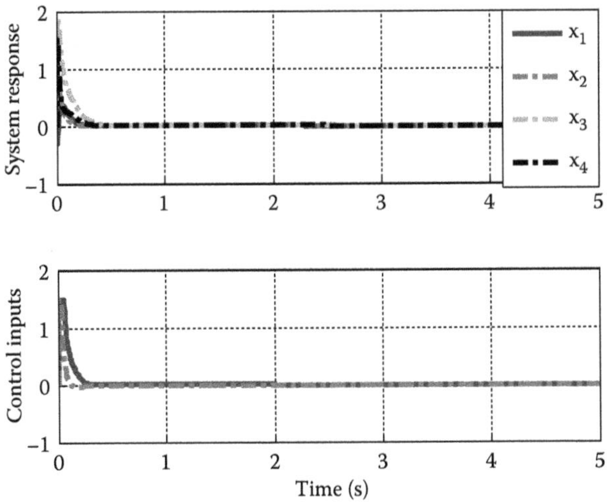

FIGURE 4.11 System response and control inputs.

First, the system response and control input are shown in Figure 4.11. Both the system states and control clearly converge close enough to the origin within finite time, which illustrates the stability of the proposed design scheme. Next, the quantization errors for the control inputs with proposed dynamic quantizer are shown in Figure 4.12a and that for traditional uniform quantizer are shown in Figure 4.12b), respectively. Compared to Figure 4.12a and b, it is clear that the quantization errors are decreasing over time instead of being bounded for the traditional uniform quantizer, illustrating the effectiveness of the proposed dynamic quantizer design.

Next, the error history in the design procedure is given in Figure 4.13. From the figure, it can be seen that the Bellman equation error eventually converges close to zero, which illustrates the fact that the optimality is indeed achieved. More importantly, the convergence of the terminal constraint error demonstrates that the terminal constraint is also satisfied with our proposed design. Finally, the convergence of critic and actor NN weights is shown in Figure 4.14. It can be observed from the results that the novel NN structure with our proposed tuning law guarantees that the NN weights converge to constants and remain bounded, as desired. This illustrates the feasibility of NN approximation for time-varying functions.

4.4 CONCLUSIONS

In this chapter, the optimal control of quantized linear NCS with unknown system dynamics is addressed. First, the proposed novel dynamic quantizer reduces the effect of saturation and quantization error through the zoom parameter update. Dynamics of the system are not needed with an adaptive action-dependent value function $V_{AD}(\boldsymbol{x}_k, \boldsymbol{u}_k, N-k)$. An additional error is defined and incorporated in the update law so that the terminal constraint for the finite horizon can be properly satisfied. An initial admissible control policy ensures the stability of the system while the value function and the kernel matrix $\boldsymbol{G}_k$ are adjusted online. All the parameters

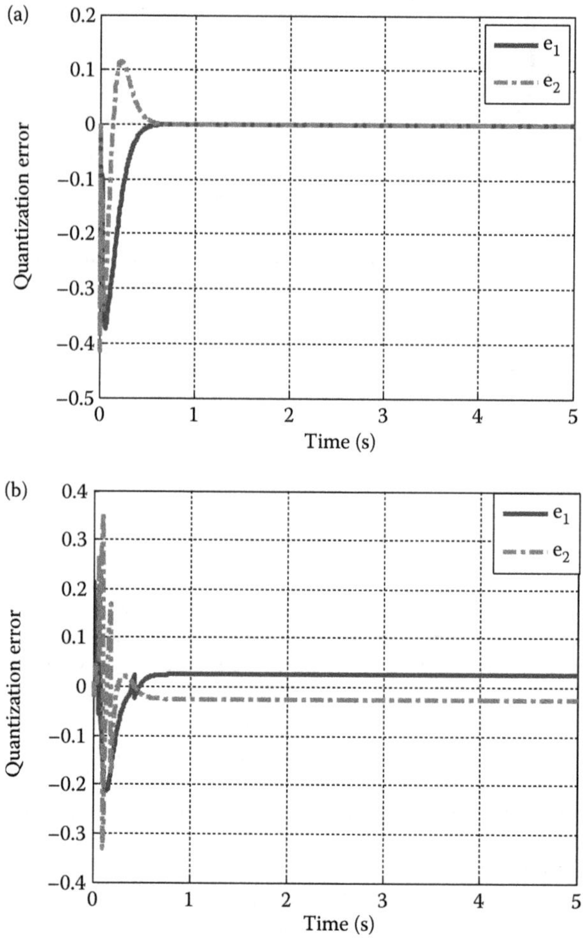

FIGURE 4.12 (a) Quantization error with dynamic quantizer. (b) Quantization error with static quantizer.

are tuned in an online and forward-in-time manner. Policy and value iterations are not needed. The stability of the overall closed-loop system is demonstrated by using Lyapunov and geometric sequence analysis when the state vector is perfectly measured and in the presence of a lossless network.

Next, the optimal control of quantized nonlinear NCS with unknown system dynamics is addressed. The dynamic quantizer effectively mitigates the quantization error for the control inputs while the NN-based Luenberger observer relaxes the need for an additional identifier and generates the state vector estimate with perfect output measurements. The time-dependency nature of the finite horizon is handled by an NN structure with constant weights and time-varying activation function. The terminal constraint is properly satisfied by minimizing an additional error term along the system trajectory. All NN weights are tuned online by using proposed

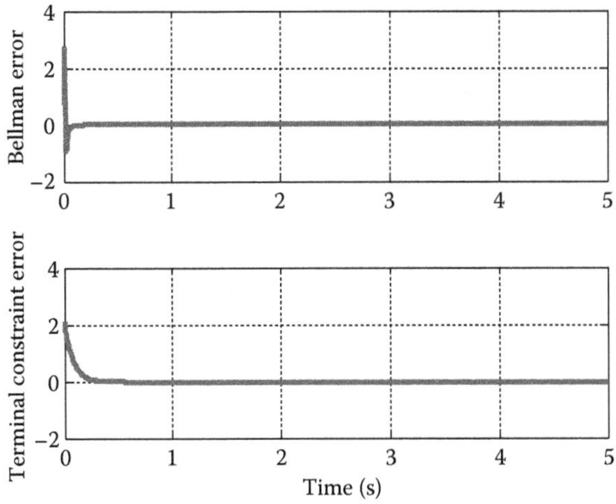

FIGURE 4.13 History of error terms.

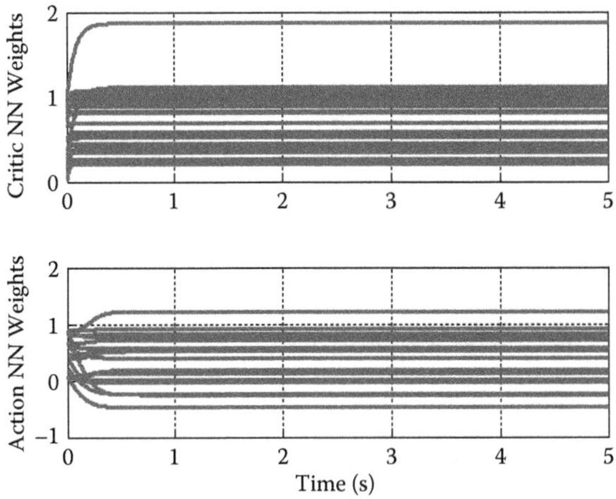

FIGURE 4.14 Convergence of critic/actor NN weight.

update laws, and Lyapunov stability theory demonstrated that the approximated control input converges close to its optimal value as time evolves. The performance of the proposed finite-time near-optimal regulator is demonstrated via simulation.

PROBLEMS

Section 4.2

Problem 4.2.1: Consider the continuous-time dynamics given for the power system given in Problem 3.2.2 with a quantizer. Discretize the system dynamics.

Select the weighting matrices and terminal constraint matrix as $\boldsymbol{Q} = 0.1\boldsymbol{I}$, $\boldsymbol{R} = 0.1\boldsymbol{I}$, and $\boldsymbol{S}_N = \boldsymbol{I}$, where $\boldsymbol{I}$ is the identity matrix with appropriate dimension. The augmented states z_k^q are generated as $z_k^q = [(\boldsymbol{x}_k^q)^T \ (\boldsymbol{u}_{qk}^q)^T]^T$ and the regression function was generated per Equation 4.55. Note that the regression function (4.55) is given by using the Kronecker product quadratic polynomial, which guarantees that the value function is positive. Select a suitable admissible control and state vector. For the dynamic quantizer design, select the parameters as $\beta = 0.9$ and $\gamma = 0.9$.

For the action-dependent value function estimator, select the design parameter as $\alpha_\theta = 0.01$. The time-dependent basis function $\varphi(N-k)$ is selected in Equation 4.25 with saturation. Note that for finite time period, $\varphi(N-k)$ is always bounded. Saturation for $\varphi(N-k)$ is to ensure the magnitude of $\varphi(N-k)$ is within a reasonable range such that the parameter estimation is computable. The initial values for $\hat{\theta}_k$ are randomly selected. Perform MATLAB simulation and redo the plots presentation Section 4.2.2.

Problem 4.2.2: Repeat Problem 3.2.3 with a quantizer. elect the design parameters listed above.

Problem 4.2.3: Repeat Problem 3.2.3 with a quantizer and different initial conditions. Select the design parameters listed above.

Section 4.3

Problem 4.3.1: Repeat Example given in Section 4.3 with various initial conditions on the state vector and weight estimates.

APPENDIX 4A

Proof of Theorem 4.1

Consider the Lyapunov candidate function as

$$L_\theta = \underset{\tau,\gamma}{E}\left\{\tilde{\theta}_k^T\tilde{\theta}_k\right\} \tag{4A.1}$$

The first difference of L_θ is given by

$$\begin{aligned}\Delta L_\theta = \underset{\tau,\gamma}{E}\left[\tilde{\theta}_{k+1}^T\tilde{\theta}_{k+1} - \tilde{\theta}_k^T\tilde{\theta}_k\right] \le \underset{\tau,\gamma}{E}\Bigg[&\tilde{\theta}_k^T\tilde{\theta}_k - 2\alpha_\theta \frac{\tilde{\theta}_k^T\Delta\xi(z_k^q,k)\Delta\xi^T(z_k^q,k)\tilde{\theta}_k}{\Delta\xi^T(z_k^q,k)\Delta\xi(z_k^q,k)+1} \\ &- 2\alpha_\theta \frac{\tilde{\theta}_k^T\varphi(0)\varphi^T(0)\tilde{\theta}_k}{\left\|\sigma_N\right\|^2+1} + \Bigg(\alpha_\theta \frac{\Delta\xi(z_k^q,k)\Delta\xi^T(z_k^q,k)\tilde{\theta}_k}{\Delta\xi^T(\mathbf{z}_k^q,k)\Delta\xi(\mathbf{z}_k^q,k)+1} \\ &+ \alpha_\theta \frac{\varphi(0)\varphi^T(0)\tilde{\theta}_k}{\left\|\sigma_N\right\|^2+1}\Bigg)^2 - \tilde{\theta}_k^T\tilde{\theta}_k\Bigg]\end{aligned}$$

$$\leq -2\alpha_\theta \underset{\tau,\gamma}{E}\left[\frac{\tilde{\theta}_k^T \Delta\xi(z_k^q,k)\Delta\xi^T(z_k^q,k)\tilde{\theta}_k}{\Delta\xi^T(z_k^q,k)\Delta\xi(z_k^q,k)+1} - 2\alpha_\theta \frac{\tilde{\theta}_k^T \varphi(0)\varphi^T(0)\tilde{\theta}_k}{\|\varphi(0)\|^2+1}\right.$$

$$+2\alpha_\theta^2 \frac{\tilde{\theta}_k^T \Delta\xi(z_k^q,k)\Delta\xi^T(z_k^q,k)\tilde{\theta}_k}{\Delta\xi^T(z_k^q,k)\Delta\xi(z_k^q,k)+1} + 2\alpha_\theta^2 \frac{\tilde{\theta}_k^T \varphi(0)\varphi^T(0)\tilde{\theta}_k}{\|\varphi(0)\|^2+1}$$

$$\leq -2\alpha_\theta(1-\alpha_\theta)\left(\frac{\Delta\xi_{\min}^2}{1+\Delta\xi_{\min}^2} + \frac{\|\varphi(0)\|^2}{\|\varphi(0)\|^2+1}\right)\underset{\tau,\gamma}{E}\left\|\tilde{\theta}_k\right\|^2$$

Define $\zeta = 2\alpha_\theta(1-\alpha_\theta)\left(\frac{\Delta\xi_{\min}^2}{1+\Delta\xi_{\min}^2} + \frac{\|\varphi(0)\|^2}{\|\varphi(0)\|^2+1}\right)$. Note that $0<\zeta<1$ with $0<\alpha_\theta < 1/4$. Also, we have $0 < \Delta\xi_{\min}^2 \leq \|\Delta\xi(z_k^q,k)\|^2$ due to the PE condition; then,

$$\Delta L_\theta \leq -2\alpha_\theta(1-\alpha_\theta)\left(\frac{\Delta\xi_{\min}^2}{1+\Delta\xi_{\min}^2} + \frac{\|\varphi(0)\|^2}{\|\varphi(0)\|^2+1}\right)\underset{\tau,\gamma}{E}\left\|\tilde{\theta}_k\right\|^2 \equiv -\zeta \underset{\tau,\gamma}{E}\left\|\tilde{\theta}_k\right\|^2 \tag{4A.2}$$

Therefore, the first difference of Lyapunov function ΔL_θ is negative definite while Lypaunov function L_θ is positive definite. Moreover, using standard Lyapunov theory and geometric sequence theory, within finite horizon, the parameter estimation error will be bounded in the mean with bounds depending upon initial condition $B_{\tilde{\theta},0}$ with $\underset{\tau,\gamma}{E}\|\tilde{\theta}_0\|^2 \leq B_{\tilde{\theta},0}$ and terminal time NT_s, that is,

$$\underset{\tau,\gamma}{E}\left\|\tilde{\theta}_k\right\|^2 \leq (1-\zeta)^k B_{\tilde{\theta},0}, \quad \forall k = 0,1,\ldots,N \tag{4A.3}$$

Furthermore, when $N \to \infty$, the estimation error $\underset{\tau,\gamma}{E}\{\tilde{\theta}_k\}$ will converge to zero asymptotically in the mean.

APPENDIX 4B

Proof of Lemma 4.3

Recall from the quantizer design, for the state quantization, the quantization error is always bounded by $e_{Mx,k}$, as shown in Equation 4.31. Therefore, instead of dealing with the quantization error directly, we focus on the analysis of quantization error bound. Recalling from Equation 4.2, we have

$$\frac{e_{Mx,k+1}^2}{e_{Mx,k}^2} = \frac{\mu_{x,k+1}^2 \Delta_{x,k+1}^2}{\mu_{x,k}^2 \Delta_{x,k}^2} = \frac{\|x_{k+1}\|^2}{\|x_k\|^2} \tag{4B.1}$$

Substituting the system dynamics Equation 4.5 into Equation 4B.1 yields

$$\frac{e_{Mx,k+1}^2}{e_{Mx,k}^2} = \frac{\left\| \boldsymbol{A}\boldsymbol{x}_k + \boldsymbol{B}\boldsymbol{u}_{qk} + \boldsymbol{B}\boldsymbol{e}_{uk} \right\|^2}{\left\| \boldsymbol{x}_k \right\|^2} = \frac{\left\| \boldsymbol{A}\boldsymbol{x}_k + \boldsymbol{B}\hat{\boldsymbol{K}}_k\boldsymbol{x}_k^q + \boldsymbol{B}\boldsymbol{e}_{u,k} \right\|^2}{\left\| \boldsymbol{x}_k \right\|^2}$$
$$= \frac{\left\| \boldsymbol{A}\boldsymbol{x}_k + \boldsymbol{B}\boldsymbol{K}_k^*\boldsymbol{x}_k + \boldsymbol{B}\tilde{\boldsymbol{K}}_k\boldsymbol{x}_k - \boldsymbol{B}\hat{\boldsymbol{K}}_k\boldsymbol{e}_{x,k} + \boldsymbol{B}\boldsymbol{e}_{u,k} \right\|^2}{\left\| \boldsymbol{x}_k \right\|^2} \tag{4B.2}$$

with $\boldsymbol{K}_k^*$ the true Kalman gain satisfying $\left\| \boldsymbol{K}_k^* \right\| \le K_M$, and $\tilde{\boldsymbol{K}}_k = \boldsymbol{K}_k^* - \hat{\boldsymbol{K}}_k$ being the Kalman gain error.

Applying the Cauchy–Schwartz inequality and using Lemma 4.1, Equation 4B.2 can be further written as

$$\frac{e_{Mx,k+1}^2}{e_{Mx,k}^2} \le \frac{3\left\| \boldsymbol{A}\boldsymbol{x}_k + \boldsymbol{B}\boldsymbol{K}_k^*\boldsymbol{x}_k \right\|^2}{\left\| \boldsymbol{x}_k \right\|^2} + \frac{3\left\| \boldsymbol{B}\tilde{\boldsymbol{K}}_k\boldsymbol{x}_k - \boldsymbol{B}\hat{\boldsymbol{K}}_k\boldsymbol{e}_{x,k} \right\|^2}{\left\| \boldsymbol{x}_k \right\|^2} + \frac{3\left\| \boldsymbol{B}\boldsymbol{e}_{u,k} \right\|^2}{\left\| \boldsymbol{x}_k \right\|^2}$$
$$\le 3\rho + \frac{6\left\| \boldsymbol{B}\tilde{\boldsymbol{K}}_k\boldsymbol{x}_k \right\|^2}{\left\| \boldsymbol{x}_k \right\|^2} + \frac{6\left\| \boldsymbol{B}\hat{\boldsymbol{K}}_k\boldsymbol{e}_{x,k} \right\|^2}{\left\| \boldsymbol{x}_k \right\|^2} + \frac{3\left\| \boldsymbol{B}\boldsymbol{e}_{u,k} \right\|^2}{\left\| \boldsymbol{x}_k \right\|^2} \tag{4B.3}$$

Recall from Equation 4.22 and the definition of the adaptive estimator that we have $\boldsymbol{K}_k^* = (\boldsymbol{G}_k^{uu})^{-1}\boldsymbol{G}_k^{ux} = f(\boldsymbol{g}_k)$ and similarly $\hat{\boldsymbol{K}}_k = (\hat{\boldsymbol{G}}_k^{uu})^{-1}\hat{\boldsymbol{G}}_k^{ux} = f(\hat{\boldsymbol{g}}_k)$; then the Kalman gain error can be represented as

$$\left\| \tilde{\boldsymbol{K}}_k \right\| = \left\| f(\boldsymbol{g}_k) - f(\hat{\boldsymbol{g}}_k) \right\| \le L_f \left\| \tilde{\boldsymbol{g}}_k \right\|$$
$$= L_f \left\| \tilde{\theta}_k \varphi(N-k) \right\| \le L_f \varphi_{\max} \left\| \tilde{\theta}_k \right\| \equiv \varsigma_K \left\| \tilde{\theta}_k \right\| \tag{4B.4}$$

where L_f is a positive Lipchitz constant, $\varphi_{\max}$ always exists since the time of interest is finite and hence $\varphi(N-k)$ is always bounded, and $\varsigma_K = L_f\varphi_{\max}$. Hence, Equation 4B.3 becomes

$$\frac{e_{Mx,k+1}^2}{e_{Mx,k}^2} \le 3\rho + 6B_M^2 \left\| \tilde{\boldsymbol{K}}_k \right\|^2 + \frac{6B_M^2 \left\| \hat{\boldsymbol{K}}_k \right\|^2}{2^{2R}} + \frac{3B_M^2 \left\| \boldsymbol{e}_{u,k} \right\|^2}{\left\| \boldsymbol{x}_k \right\|^2}$$
$$\le 3\rho + 6B_M^2 L_f^2 \varphi_{\max}^2 \left\| \tilde{\theta}_k \right\|^2 + \frac{6B_M^2 \left\| \hat{\boldsymbol{K}}_k \right\|^2}{2^{2R}} + \frac{3B_M^2 \left\| \boldsymbol{e}_{u,k} \right\|^2}{\left\| \boldsymbol{x}_k \right\|^2} \tag{4B.5}$$

Furthermore, since $\left\| \boldsymbol{K}_k^* \right\| \le K_M$, we have

$$\frac{\left\| \boldsymbol{e}_{uk} \right\|^2}{\left\| \boldsymbol{x}_k \right\|^2} \le \frac{e_{Mu,k}^2}{\left\| \boldsymbol{x}_k \right\|^2} \le \frac{1}{2^{2R}} \frac{\left\| \boldsymbol{u}_{qk} \right\|^2}{\left\| \boldsymbol{x}_k \right\|^2} \le \frac{1}{2^{2R}} \frac{\left\| \hat{\boldsymbol{K}}_k \boldsymbol{x}_k^q \right\|^2}{\left\| \boldsymbol{x}_k \right\|^2} \le \frac{1}{2^{2R}} \frac{\left\| \hat{\boldsymbol{K}}_k \boldsymbol{x}_k + \hat{\boldsymbol{K}}_k \boldsymbol{e}_{x,k} \right\|^2}{\left\| \boldsymbol{x}_k \right\|^2}$$

$$\le \frac{1}{2^{2R}} \frac{\left\| \tilde{\boldsymbol{K}}_k \boldsymbol{x}_k + \hat{\boldsymbol{K}}_k \boldsymbol{e}_{x,k} + \boldsymbol{K}_k^* \boldsymbol{x}_k \right\|^2}{\left\| \boldsymbol{x}_k \right\|^2} \tag{4B.6}$$

$$\le \frac{1}{2^{2R}} \left(3B_M^2 L_f^2 \varphi_{\max}^2 \left\| \tilde{\theta}_k \right\|^2 + \frac{3\left\| \hat{\boldsymbol{K}}_k \right\|^2}{2^{2R}} + 3\boldsymbol{K}_M^2 \right)$$

Therefore, Equation 4B.5 becomes

$$\frac{e_{Mx,k+1}^2}{e_{Mx,k}^2} \le 3\rho + 6B_M^2 L_f^2 \varphi_{\max}^2 \left\| \tilde{\theta}_k \right\|^2 + \frac{6B_M^2 \left\| \hat{K}_k \right\|^2}{2^{2R}}$$

$$+ \frac{3B_M^2}{2^{2R}} \left(3B_M^2 L_f^2 \varphi_{\max}^2 \left\| \tilde{\theta}_k \right\|^2 + \frac{3\left\| \hat{K}_k \right\|^2}{2^{2R}} + 3K_M^2 \right)$$

$$\le 3\rho + B_M^2 L_f^2 \varphi_{\max}^2 \left(6 + \frac{9}{2^{2R}} \right) \left\| \tilde{\theta}_k \right\|^2 + \frac{B_M^2}{2^{2R}} \left(6 + \frac{9}{2^{2R}} \right) \left\| \hat{K}_k \right\|^2 + \frac{9B_M^2 K_M^2}{2^{2R}}$$

Hence, for the quantizer, there exists a finite number of bits R_f such that for all $R \ge R_f$,

$$\frac{B_M^2}{2^{2R_f}} \left(6 + \frac{9}{2^{2R_f}} \right) \left\| \hat{\boldsymbol{K}}_k \right\|^2 + \frac{9B_M^2 K_M^2}{2^{2R_f}} \le \frac{3 - 3\rho}{2} \tag{4B.7}$$

Therefore, Equation 4B.5 can be written as

$$\frac{e_{Mx,k+1}^2}{e_{Mx,k}^2} \le 3\rho + B_M^2 L_f^2 \varphi_{\max}^2 \left(6 + \frac{9}{2^{2R_f}} \right) \left\| \tilde{\theta}_k \right\|^2 + \frac{(1 - 3\rho)}{2}$$

$$\le \frac{(1 + 3\rho)}{2} + B_M^2 L_f^2 \varphi_{\max}^2 \left(6 + \frac{9}{2^{2R_f}} \right) \left\| \tilde{\theta}_k \right\|^2 \tag{4B.8}$$

Recall from Theorem 4.1, since $0 < \alpha_\theta < 1/4$, thus $0 < \zeta < 1$, which further implies $\|\tilde{\theta}_k\|^2 \le (1-\zeta)^k \|\tilde{\theta}_0\|^2$. Hence, Equation 4B.8 becomes

$$\frac{e^2_{Mx,k+1}}{e^2_{Mx,k}} \le \frac{(1+3\rho)}{2} + B_M^2 L_f^2 \varphi_{\max}^2 \left(6 + \frac{9}{2^{2R_f}}\right)(1-\zeta)^k \left\|\tilde{\theta}_0\right\|^2 \tag{4B.9}$$

Therefore, there exists a finite number k_f such that for all $k > k_f$, we have $B_M^2 L_f^2 \varphi_{\max}^2 (6 + 9/2^{2R_f})(1-\zeta)^{k_f} \|\tilde{\theta}_0\|^2 < (1-3\rho)/2$. Hence,

$$\frac{e^2_{Mx,k+1}}{e^2_{Mx,k}} \equiv \eta \le \frac{(1+3\rho)}{2} + \frac{(1-3\rho)}{2} < 1 \tag{4B.10}$$

According to Equation 4B.10, within finite horizon, the quantization error for system state is *bounded* in the mean with bound depending upon the initial quantization error bound $e_{Mx,0}$ and terminal time, that is,

$$e^2_{Mx,k} \le (1-\eta)^k e^2_{Mx,0}, \quad \forall k = 0,1,\ldots,N \tag{4B.11}$$

Further, the quantization error bound for the system states $e_{Mx,k}$ converges to zero asymptotically as $N \to \infty$. Since the quantization error never exceeds the bound, the state quantization error also converges to zero as $N \to \infty$.

APPENDIX 4C

Proof of Lemma 4.4

Recall from Equation 4.32, the certainty-equivalent control input is given as

$$\boldsymbol{u}_{qk} = \underset{\tau,\gamma}{E}\left[\hat{K}_k x_k^q\right] = \underset{\tau,\gamma}{E}\left[K_k^* x_k^q - \tilde{K}_k x_k^q\right] \tag{4C.1}$$

where $\tilde{\boldsymbol{K}}_k = \boldsymbol{K}_k^* - \hat{\boldsymbol{K}}_k$ is the Kalman gain error. Similar to the state quantization, we have the quantization error bound for the control input as

$$\begin{aligned} e_{Mu,k} = \frac{\left\|\boldsymbol{u}_{qk}\right\|}{2^R} &\le \frac{K_M}{2^R} \underset{\tau,\gamma}{E}\left\|\boldsymbol{x}_k\right\| + \frac{K_M}{2^R} \underset{\tau,\gamma}{E}\left\|\boldsymbol{e}_{x,k}\right\| \\ &+ \frac{L_f \varphi_{\max}}{2^R} \underset{\tau,\gamma}{E}\left\|\tilde{\theta}_k\right\|\left\|\boldsymbol{x}_k\right\| + \frac{L_f \varphi_{\max}}{2^R} \underset{\tau,\gamma}{E}\left\|\tilde{\theta}_k\right\|\left\|\boldsymbol{e}_{x,k}\right\| \\ &\le \left(1 + \frac{1}{2^R}\right) K_M e_{Mx,k} + \left(1 + \frac{1}{2^R}\right) L_f \varphi_{\max} \underset{\tau,\gamma}{E}\left\|\tilde{\theta}_k\right\| e_{Mx,k} \equiv e_{MMu,k} \end{aligned} \tag{4C.2}$$

Define the Lyapunov candidate function as $L(e_{MMu,k}) = e_{MMu,k}^2$. The first difference of $L(e_{MMu,k})$ is given by

$$\begin{aligned}\Delta L(e_{MMu,k}) &= e_{MMu,k+1}^2 - e_{MMu,k}^2 \\ &= \left(1+\frac{2}{2^R}\right)^2 K_M^2 e_{Mx,k+1}^2 + \left(1+\frac{1}{2^R}\right) L_f^2 \varphi_{\max}^2 \left\|\tilde{\theta}_{k+1}\right\|^2 e_{Mx,k+1}^2 \\ &\quad - \left(1+\frac{1}{2^R}\right)^2 K_M^2 e_{Mx,k}^2 - \left(1+\frac{1}{2^R}\right) L_f^2 \varphi_{\max}^2 \left\|\tilde{\theta}_k\right\|^2 e_{Mx,k}^2 \\ &\le -(1-\eta^2)\left(1+\frac{1}{2^R}\right)^2 K_M^2 e_{Mx,k}^2 \\ &\quad -(1-\zeta^2\eta^2)\left(1+\frac{1}{2^R}\right)^2 L_f^2 \varphi_{\max}^2 \left\|\tilde{\theta}_k\right\|^2 e_{Mx,k}^2 \le \delta e_{MMu,k}^2\end{aligned} \tag{4C.3}$$

where $0 < \delta < (1/2)\min\{1-\eta^2, 1-\zeta^2\eta^2\} < 1$. According to Equation 4C.3, within finite horizon, the quantization error bound for control input is bounded in the mean with the bound depending upon initial quantization error bound $e_{MMu,0}$ and terminal time, that is,

$$e_{MMu,k}^2 \le (1-\delta)^k e_{MMu,0}^2, \qquad \forall k = 0,1,\ldots,N \tag{4C.4}$$

Moreover, since the first difference of Lyapunov function $\Delta L(e_{MMu,k})$ is negative definite while Lypaunov function $L(e_{MMu,k})$ is positive definite, we have $e_{MMu,k} \to 0$ as $N \to \infty$. Since, $\|e_{u,k}\| \le e_{Mu,k} \le e_{MMu,k}$, $\|e_{u,k}\| \to 0$ as $N \to \infty$.

APPENDIX 4D

Proof of Theorem 4.2

Consider the Lyapunov candidate function as

$$\begin{aligned}L &= \Lambda_5 L(\boldsymbol{x}_k) + \Lambda_1\Lambda_5 L(\tilde{\theta}_k) + \Lambda_2\Lambda_5(1-\zeta^2\eta^2) L(\tilde{\theta}_k, e_{Mx,k}) \\ &\quad + \Lambda_3\Lambda_5 L(e_{Mx,k}) + \Lambda_4\Lambda_5 L(e_{Mu,k})\end{aligned} \tag{4D.1}$$

where $L(\boldsymbol{x}_k) = \boldsymbol{x}_k^{\mathrm{T}}\boldsymbol{x}_k$, $L(\tilde{\theta}_k, e_{Mx,k}) = \|\tilde{\theta}_k\|^2 e_{Mx,k}^2$, $L(e_{Mx,k}) = e_{Mx,k}^2$, $L(e_{Mu,k}) = e_{Mu,k}^2$, $\Lambda_1 = 24B_M^2 L_f^2 \varphi_{\max}^2/(1-\zeta)$, $\Lambda_2 = 24B_M^2 L_f^2 \varphi_{\max}^2/(1-\zeta^2\eta^2)$, $\Lambda_3 = 12B_M^2 K_M^2/(1-\eta^2)$ and $\Lambda_4 = 12B_M^2/((1-\eta^2)(1+1/2^R)^2 K_M^2)$, $\Lambda_5 = \varpi/\max\{12B_M^2 L_f^2\varphi_{\max}^2\Theta_M^2\zeta_M^2, 6B_M^2K_M^2, 3B_M^2\}$ with $0 < \varpi < \min\left\{1, \max\{12B_M^2 L_f^2\varphi_{\max}^2\Theta_M^2\zeta_M^2, 6B_M^2K_M^2, 3B_M^2\}\right\} < 1$.

Next, consider each term in Equation 4D.1 individually. Applying the Cauchy–Schwartz inequality and recalling Lemma 4.1, we have

$$\Delta L(\boldsymbol{x}_k) = \underset{\tau,\gamma}{E}\left[\boldsymbol{x}_{k+1}^T\boldsymbol{x}_{k+1} - \boldsymbol{x}_k^T\boldsymbol{x}_k\right] = \underset{\tau,\gamma}{E}\left\|\left\|\boldsymbol{x}_{k+1}\right\|^2 - \left\|\boldsymbol{x}_k\right\|^2\right\|$$

$$= \underset{\tau,\gamma}{E}\left[\left\|\boldsymbol{A}\boldsymbol{x}_k + \boldsymbol{B}\boldsymbol{u}_{qk} + \boldsymbol{B}\boldsymbol{e}_{u,k}\right\|^2 - \left\|\boldsymbol{x}_k\right\|^2\right]$$

$$= \underset{\tau,\gamma}{E}\left\{\left\|\boldsymbol{A}\boldsymbol{x}_k + \boldsymbol{B}\boldsymbol{K}_k^*\boldsymbol{x}_k + \boldsymbol{B}\hat{\boldsymbol{K}}_k\boldsymbol{x}_k^q - \boldsymbol{B}\boldsymbol{K}_k^*\boldsymbol{x}_k + \boldsymbol{B}\boldsymbol{e}_{u,k}\right\|^2 - \left\|\boldsymbol{x}_k\right\|^2\right\}$$

$$\le 3\underset{\tau,\gamma}{E}\left\|\boldsymbol{A}\boldsymbol{x}_k + \boldsymbol{B}\boldsymbol{K}_k^*\boldsymbol{x}_k\right\|^2 + 6B_M^2\underset{\tau,\gamma}{E}\left\|\tilde{\boldsymbol{K}}_k\right\|^2\left\|\boldsymbol{x}_k^q\right\|^2 + 6B_M^2K_M^2\underset{\tau,\gamma}{E}\left\|\boldsymbol{e}_{\boldsymbol{x},k}\right\|^2$$

$$+ 3B_M^2\underset{\tau,\gamma}{E}\left\|\boldsymbol{e}_{\boldsymbol{u}k}\right\|^2 - \underset{\tau,\gamma}{E}\left\|\boldsymbol{x}_k\right\|^2$$

$$\le -(1-3\rho)\underset{\tau,\gamma}{E}\left\|\boldsymbol{x}_k\right\|^2 + 12B_M^2L_f^2\varphi_{\max}^2\underset{\tau,\gamma}{E}\left\|\tilde{\theta}_k\right\|^2\Theta_M^2\underset{\tau,\gamma}{E}\left\|\xi_{k-1}\right\|^2$$

$$+ 12B_M^2L_f^2\varphi_{\max}^2\underset{\tau,\gamma}{E}\left\|\tilde{\theta}_k\right\|^2 e_{M\boldsymbol{x},k}^2 + 6B_M^2K_M^2e_{M\boldsymbol{x},k}^2 + 3B_M^2e_{M\boldsymbol{u},k}^2$$

$$\le -(1-3\rho)\underset{\tau,\gamma}{E}\left\|\boldsymbol{x}_k\right\|^2 + 12B_M^2L_f^2\varphi_{\max}^2\underset{\tau,\gamma}{E}\left\|\tilde{\theta}_k\right\|^2\Theta_M^2\xi_M^2$$

$$+ 12B_M^2L_f^2\varphi_{\max}^2\underset{\tau,\gamma}{E}\left\|\tilde{\theta}_k\right\|^2 e_{M\boldsymbol{x},k}^2 + 6B_M^2K_M^2e_{M\boldsymbol{x},k}^2 + 3B_M^2e_{M\boldsymbol{u},k}^2$$

where $\|\Theta\| = \|[\boldsymbol{A}\ \ \boldsymbol{B}]\| \le \Theta_M$ and the history information satisfying $\|\xi_{k-1}\| = \|[\boldsymbol{x}_{k-1}^T\ \ \boldsymbol{u}_{qk-1}^T]^T\| \le \xi_M$. Next, recalling from Theorem 4.1, Lemma 4.3, and Lemma 4.4, the overall difference of the Lyapunov candidate function is given by

$$\Delta L = \Lambda_5\Delta L(\boldsymbol{x}_k) + \Lambda_1\Lambda_5\Delta L(\tilde{\theta}_k) + \Lambda_3\Lambda_5\Delta L(e_{M\boldsymbol{x},k})$$

$$+ \Lambda_2\Lambda_5(1-\zeta^2\eta^2)\Delta L(\tilde{\theta}_k, e_{M\boldsymbol{x},k}) + \Lambda_4\Lambda_5\Delta L(e_{M\boldsymbol{u},k})$$

$$\le -\Lambda_5(1-3\rho)\underset{\tau,\gamma}{E}\left\|\boldsymbol{x}_k\right\|^2 + \Lambda_5 12B_M^2L_f^2\varphi_{\max}^2\underset{\tau,\gamma}{E}\left\|\tilde{\theta}_k\right\|^2\Theta_M^2\xi_M^2$$

$$+ \Lambda_5 12B_M^2L_f^2\varphi_{\max}^2\underset{\tau,\gamma}{E}\left\|\tilde{\theta}_k\right\|^2 e_{M\boldsymbol{x},k}^2 + \Lambda_5 6B_M^2K_M^2e_{M\boldsymbol{x},k}^2 + 3\Lambda_5B_M^2e_{M\boldsymbol{u},k}^2$$

$$- \Lambda_1\Lambda_5(1-\zeta)\left\|\tilde{\theta}_k\right\|^2 - \Lambda_4\Lambda_5(1-\eta^2)\left(\frac{1}{2^R}\right)^2 K_M^2e_{M\boldsymbol{x},k}^2$$

$$- \Lambda_2\Lambda_5(1-\zeta^2\eta^2)\underset{\tau,\gamma}{E}\left\|\tilde{\theta}_k\right\|^2 e_{M\boldsymbol{x},k}^2 - \Lambda_1\Lambda_5(1-\eta^2)\underset{\tau,\gamma}{E}\left\|\tilde{\theta}_k\right\|^2$$

$$- \Lambda_4\Lambda_5(1-\zeta^2\eta^2)\left(\frac{1}{2^R}\right)^2 L_f^2\varphi_{\max}^2\underset{\tau,\gamma}{E}\left\|\tilde{\theta}_k\right\|^2 e_{M\boldsymbol{x},k}^2$$

$$\le -\Lambda_5(1-3\rho)\underset{\tau,\gamma}{E}\left\|\boldsymbol{x}_k\right\|^2 - \varpi\underset{\tau,\gamma}{E}\left\|\tilde{\theta}_k\right\|^2 - \varpi e_{M\boldsymbol{x},k}^2 - \varpi e_{M\boldsymbol{u},k}^2$$

where $0 < \rho < 1/3$, $0 < \Lambda_5 < 1$, and $0 < \varpi < 1$.

Therefore, the first difference of Lyapunov function ΔL is negative definite while Lypaunov function L is positive definite. Moreover, using standard Lyapunov theory and geometric sequence theory, within finite horizon, the system states, parameter estimation error, state quantization error bound, and control input quantization error bound will be bounded in the mean with bounds depending upon initial conditions $B_{x,0}, B_{\tilde{\theta},0}, e_{Mx,0}^2, e_{Mu,0}^2$ with $\|\boldsymbol{x}(0)\|^2 \le B_{x,0}$, $\|\tilde{\theta}_0\|^2 \le B_{\tilde{\theta},0}$, $\|\boldsymbol{e}_{x,0}\|^2 \le e_{Mx,0}^2$, $\|\boldsymbol{e}_{u,0}\|^2 \le e_{Mu,0}^2$ and terminal time, that is,

$$\underset{\tau,\gamma}{E}\|\boldsymbol{x}_k\|^2 \le (1-(1-3\rho)\Lambda_5)^k B_{x,0} \equiv B_{x,k}, \quad \forall k = 0,1,\ldots,N$$

$$\text{or } \underset{\tau,\gamma}{E}\|\tilde{\theta}_k\|^2 \le (1-\varpi)^k B_{\tilde{\theta},0} \equiv B_{\tilde{\theta},k}, \quad \forall k = 0,1,\ldots,N$$

$$\text{or } \underset{\tau,\gamma}{E}\|\boldsymbol{e}_{x,k}\|^2 \le (1-\varpi)^k e_{Mx,0}^2 \equiv B_{e_x,k}, \quad \forall k = 0,1,\ldots,N$$

$$\text{or } \underset{\tau,\gamma}{E}\|\boldsymbol{e}_{u,k}\|^2 \le (1-\varpi)^k e_{Mu,0}^2 \equiv B_{e_u,k}, \quad \forall k = 0,1,\ldots,N \tag{4D.2}$$

Further, since $0 < \rho < 1/3$ and $0 < \varpi < 1$, the bounds in Equation 4D.2 are monotonically decreasing as k increases. When $N \to \infty$, all the bounds tend to zero and the stochastic asymptotic stability in the mean square of the closed-loop system is achieved.

Eventually, while time goes to fixed final time NT_s, we have the upper bound for $\hat{\boldsymbol{u}}_{qk}^q - \boldsymbol{u}_k^*$ as

$$\begin{aligned}
\underset{\tau,\gamma}{E}\|\hat{\boldsymbol{u}}_{qk}^q - \boldsymbol{u}_k^*\| &= \underset{\tau,\gamma}{E}\|\hat{\boldsymbol{K}}_k\boldsymbol{x}_k^q + \boldsymbol{e}_{u,k} - \boldsymbol{K}_k^*\boldsymbol{x}_k\| = \underset{\tau,\gamma}{E}\|\tilde{\boldsymbol{K}}_k\boldsymbol{x}_k + \boldsymbol{K}_k^*\boldsymbol{e}_{x,k} - \tilde{\boldsymbol{K}}_k\boldsymbol{e}_{x,k} + \boldsymbol{e}_{u,k}\| \\
&\le \varsigma_K \underset{\tau,\gamma}{E}\|\tilde{\theta}_k\|\|\boldsymbol{x}_k\| + \varsigma_K \underset{\tau,\gamma}{E}\|\tilde{\theta}_k\|\|\boldsymbol{e}_{x,k}\| + K_M \underset{\tau,\gamma}{E}\|\boldsymbol{e}_{x,k}\| + \underset{\tau,\gamma}{E}\|\boldsymbol{e}_{u,k}\| \\
&\le \varsigma_K\sqrt{B_{\tilde{\theta},k}B_{x,k}} + (\varsigma_K\sqrt{B_{\tilde{\theta},k}} + K_M)\sqrt{B_{e_x,k}} + \sqrt{B_{e_u,k}} \equiv \varepsilon_{us}
\end{aligned} \tag{4D.3}$$

where $B_{\tilde{\theta},k}, B_{x,k}, B_{e_x,k}, B_{e_u,k}$ are given in Equation 4D.2. It is important to note that all the bounds will converge to zero, that is, $B_{\tilde{\theta},k} \to 0, B_{x,k} \to 0, B_{e_x,k} \to 0, B_{e_u,k} \to 0$, when $N \to \infty$. Moreover, the estimated control input will tend to optimal control, $\underset{\tau,\gamma}{E}[\hat{\boldsymbol{u}}_{qk}^q] \to \underset{\tau,\gamma}{E}[\boldsymbol{u}_k^*]$, since $\varepsilon_{us} \to 0$ when $N \to \infty$.

REFERENCES

Abu-Khalaf, M. and Lewis, F.L. 2005. Near optimal control laws for nonlinear systems with saturating actuators using a neural network HJB approach. *Automatica*, 41, 779–791.

Beard, R.W. 1995. Improving the closed-loop performance of nonlinear systems. PhD dissertation, Electrical Engineering Department, Rensselaer Polytechnic Institute, USA.

Bertsekas, D.P. 2005. Dynamic programming and suboptimal control: A survey from ADP to MPC. *European Journal of Control*, 11, 310–334.

Bradtke, S.J. and Ydstie, B.E. 1994. Adaptive linear quadratic control using policy iteration. *Proceedings of the American Control Conference*, Baltimore, MD, pp. 3475–3479.

Brockett, R.W. and Liberzon, D. 2000. Quantized feedback stabilization of linear systems. *IEEE Transactions on Automatic Control*, 45, 1279–1289.

Chen, C.T. 2012. *Linear System Theory and Design*. Oxford University Press, New York.

Chen, Z. and Jagannathan, S. 2008. Generalized Hamilton–Jacobi–Bellman formulation based neural network control of affine nonlinear discrete-time systems. *IEEE Transactions on Neural Networks*, 19, 90–106.

Cybenko, G. 1989. Approximation by superpositions of a sigmoidal function. *Mathematics of Control, Signals, and Systems*, 2, 303–314.

Delchamps, D.F. 1990. Stabilizing a linear system with quantized state feedback. *IEEE Transactions on Automatic Control*, 35, 916–924.

Dierks, T. and Jagannathan, S. 2012. Online optimal control of affine nonlinear discrete-time systems with unknown internal dynamics by using time-based policy update. *IEEE Transactions on Neural Networks and Learning Systems*, 23, 1118–1129.

Elia, N. and Mitter, S.K. 2001. Stabilization of linear systems with limited information. *IEEE Transactions on Automatic Control*, 46, 1384–1400.

Green, M. and Moore, J.B. 1986. Persistency of excitation in linear systems. *Systems and Control Letters*, 7, 351–360.

Heemels, H., Teel A.R., Wouw, N., and Nesic, D. 2010. Networked control systems with communication constraints: Tradeoffs between transmission intervals, delays and performance. *IEEE Transactions on Automatic Control*, 55, 1781–1796.

Heydari, A. and Balakrishnan, S.N. 2013. Finite-horizon control-constrained nonlinear optimal control using single network adaptive critics. *IEEE Transactions on Neural Networks and Learning Systems*, 24, 145–157.

Khalil, H.K. and Praly, L. 2014. High-gain observers in nonlinear feedback control. *International Journal of Robust and Nonlinear Control*, 24, 993–1015.

Lewis, F.L., Jagannathan, S., and Yesildirek, A. 1999. *Neural Network Control of Robot Manipulators and Nonlinear Systems*. CRC Press/Taylor & Francis Group, UK.

Liberzon, D. 2003. Hybrid feedback stabilization of systems with quantized signals. *Automatica*, 39, 1543–1554.

Lyshevski, S.E. 1998. Optimal control of nonlinear continuous-time systems: Design of bounded controllers via generalized nonquadratic functionals. *Proceedings of the American Control Conference*, USA, pp. 205–209.

Narendra, K.S. and Annaswamy, A.M. 1989. *Stable Adaptive Systems*. Prentice-Hall, Englewood Cliffs, NJ.

Narendra, K.S. and Parthasarathy, K. 1990. Identification and control of dynamical systems using neural networks. *IEEE Transactions on Neural Networks*, 1, 4–27.

Saberi, A., Lin, Z., and Teel, A. 1996. Control of linear systems with saturating actuators. *IEEE Transactions on Automatic Control*, 41, 368–378.

Sandberg, I.W. 1998. Notes on uniform approximation of time-varying systems on finite time intervals. *IEEE Transactions on Circuits and Systems–I, Fundamental Theory and Application*, 45, 863–865.

Sarangapani, J. 2006. *Neural Network Control of Nonlinear Discrete-Time Systems*, CRC Press, Florida, USA.

Si, J., Barto, A.G., Powell, W.B., and Wunsch, D. 2004. *Handbook of Learning and Approximate Dynamic Programming*. Wiley, New York.

Slotine, J.E. and Li, W. 1991. *Applied Nonlinear Control*, Prentice-Hall, Englewood Cliffs, NJ.

Stengel, R.F. 1986. *Stochastic Optimal Control: Theory and Application*. Wiley-Interscience, New York.

Sussmann, H, Sontag, E.D., and Yang, Y. 1994. A general result on the stabilization of linear systems using bounded controls. *IEEE Transactions on Automatic Control*, 39, 2411–2425.

Tse, D. and Viswanath, P. 2005. *Fundamentals of Wireless Communication*, Cambridge University Press, Cambridge, UK.

Wang, F.Y., Jin, N., Liu, D., and Wei, Q. 2011. Adaptive dynamic programming for finite-horizon optimal control of discrete-time nonlinear systems with $\varepsilon = 0$-error bound. *IEEE Transactions on Neural Networks*, 22, 24–36.

Watkins, C. 1989. Learning from delayed rewards. PhD dissertation, Cambridge University, England.

Werbos, P.J. 1983. A menu of designs for reinforcement learning over time. *Journal of Neural Networks Control*, 3, 835–846.

Xu, H. and Jagannathan, S. 2013. Stochastic optimal controller design for uncertain nonlinear networked control system via neuro dynamic programming. *IEEE Transactions on Neural Networks and Learning Systems*, 24, 471–484.

Zhao, Q., Xu, H., and Jagannathan, S. 2012a. Optimal adaptive controller scheme for uncertain quantized linear discrete-time system. *Proceedings of the 51st IEEE Conference on Decision and Control*, Maui, Hawaii, USA, pp. 6132–6137.

Zhao, Q., Xu, H., and Jagannathan, S. 2012b. Adaptive dynamic programming-based-state quantized networked control system without value and/or policy iterations. *Proceedings of the International Joint Conference on Neural Networks (IJCNN)*, Brisbane, Australia, pp. 1–7.

Zhao, Q., Xu, H., and Jagannathan, S. 2013. Finite-horizon optimal control design for uncertain linear discrete-time systems. *Proceedings of the IEEE Symposium on Approximate Dynamic Programming and Reinforcement Learning (ADPRL)*, Singapore.

Zhao, Q., Xu, H., and Jagannathan, S. 2015a. Optimal control of uncertain quantized linear discrete-time systems. *International Journal of Adaptive Control and Signal Processing*, 29(3), 325–345, 2015.

Zhao, Q., Xu, H., and Jagannathan, S. 2015b. Neural network-based finite-horizon optimal control of quantized uncertain affine nonlinear discrete-time systems. *IEEE Transactions on Neural Networks and Learning Systems*, 26(3), 486–499, 2015.

5 Optimal Control of Uncertain Linear Networked Control Systems in Input–Output Form with Disturbance Inputs

Feedback systems with control loops closed through a real-time network are referred to as NCS (Halevi and Ray 1988; Yang et al. 2011) as described in the previous chapters. The feedback system can be a linear (see Chapter 3) or a nonlinear system (Chapter 6).

Recently, Walsh et al. (1999a,b) and Lian et al. (2001) considered the stability and performance of an NCS with constant delays. Krtolica et al. (1994) analyzed the stability of an NCS with random network-induced delays while Wu and Chen (2007) studied the stability of an NCS with packet losses. Eventually, Zhang et al. (2001) conducted the stability analysis of an NCS with network delays and packet losses and derived a stability region. These designs did not consider deterministic disturbance inputs.

Lian et al. (2003) introduced the optimal controller design (Stengel 1986) for an NCS without taking into account the disturbances. By using the stochastic optimal control theory (Åstrom 1970; Stengel 1986; Chen and Guo 1991), Nilsson et al. (1998) introduced the optimal and suboptimal control design for a linear NCS with known network-induced delays. In addition, these designs did not consider the optimal design for an unknown NCS under worst-case disturbance, referred to as NCS quadratic zero-sum games (Basar and Olsder 1995). Meanwhile, widely known output feedback design has not been developed for both known and unknown NCS quadratic zero-sum games.

Recently, Al-Tamimi et al. (2007) employed Q-learning (Watkins, 1989) and ADP (Werbos 1999) to solve the optimal strategy for uncertain discrete-time linear time-invariant system quadratic zero-sum games. Though the value and policy iteration-based approach works forward-in-time for optimal control, it requires a large number of iterations within a sampling interval for convergence, which can be a bottleneck for real-time control, including the NCS. Moreover, convergence of the algorithm is introduced while the stability of the overall system is not demonstrated. The network imperfections, such as delays and packet losses, can make the optimal design more involved (Al-Tamimi et al. 2007) and cause instability (Zhang et al. 2001) if they are not properly accounted for.

In the previous chapters, the disturbance inputs are not included. In this chapter, a time-driven ADP approach (Xu et al. 2014; Dierks and Jagannathan, 2012) is undertaken to obtain stochastic optimal regulation of uncertain linear NCS under worst-case disturbances, that is, NCS quadratic zero-sum games in input–output form with network imperfections (e.g., network-induced delays and packet losses) by solving the Bellman equation (Wonham 1968) online and in forward-in-time manner. The output is considered available without any noise in the measurements.

First, a known linear time-invariant system with deterministic disturbances enclosed by a communication network is represented as an uncertain stochastic linear discrete-time system. Given that the original linear continuous-time system is controllable and observable, it is proven that the stochastic linear discrete-time system is also controllable and observable when network imperfections are incorporated. Next, the uncertain stochastic linear discrete-time system is further rewritten in the input–output form for the controller design.

Subsequently, by using an initial admissible control, the certainty-equivalent value function is estimated adaptively and online (Al-Tamimi et al. 2007; Lewis and Vamvoudakis 2011; Yang et al., 2011) while its unknown parameters are tuned by using a novel update law since solving the stochastic game-theoretic Riccati equation (GRE) in a traditional manner requires information about system matrices. Then, by using ADP and the zero-sum game-theoretical formulation, the optimal control policies that optimize the cost function under worst-case disturbances are obtained based on the information provided by the estimated value function. Thus the proposed time-based ADP scheme is not only developed for NCS quadratic zero-sum games but also relaxes the need for system dynamics and information on network-induced delay and packet losses while rendering the optimal solution without using value and policy iterations. Finally, the overall stability of the closed-loop system is demonstrated by using Lyapunov theory.

The importance of this chapter stems from the fact that a game-theoretic adaptive system is proposed to create controllers for NCS quadratic zero-sum games that learn to coexist with an L_2-gain worst-case disturbance signal (Basar and Olsder 1995). In the control system design, this problem is defined as a two-player game that corresponds to the well-known H_∞ control. Next, background information is introduced.

First, Section 5.1 presents the traditional two-player zero-sum game design by using the GRE-based solution. Subsequently, Section 5.2 develops the infinite-horizon optimal adaptive design for LNCS quadratic zero-sum games (Basar and Olsder 1995).

5.1 TRADITIONAL TWO-PLAYER ZERO-SUM GAME DESIGN AND GAME-THEORETIC RICCATI EQUATION-BASED SOLUTION

Consider a controllable and observable stochastic linear two-player zero-sum game whose state-dependent dynamics are given by

$$x_{k+1} = A_k x_{d,k} + B_k u_k + D_k d_k + w_k \tag{5.1}$$

where x_k, u_k, d_k, and w_k represent system state, control and disturbance inputs, and zero-mean Gaussian process noise, respectively, while $A_k \in \Re^{n\times n}$, $B_k \in \Re^{n\times m}$, $D_k \in \Re^{n\times l}$ denote deterministic system matrices. It is easy to verify that linear two-player zero-sum games have a unique equilibrium point, $x = 0$, on a set Ω in the absence of process noise while the system states are measurable. According to these conditions, under perfect state measurements, the stochastic optimal strategy that optimize the certainty-equivalent stochastic cost function $J_k = \underset{\tau,\gamma}{E} \sum_{i=k}^{\infty}(x_i^T Q x_i + u_i^T R u_i - d_i^T S d_i)$ for linear two-player zero-sum games (5.1) can be derived as (Al-Tamimi et al. 2007) $u_k^* = -K_k x_k$, $d_k^* = -L_k x_k$ where K_k, L_k denote Kalman gains for the control inputs and worst-case disturbance signals, respectively.

If we assume that there exists a solution to the GRE, that is strictly feedback stabilizing, and then it can be shown (Basar and Olsder 1995) that policies attain a saddle-point equilibrium (Al-Tamimi et al. 2007), which implies that minimax is equal to maximin, in the restricted class of feedback stabilizing policies. Assuming that the game has a value and is solvable, then it is known that stochastic cost function is quadratic and given by

$$J_k = \underset{\tau,\gamma}{E}(x_k^T P_k x_k) \tag{5.2}$$

where matrix $P_k \geq 0$ is a solution to the certainty-equivalent SGRE (Lu et al. 2004). The optimal action-dependent value function of linear two-player zero-sum games is now defined to be

$$\begin{aligned} V(x_k,u_k,d_k) &= \underset{\tau,\gamma}{E}\{x_k^T Q x_k + u_k^T R u_k - d_k^T S d_k + J_{k+1}\} \\ &= \underset{\tau,\gamma}{E}\{r(x_k,u_k,d_k) + J_{k+1}\} = \underset{\tau,\gamma}{E}\{r(x_k,u_k,d_k) + x_{k+1}^T P_{k+1} x_{k+1}\} \end{aligned} \tag{5.3}$$

where $r(x_k,u_k,d_k) = x_k^T Q x_k + u_k^T R u_k - d_k^T S d_k$.

Since stochastic optimal control inputs and worst-case disturbance signals, u_k^*, d_k^*, are dependent on augment measured system state x_k, which is known at time k, the certainty-equivalent value function can be expressed as

$$V(x_k,u_k,d_k) = [x_k^T \; u_k^T \; d_k^T]^T \underset{\tau,\gamma}{E}(H_k)[x_k^T \; u_k^T \; d_k^T]$$

Then, using the Bellman equation and cost function, we can get

$$\underset{\tau,\gamma}{E}\{x_k^T P_k x_k\} = \underset{\tau,\gamma}{E}\{r(x_k,u_k,d_k) + x_{k+1}^T P_{k+1} x_{k+1}\} \tag{5.4}$$

Then, using optimal theory (Chen and Guo 1991), for zero-sum game (Basar and Olsder 1995), gain matrix associated with optimal control policy and worst-case disturbances can be expressed as

$$\underset{\tau,\gamma}{E}[K_k] = -\left(R + \underset{\tau,\gamma}{E}(B_k^T P_{k+1} B_k) - \underset{\tau,\gamma}{E}(B_k^T P_{k+1} D_k)\left(\underset{\tau,\gamma}{E}(D_k^T P_{k+1} D_k) - S\right)^{-1}\right.$$
$$\times \underset{\tau,\gamma}{E}(D_k^T P_{k+1} B_k)\Bigg)^{-1}\left(\underset{\tau,\gamma}{E}(B_k^T P_{k+1} D_k)\left(\underset{\tau,\gamma}{E}(D_k^T P_{k+1} D_k) - S\right)^{-1}\right. \tag{5.5}$$
$$\left.\times \underset{\tau,\gamma}{E}(D_k^T P_{k+1} A_k) - \underset{\tau,\gamma}{E}(B_k^T P_{k+1} A_k)\right)$$

and

$$\underset{\tau,\gamma}{E}[L_k] = -\left(\underset{\tau,\gamma}{E}(D_k^T P_{k+1} D_k) - S - \underset{\tau,\gamma}{E}(D_k^T P_{k+1} B_k)\left(\underset{\tau,\gamma}{E}(B_k^T P_{k+1} B_k) + R\right)^{-1}\right.$$
$$\times \underset{\tau,\gamma}{E}(B_k^T P_{k+1} D_k)\Bigg)^{-1}\left(\underset{\tau,\gamma}{E}(D_k^T P_{k+1} B_k)\,\underset{\tau,\gamma}{E}(B_k^T P_{k+1} B_k) + R\right)^{-1} \tag{5.6}$$
$$\times \underset{\tau,\gamma}{E}(B_k^T P_{k+1} A_k) - \underset{\tau,\gamma}{E}(D_k^T P_{k+1} A_k)$$

The solution of the certainty-equivalent SGRE equation and the control policies evolve in a backward-in-time manner, provided the system matrices are known beforehand. When the system matrices are uncertain, the SGRE solution cannot be found. Additionally, generating the control policy in a forward-in-time manner has significant practical value for hardware implementation, which is not possible with traditional optimal control techniques. For LNCS two-player zero-sum games, the system matrices are times-varying and uncertain due to network imperfections necessitating an optimal adaptive approach.

5.2 INFINITE-HORIZON OPTIMAL ADAPTIVE DESIGN

In this section, we use the idea of ADP (Watkins 1989; Werbos 1999) and the concept of adaptive estimation of value function to develop stochastic optimal strategy for NCS quadratic zero-sum games with uncertain linear time-varying system dynamics that change slowly in comparison with the sampling interval due to communication imperfections such as network-induced delays and packet losses. A perfect measurement is considered. Thus, in this section, first, the background of LNCS quadratic zero-sum games is introduced. Then, we introduce an adaptive estimation scheme to obtain the unknown value function for NCS quadratic zero-sum games in input–output form with network imperfections. Subsequently, a model-free online tuning of the parameters based on adaptive estimation of cost and value function and the ADP algorithm will be proposed. Eventually, the convergence proof is given.

5.2.1 Background

5.2.1.1 LNCS Quadratic Zero-Sum Games

The basic structure of NCS considered in this chapter is shown as Figure 3.1 where the feedback control loop is closed over a communication network. Consider the following linear time-invariant system with communication imperfections (i.e., network-induced delays and packet losses) described by

$$
\begin{aligned}
\dot{x}(t) &= Ax(t) + \gamma(t)Bu(t-\tau(t)) + \gamma(t)Dd(t-\tau(t)) \\
y(t) &= Cx(t)
\end{aligned}
\tag{5.7}
$$

where

$$
\gamma(t) = \begin{cases} \mathbf{I}^{n\times n} & \text{if the control input is received at time } t \\ \mathbf{0}^{n\times n} & \text{if the control input is lost at time } t \end{cases}
$$

where $x(t) \in \Re^n$, $y(t) \in \Re^r$, $u(t) \in \Re^m$, $d(t) \in \Re^l$ represent the state, output, control, and disturbance input vectors, respectively, and $A \in \Re^{n\times n}$, $B \in \Re^{n\times m}$, $C \in \Re^{r\times n}$, $D \in \Re^{n\times l}$ denote the system matrices. Recalling Assumption 3.1, it can be deduced that the sum of network-induced delays is bounded above such that $\tau(t) = \tau_{sc}(t) + \tau_{ca}(t) < bT_s$, where b represents the delay bound while T_s is the sampling interval.

Recalling Section 3.2.1, integration of Equation 5.7 over a sampling interval $[kT_s,(k+1)T_s]\forall k$ yields

$$
\begin{aligned}
x_{k+1} &= A_s x_k + \sum_{i=0}^{b} \gamma_{k-i} B_i^k u_{k-i} + \sum_{i=0}^{b} \gamma_{k-i} D_i^k d_{k-i} \\
y_k &= Cx_k
\end{aligned}
\tag{5.8}
$$

where

$$
x_k = x(kT), A_s = e^{AT}, \quad B_0^k = \int_{\tau_0^k}^{T} e^{A(T-s)}\, ds\, B \cdot \mathbf{1}(T-\tau_0^k)
$$

$$
B_i^k = \int_{\tau_i^k - iT}^{\tau_{i-1}^k-(i-1)T} e^{A(T-s)}\, dsB \cdot \delta(T+\tau_{i-1}^k - \tau_i^k)\cdot \delta(\tau_i^k - iT) \quad \forall i = 1,2,\ldots,b
$$

$$
D_i^k = \int_{\tau_i^k - iT}^{\tau_{i-1}^k-(i-1)T} e^{A(T-s)}\, ds\, D \cdot \delta(T+\tau_{i-1}^k - \tau_i^k)\cdot \delta(\tau_i^k - iT) \quad \forall i = 1,2,\ldots,b
$$

$$
D_0^i = \int_{\tau_0^k}^{T} e^{A(T-s)}\, ds\, D \cdot \delta((k+1)T - \tau_0^k); \quad \delta(x) = \begin{cases} 1, & x \ge 0 \\ 0, & x < 0 \end{cases}
$$

and

$$\gamma_{k-i} = \begin{cases} 1, & \text{if } u_{k-i} \text{ was received during } [kT_s,(k+1)T_s) \\ 0, & \text{if } u_{k-i} \text{ was lost during } [kT_s,(k+1)T_s) \end{cases}$$

By using a new augment state variable $z_k = [x_k^T \; u_{k-1}^T \ldots u_{k-b}^T \; d_{k-1}^T \ldots d_{k-b}^T]^T$ which includes the current system state, previous control, and disturbance inputs, Equation 5.8 can be expressed as a linear time-varying stochastic discrete-time system described by

$$z_{k+1} = A_{zk}z_k + B_{zk}u_k + D_{zk}d_k, \quad y_k^n = C_z z_k \tag{5.9}$$

where the system matrices are a function of the unknown random delays, and packet losses, which are given by

$$A_{zk} = \begin{bmatrix} A_s & \gamma_{k-1}B_1^k & \cdots & \gamma_{k-i}B_i^k & \cdots & \gamma_{k-b}B_b^k & \gamma_{k-1}D_1^k & \cdots & \gamma_{k-i}D_i^k & \cdots & \gamma_{k-b}D_b^k \\ 0 & 0 & \cdots & \cdots & 0 & 0 & 0 & 0 & \cdots & 0 & 0 \\ 0 & I_m & 0 & \cdots & 0 & 0 & 0 & 0 & \cdots & 0 & 0 \\ 0 & 0 & I_m & \cdots & 0 & 0 & 0 & 0 & \cdots & 0 & 0 \\ \vdots & \vdots & & \ddots & & \vdots & \vdots & 0 & \cdots & \vdots & \vdots \\ 0 & 0 & \cdots & \cdots & I_m & 0 & 0 & 0 & \cdots & 0 & 0 \\ 0 & 0 & \cdots & \cdots & 0 & 0 & 0 & 0 & \cdots & 0 & 0 \\ 0 & 0 & \cdots & \cdots & 0 & 0 & I_l & 0 & \cdots & 0 & 0 \\ 0 & 0 & \cdots & \cdots & 0 & 0 & 0 & I_l & \cdots & 0 & 0 \\ \vdots & & \ddots & & \vdots & \vdots & \vdots & & \ddots & & \vdots \\ 0 & 0 & \cdots & \cdots & 0 & 0 & 0 & 0 & \cdots & I_l & 0 \end{bmatrix}$$

and

$$B_{zk} = [(\gamma_k B_0^k)^T \; I_m \; 0 \ldots 0]^T, \quad C_z = diag\{C \; I_m \; I_m \ldots I_l\}$$

$$D_{zk} = [(\gamma_k D_0^k)^T \cdots\cdots I_l \; 0 \cdots 0]^T$$

and

$$y_k^n = [y_k^T \; u_{k-1}^T \ldots u_{k-b}^T \; d_{k-1}^T \ldots d_{k-b}^T]^T$$

where I_m, I_l are $m \times m$ and $l \times l$ identity matrices. Note that using augment state z_k will not affect the optimal control policy design since it only incorporates system history information while Equation 5.9 is equivalent to Equation 5.8.

From Equations 5.8 and 5.9, it is clear that when the network imperfections are included, the time-invariant linear continuous-time system with known system dynamics (5.8) becomes an uncertain stochastic time-varying system. Moreover, when communication network changes are considered more slowly (Yang et al. 2011) when compared to the sampling rate, the NCS system description (5.9) can be considered as a linear but slowly time-varying stochastic discrete-time system. The network-induced random delays and packet losses are not accurately known beforehand except for their upper bounds and thus the NCS dynamics become uncertain and stochastic in nature. Meanwhile, the controllability and observability properties of the stochastic linear discrete-time system are now studied in the lemma stated next.

Lemma 5.1

Given the origin linear system (A,B) is controllable and (A,C) is observable, the linear discrete-time system (5.9) (A_{zk},B_{zk}) with augmented state is also controllable and (A_{zk},C_z) observable.

Proof: Refer to Appendix 5A.

On the other hand, it is important to note that network imperfections will affect the stability of the closed-loop system, which is stated next.

Lemma 5.2

Given a controllable LNCS zero-sum game with matrices (A_{zk},B_{zk}), the eigenvalues of the closed-loop system and overall stability are dependent upon the expected values of network-induced delay and packet losses.

Proof: Refer to Appendix 5B.

Remark 5.1

The network-induced random delays and packet losses affect the closed-loop poles in an interesting way. For a fixed feedback gain matrix, an increase in the mean value of the random delay and/or packet losses can move the poles closer to the unit disk and eventually become unstable as expected. Normally, it is not possible to know the actual value of the random delays and packet losses over time.

In the next section, the system description given in Equation 5.9 will be transformed into the input–output form for optimal controller design since the states are considered unavailable.

5.2.1.2 LNCS Quadratic Zero-Sum Games in Input–Output Form

According to Lemma 5.1, (A_{zk},B_{zk}) is controllable and (A_{zk},C_z) is observable with the observability index N, and Equation 5.9 can be expressed as

$$\begin{aligned} z_k &= \prod_{i=k-N}^{k-1} A_{zi} z_{k-N} + T^c_{k-1,k-N}\bar{u}_{k-1,k-N} + T^d_{k-1,k-N}\bar{d}_{k-1,k-N} \\ \bar{y}_{k-1,k-N} &= G_{k-1,k-N} z_{k-N} + M^c_{k-2,k-N}\bar{u}_{k-2,k-N} + M^d_{k-2,k-N}\bar{d}_{k-2,k-N} \end{aligned} \tag{5.10}$$

where

$$T^{c}_{k-1,k-N} = \left[B_{zk-1}\ A_{zk-1}B_{zk-2} \cdots \prod_{i=k-N+1}^{k-1} A_{zi}B_{zk-N} \right],$$

$$G_{k-1,k-N} = \left[\left(C_z \prod_{i=k-N}^{k-2} A_{zi} \right)^T \left(C_z \prod_{i=k-N}^{k-3} A_{zi} \right)^T \cdots C_z^T \right]^T$$

are the controllability and observability matrices, respectively,

$$T^{d}_{k-1,k-N} = \left[D_{zk-1}\ A_{zk-1}D_{zk-2} \cdots \prod_{i=k-N+1}^{k-1} A_{zi}D_{zk-N} \right],$$

$$\overline{u}_{k-1,k-N} = [u_{k-1}^T\ u_{k-2}^T \ldots u_{k-N}^T]^T, \overline{d}_{k-1,k-N} = [d_{k-1}^T\ d_{k-2}^T \ldots d_{k-N}^T]^T,$$

$$\overline{y}_{k-1,k-N} = [y_{k-1}^T\ y_{k-2}^T \ldots y_{k-N}^T]^T$$

are historical augment control policies, disturbances, and system outputs, respectively, and

$$M^{c}_{k-2,k-N} = \begin{bmatrix} C_z B_{zk-2} & C_z A_{zk-2} B_{zk-3} & \cdots & C_z \prod_{i=k-N+1}^{k-2} A_{zi} B_{zk-N} \\ 0 & C_z B_{zk-3} & \cdots & C_z \prod_{i=k-N+1}^{k-3} A_{zi} B_{zk-N} \\ \vdots & \vdots & \ddots & \vdots \\ 0 & 0 & \cdots & C_z B_{zk-N} \\ 0 & 0 & \cdots & 0 \end{bmatrix}$$

and

$$M^{d}_{k-2,k-N} = \begin{bmatrix} C_z D_{zk-2} & C_z A_{zk-2} D_{zk-3} & \cdots & C_z \prod_{i=k-N+1}^{k-2} A_{zi} D_{zk-N} \\ 0 & C_z D_{zk-3} & \cdots & C_z \prod_{i=k-N+1}^{k-3} A_{zi} D_{zk-N} \\ \vdots & \vdots & \ddots & \vdots \\ 0 & 0 & \cdots & C_z D_{zk-N} \\ 0 & 0 & \cdots & 0 \end{bmatrix}$$

According to controllability and observability definitions, the observability matrix, $G_{k-1,k-N}$, has full rank and is invertible. Therefore, the system state can be expressed by using measured outputs, controller, and disturbance inputs as

$$z_{k-N}=G_{k-1,k-N}^{-1}\bar{y}_{k-1,k-N}-G_{k-1,k-N}^{-1}M_{k-2,k-N}^{c}\bar{u}_{k-2,k-N}-G_{k-1,k-N}^{-1}M_{k-2,k-N}^{d}\bar{d}_{k-2,k-N}$$

In the other words,

$$z_k=V_{k-1}\mathbf{y}_{k-1}^{c}+B_{zk-1}u_{k-1}+D_{zk-1}d_{k-1} \tag{5.11}$$

where $\mathbf{y}_k^c=[\bar{y}_{k,k-N+1}^T\ \bar{u}_{k-1,k-N+1}^T\ \bar{d}_{k-1,k-N+1}^T]^T$ is augment measured output and

$$V_{k-1}=[V_{1,k-1}\ V_{2,k-1}\ V_{3,k-1}]=\left[\prod_{i=k-N}^{k-1}A_{zi}G_{k-1,k-N}^{-1}\left(A_{zk-1}T_{k-2,k-N}^{c}-\prod_{i=k-N}^{k-1}A_{zi}G_{k-1,k-N}^{-1}M_{k-2,k-N}^{c}\right)\right.$$
$$\left.\left(A_{zk-1}T_{k-2,k-N}^{d}-\prod_{i=k-N}^{k-1}A_{zi}G_{k-1,k-N}^{-1}M_{k-2,k-N}^{d}\right)\right]$$

According to Equations 5.9 and 5.10, linear NCS quadratic zero-sum game can be represented with measured inputs and output as

$$\mathbf{y}_{k+1}^{c}=Q_k\mathbf{y}_k^{c}+F_ku_k+W_kd_k \tag{5.12}$$

where $F_k=[(C_zB_{zk})^T\ I_m\ldots 0]^T$, $W_k=[(C_zD_{zk})^T 0\ldots I_l\ldots 0]^T$, $\mathbf{y}_{k+1}^c$ denotes augmented measured output at time $(k+1)T_s$,and

$$Q_k=\begin{bmatrix} C_zV_{1,k-1} & & C_zV_{2,k-1} & & C_zV_{3,k-1}\\ 0 & \cdots & 0 & \cdots & 0\\ 0 & I_{m\times(N-2)} & 0 & \cdots & 0\\ 0 & \cdots & 0 & \cdots & 0\\ 0 & \cdots & 0 & I_{l\times(N-2)} & 0 \end{bmatrix}$$

When compared to Equation 5.9, the linear NCS quadratic zero-sum game is expressed in the input–output form given by Equation 5.12, which will be utilized in the development of stochastic adaptive optimal strategies. Therefore, in this chapter, based on optimal control theory (Stengel 1986), the certainty-equivalent stochastic cost function can be defined as

$$J_k=\underset{\tau,\gamma}{E}\left[\sum_{i=k}^{\infty}(x_i^TGx_i+u_i^TRu_i-d_i^TSd_i)\right]\quad \forall k=0,1,2,\ldots \tag{5.13}$$

where u_i, d_i are control inputs and disturbances, respectively, G is a symmetric positive semidefinite matrix, R is a symmetric positive definite matrix, S is a symmetric positive definite matrix defined equal to the square of upper bound γ on the desired L_2 gain disturbance attenuation (i.e., $S = \gamma^2 I$, I is an identity matrix) (Basar and Olsder 1995; Al-Tamimi et al. 2007), and $\underset{\tau,\gamma}{E}(\cdot)$ is the expectation operator of $\sum_{i=k}^{\infty}(x_i^T Q x_i + u_i^T R u_i - d_i^T S d_i)$ based on network imperfections (i.e., network-induced delays and packet losses) at each time interval. After redefining the augment state variable z_k, the original stochastic cost function, Equation 5.13 can be expressed with the augment measured outputs with perfect measurements as

$$J_k = \underset{\tau,\gamma}{E}\left[\sum_{i=k}^{\infty}(y_i^{cT} G_y y_i^c + u_i^T R_y u_i - d_i^T S_y d_i)\right] \quad \forall k = 0,1,2,\ldots \tag{5.14}$$

where

$$G_y = \begin{bmatrix} Q & 0 & \cdots & 0 \\ 0 & 0 & \cdots & 0 \\ \vdots & & \ddots & \vdots \\ 0 & 0 & \cdots & 0 \end{bmatrix}, \quad R_y = R, \text{ and } S_y = S$$

Note that G_y is still a symmetric positive semidefinite matrix while R_y, S_y are symmetric positive definite matrices, respectively.

5.2.2 Stochastic Value Function

In this section, we formulate Bellman's optimality principle for NCS quadratic zero-sum games by using the concept of ADP under network imperfections described by Equation 5.9. It is easy to verify that NCS quadratic zero-sum games has a unique equilibrium point, $\mathbf{y} = 0$, on a set Ω while the system outputs are measurable. According to these conditions, the stochastic optimal strategy that optimizes the stochastic cost function J_k for NCS system (5.9) can be derived from (Basar and Olsder 1995; Al-Tamimi et al. 2007) $u_k^* = -K_k y_k$, $d_k^* = -L_k y_k$ with K_k, L_k being the optimal Kalman gains for the control inputs and worst-case disturbance signals, respectively.

If we assume that there is a solution to the stochastic game-theoretic Riccati equation (SGRE) that is strictly feedback stabilizing, then it can be shown (Basar and Olsder 1995) that policies attain a saddle-point equilibrium (Al-Tamim et al. 2007), which implies that minimax is equal to maximin, in the restricted class of feedback stabilizing policies. Assuming that game has a value and is solvable, then it is known that the certainty-equivalent value function is quadratic in augment measured outputs (5.11) and is given by

$$J_k = \underset{\tau,\gamma}{E}(y_k^{cT} P_k y_k^c) \tag{5.15}$$

where matrix $P_k \geq 0$ is a solution to the certainty-equivalent SGRE (Lu et al. 2004). The certainty-equivalent optimal action-dependent value function of LNCS quadratic zero-sum games is now defined to be

$$\begin{aligned} V(y_k^c, u_k, d_k) &= \underset{\tau,\gamma}{E}\left\{ y_k^{cT} G_y y_k^c + u_k^T R_y u_k - d_k^T S_y d_k + J_{k+1} \right\} \\ &= \underset{\tau,\gamma}{E}\left\{ r(y_k^c, u_k, d_k) + J_{k+1} \right\} = \underset{\tau,\gamma}{E}\left\{ [y_k^{cT}\ u_k^T\ d_k^T] H_k [y_k^{cT}\ u_k^T\ d_k^T]^T \right\} \end{aligned} \tag{5.16}$$

where

$$r(y_k^c, u_k, d_k) = y_k^{cT} G_y y_k^c + u_k^T R_y u_k - d_k^T S_y d_k$$

Since stochastic optimal control inputs and worst-case disturbance signals, u_k^*, d_k^*, are dependent on augment measured outputs y_k^c, which is known at time k, the value function can be expressed as

$$V(y_k^c, u_k, d_k) = [y_k^{cT}\ u_k^T\ d_k^T]^T \underset{\tau,\gamma}{E}(H_k)[y_k^{cT}\ u_k^T\ d_k^T]$$

Then, using the Bellman equation and cost function, we can get

$$\begin{aligned} \begin{bmatrix} y_k^c \\ u_k \\ d_k \end{bmatrix}^T \underset{\tau,\gamma}{E}(H_k) \begin{bmatrix} y_k^c \\ u_k \\ d_k \end{bmatrix} &= \underset{\tau,\gamma}{E}\{ r(y_k^c, u_k, d_k) + J_{k+1} \} \\ &= \begin{bmatrix} y_k^c \\ u_k \\ d_k \end{bmatrix}^T \begin{bmatrix} G_y + \underset{\tau,\gamma}{E}(Q_k^T P_{k+1} Q_k) & \underset{\tau,\gamma}{E}(Q_k^T P_{k+1} F_k) & \underset{\tau,\gamma}{E}(Q_k^T P_{k+1} W_k) \\ \underset{\tau,\gamma}{E}(F_k^T P_{k+1} Q_k) & R_y + \underset{\tau,\gamma}{E}(F_k^T P_{k+1} F_k) & \underset{\tau,\gamma}{E}(F_k^T P_{k+1} W_k) \\ \underset{\tau,\gamma}{E}(W_k^T P_{k+1} Q_k) & \underset{\tau,\gamma}{E}(W_k^T P_{k+1} F_k) & \underset{\tau,\gamma}{E}(W_k^T P_{k+1} W_k) - S_y \end{bmatrix} \\ &\quad \times \begin{bmatrix} y_k^c \\ u_k \\ d_k \end{bmatrix} \end{aligned} \tag{5.17}$$

Therefore, $\underset{\tau,\gamma}{E}(H_k)$ can be written as

$$\bar{H}_k = \underset{\tau,\gamma}{E}(H_k) = \begin{bmatrix} \bar{H}_k^{zz} & \bar{H}_k^{zu} & \bar{H}_k^{zd} \\ \bar{H}_k^{uz} & \bar{H}_k^{uu} & \bar{H}_k^{ud} \\ \bar{H}_k^{dz} & \bar{H}_k^{du} & \bar{H}_k^{dd} \end{bmatrix}$$

$$= \begin{bmatrix} G_y + \underset{\tau,\gamma}{E}(Q_k^T P_{k+1} Q_k) & \underset{\tau,\gamma}{E}(Q_k^T P_{k+1} F_k) & \underset{\tau,\gamma}{E}(Q_k^T P_{k+1} W_k) \\ \underset{\tau,\gamma}{E}(F_k^T P_{k+1} Q_k) & R_y + \underset{\tau,\gamma}{E}(F_k^T P_{k+1} F_k) & \underset{\tau,\gamma}{E}(F_k^T P_{k+1} W_k) \\ \underset{\tau,\gamma}{E}(W_k^T P_{k+1} Q_k) & \underset{\tau,\gamma}{E}(W_k^T P_{k+1} F_k) & \underset{\tau,\gamma}{E}(W_k^T P_{k+1} W_k) - S_y \end{bmatrix} \tag{5.18}$$

Then, using stochastic optimal control theory (Stengel 1986; Chen and Guo, 1991), for zero-sum game (Al-Tamim et al. 2007), the gain matrix associated with optimal control policies and worst-case disturbances can be expressed in terms of $\bar{H}_k$ as

$$\begin{aligned} \underset{\tau,\gamma}{E}[K_k] &= -(R_y + \underset{\tau,\gamma}{E}(F_k^T P_{k+1} F_k) - \underset{\tau,\gamma}{E}(F_k^T P_{k+1} W_k)\left(\underset{\tau,\gamma}{E}(W_k^T P_{k+1} W_k) - S_y\right)^{-1} \\ &\quad \times \underset{\tau,\gamma}{E}(W_k^T P_{k+1} F_k))^{-1}(\underset{\tau,\gamma}{E}(F_k^T P_{k+1} W_k)\left(\underset{\tau,\gamma}{E}(W_k^T P_{k+1} W_k) - S_y\right)^{-1} \\ &\quad \times \underset{\tau,\gamma}{E}(W_k^T P_{k+1} Q_k) - \underset{\tau,\gamma}{E}(F_k^T P_{k+1} Q_k)) \\ &= -(\bar{H}_k^{uu} - \bar{H}_k^{ud}(\bar{H}_k^{dd})^{-1}\bar{H}_k^{du})^{-1}(\bar{H}_k^{ud}(\bar{H}_k^{dd})^{-1}\bar{H}_k^{dz} - \bar{H}_k^{uz}) \end{aligned} \tag{5.19}$$

and

$$\begin{aligned} \underset{\tau,\gamma}{E}[L_k] &= -(\underset{\tau,\gamma}{E}(W_k^T P_{k+1} W_k) - S_y - \underset{\tau,\gamma}{E}(W_k^T P_{k+1} F_k)\left(\underset{\tau,\gamma}{E}(F_k^T P_{k+1} F_k) + R_y\right)^{-1} \\ &\quad \times \underset{\tau,\gamma}{E}(F_k^T P_{k+1} W_k))^{-1}(\underset{\tau,\gamma}{E}(W_k^T P_{k+1} F_k)\left(\underset{\tau,\gamma}{E}(F_k^T P_{k+1} F_k) + R_y\right)^{-1} \\ &\quad \times \underset{\tau,\gamma}{E}(F_k^T P_{k+1} Q_k) - \underset{\tau,\gamma}{E}(W_k^T P_{k+1} Q_k)) \\ &= -(\bar{H}_k^{du} - \bar{H}_k^{du}(\bar{H}_k^{uu})^{-1}\bar{H}_k^{ud})^{-1}(\bar{H}_k^{du}(\bar{H}_k^{uu})^{-1}\bar{H}_k^{uz} - \bar{H}_k^{dz}) \end{aligned} \tag{5.20}$$

Equations 5.19 and 5.20 represent time-varying gains based on the solution of the certainty-equivalent SGRE and hence some interesting observations can be stated using Equations 5.19 and 5.20. If the matrix P_k is known, then we still need the slowly time-varying system matrices to compute the optimal controller and worst-case disturbance gains. On the other hand, if the slowly time-varying matrix $\bar{H}_k$ can be estimated online without the knowledge of NCS dynamics (5.19), the NCS system matrices are not required to compute the optimal strategy gains. An adaptive estimator will be utilized to learn the time-varying matrix, $\bar{H}_k$, which in turn will be used to obtain optimal time-varying gains.

Remark 5.2

It is important to note that there are several differences between the optimal design in this chapter and Al-Tamim et al. (2007). The proposed scheme is more generic and applicable to uncertain stochastic linear system for NCS and traditional time-varying continuous and discrete-time systems whereas work by Al-Tamim et al. (2007) is applicable to time-invariant systems. Therefore, the work of Al-Tamim et al. (2007) cannot maintain the stability of an NCS due to network imperfections. Second, Al-Tamim et al. (2007) uses value iteration within each sampling interval, which in turn requires a significant number of iterations for convergence of the algorithm while the proposed scheme updates the value function and control policy once every sampling interval consistent with the standard adaptive control. Third, the proposed design is developed for an NCS in input–output form whereas the design in Al-Tamim et al. (2007) is only suitable for linear state feedback zero-sum games. Finally, this chapter derives closed-loop system stability, which is not addressed in Al-Tamim et al. (2007). Therefore, the proposed optimal strategy based on adaptive estimation of cost or value function is an online and forward-in-time approach and does not require policy and value iterations.

5.2.3 Model-Free Online Tuning

The proposed online tuning approach estimates the value function (5.17) online. Since the value function includes the $\bar{H}_k$ matrix (5.18), which can be solved, the control inputs and worst-case disturbance inputs can be obtain using augmented measured outputs, Equations 5.19 and 5.20. Next, we make the following assumption since the LNCS is a slowly linear time-varying unknown system (see Remark 5.2) and the delays are bounded above while the packet losses satisfy the Bernoulli distribution, and both of them change slowly (Kreisselmeier 1986).

Assumption 5.1

The value function $V(y_k^c, u_k, d_k)$ can be expressed as linear in the unknown parameters (LIP)—a standard assumption in adaptive control (Al-Tamim et al. 2007; Xu et al. 2014).

By using the stochastic adaptive control literature (Ioannou and Sun 1996) and Equation 5.16, the value function can be represented in vector form as

$$V(y_k^c, u_k, d_k) = w_k^T \bar{H}_k w_k = \bar{h}_k^T \bar{w}_k \tag{5.21}$$

where $\bar{h}_k = vec(\bar{H}_k), w_k = [y_k^{cT} \; u^T(z_k) \; d^T(z_k)]^T \in \Re^P$, and $\bar{w}_k = (w_{k1}^2, \ldots, w_{k1}w_{kq}, w_{k2}^2, \ldots, w_{kq-1}w_{kq}, w_{kq}^2)$ is the Kronecker product quadratic polynomial basis vector

(Xu et al. 2012). $\bar{h}_k = vec(\bar{H}_k)$ with the vector function acting on $q \times q$ matrices, thus yielding a $q(q+1)/2 \times 1$ column vector.

It is important to note that the vec(·) function is constructed by stacking the columns of the matrix into one column vector with off-diagonal elements, which can be combined as $H_{mn} + H_{nm}$. Therefore, the value function can be expressed as target unknown parameter vector multiplied by the regression function $\bar{w}_k$. The time-varying matrix $\bar{H}_k$ can be considered as slowly varying (Kreisselmeier 1986). Then it can be expressed as a time-varying target parameter vector and a known regression function $\bar{w}_k$. Now, the value function $V(\boldsymbol{y}_k^c, u_k, d_k)$ estimation will be considered.

According to the definition of the value function (5.16) and the relationship between the value function and stochastic cost function (Basar and Olsder 1995), we can use the matrix $\bar{H}_k$ in Equation 5.18 to express the stochastic cost function as

$$J_k(z) = V(\boldsymbol{y}_k^c, u_k, d_k) = w_k^T \bar{H}_k w_k = \bar{h}_k^T \bar{w}_k \tag{5.22}$$

Then the value function $V(\boldsymbol{y}_k^c, u_k, d_k)$ can be estimated by an adaptive estimator in terms of estimated parameter vector $\hat{\bar{h}}_k$ as

$$\hat{J}_k(z) = \hat{V}(\boldsymbol{y}_k^c, u_k, d_k) = \hat{\bar{h}}_k^T \bar{w}_k \tag{5.23}$$

where $w_k = [y_k^{cT} \; u^T(z_k) \; d^T(z_k)]^T$ and $\bar{w}_k$ is a Kronecker product quadratic polynomial basis vector of w_k.

It is observed that the Bellman equation can be rewritten as $J_{k+1} - J_k + \underset{\tau,\gamma}{E}[r(\boldsymbol{y}_k^c, u_k, d_k)] = 0$. This relationship, however, is not guaranteed to hold when we apply the estimated matrix $\hat{\bar{H}}_k$. Hence, using delayed values for convenience, the residual error associated with Equation 5.15 can be expressed as $\hat{J}_{k+1} - \hat{J}_k + \underset{\tau,\gamma}{E}[r(\boldsymbol{y}_k^c, u_k, d_k)] = e_{hk}$, that is,

$$\begin{aligned} e_{hk} &= \underset{\tau,\gamma}{E}[r(\boldsymbol{y}_{k-1}^c, u_{k-1}, d_{k-1})] + \hat{\bar{h}}_k^T \bar{w}_k - \hat{\bar{h}}_k^T \bar{w}_{k-1} \\ &= \underset{\tau,\gamma}{E}[r(\boldsymbol{y}_{k-1}^c, u_{k-1}, d_{k-1})] + \hat{\bar{h}}_k^T (\bar{w}_k - \bar{w}_{k-1}) = \underset{\tau,\gamma}{E}[r(\boldsymbol{y}_{k-1}^c, u_{k-1}, d_{k-1})] + \hat{\bar{h}}_k^T \Delta W_{k-1} \end{aligned} \tag{5.24}$$

where $\Delta W_{k-1} = \bar{w}_k - \bar{w}_{k-1}$.

The dynamics of Equation 5.24 are then rewritten as

$$\underset{\tau,\gamma}{E}[e_{hk+1}] = \underset{\tau,\gamma}{E}[r(\boldsymbol{y}_k^c, u_k, d_k)] + \hat{\bar{h}}_{k+1}^T \Delta W_k \tag{5.25}$$

Next, we define an auxiliary residual error vector as

$$\underset{\tau,\gamma}{E}\left(\Sigma_{hk}\right) = \underset{\tau,\gamma}{E}\left(\Gamma_{k-1}\right) + \hat{\bar{h}}_k^T \Omega_{k-1} \in \Re^{1\times(1+i)} \tag{5.26}$$

where

$$\Gamma_{k-1} = \left[r(y_{k-1}^c, u_{k-1}, d_{k-1}) r(y_{k-2}^c, u_{k-2}, d_{k-2}) \ldots r(y_{k-1-i}^c, u_{k-1-i}, d_{k-1-i}) \right]$$

and

$$\Omega_{k-1} = [\Delta W_{k-1} \; \Delta W_{k-2} \cdots \Delta W_{k-1-i}], \quad 0 < i < k-1 \in N$$

with N being the set of natural real numbers. It is important to note that Equation 5.25 indicates a time history of the previous $(i + 1)$ residual errors (5.23) recalculated by using the most recent $\hat{\bar{h}}_k$. The time history of previous residual errors allows one to overcome the need for any iterative-based value and policy update schemes while still rendering optimal control and worst-case disturbance solution. Therefore, the proposed approach can be referred to as time-based ADP.

Next, the dynamics of the auxiliary vector (5.26) are generated similar to Equation 5.25 and revealed to be

$$\underset{\tau,\gamma}{E}(\Sigma_{hk+1}) = \underset{\tau,\gamma}{E}(\Gamma_k) + \hat{\bar{h}}_{k+1}^T \Omega_k \tag{5.27}$$

Now, defining the update law of the slowly time-varying matrix $\bar{H}_k$ as

$$\hat{\bar{h}}_{k+1} = \Omega_k (\Omega_k^T \Omega_k)^{-1} \underset{\tau,\gamma}{E}\left(\alpha_h \Sigma_{hk}^T - \Gamma_k^T\right) \tag{5.28}$$

where $0 < \alpha_h < 1$. Substituting Equation 5.28 into Equation 5.27 results in

$$\underset{\tau,\gamma}{E}\left(\Sigma_{hk+1}\right) = \alpha_h \underset{\tau,\gamma}{E}\left(\Sigma_{hk}\right) \tag{5.29}$$

Remark 5.3

It is observed that the cost function J_k and adaptive estimation (5.23) will become zero only when $y_k^c = 0$. Hence, when the augment measured system outputs have converged to the origin, the value function approximation is no longer updated. It can be seen as a PE requirement for the inputs to the value function estimator wherein the augment measured system outputs must be persistently exciting long enough for the adaptive estimator to learn the optimal stochastic cost function. Therefore, exploration noise is added to the control inputs and worst-case disturbance signals in order to satisfy the PE condition (Sarangapani 2006), which is given next.

Definition 5.1: Persistence of Excitation

A stochastic vector $\beta_k \in \Re^P$ is said to be PE if there exist positive constants δ, α, and $k_0 \geq 1$, such that $\sum_{k=k_0}^{k_0+\delta} E[\beta_k \beta_k^T)] \geq \alpha \boldsymbol{I}$, where $\boldsymbol{I}$ is the identity matrix.

Lemma 5.3

Persistence of excitations of vector ΔW_k (5.25) and Ω_k can be satisfied by adding exploration noise.

Proof: Refer to Xu et al. (2012).

Now define the parameter estimation error as $\tilde{\bar{h}}_k = \bar{h}_k - \hat{\bar{h}}_k$. Rewrite the Bellman equation using an adaptive estimation with target parameters (5.21) revealing $r(y_k^c, u_k, d_k) + \bar{h}_{k+1}^T \bar{w}_{k+1} = \bar{h}_{k+1}^T \bar{w}_k$, which can be expressed as

$$\underset{\tau,\gamma}{E}[r(y_k^c, u_k, d_k)] = \bar{h}_{k+1}^T \bar{w}_k - \bar{h}_{k+1}^T \bar{w}_{k+1} = -\bar{h}_{k+1}^T \Delta W_k \tag{5.30}$$

Substituting $r(y_k^c, u_k, d_k)$ into Equation 5.25 and utilizing Equation 5.24 with $e_{hk+1} = \alpha_h e_{hk}$ from Equation 5.22 yields

$$\tilde{\bar{h}}_{k+1}^T \Delta W_k = -\alpha_h \underset{\tau,\gamma}{E}\{r(y_{k-1}^c, u_{k-1}, d_{k-1})\} - \alpha_h \hat{\bar{h}}_k^T \Delta W_{k-1} \tag{5.31}$$

Using the similar method as $r(y_k^c, u_k, d_k)$, we can form $r(y_{k-1}^c, u_{k-1}, d_{k-1})$, and substituting this into Equation 5.31, we have

$$\tilde{\bar{h}}_{k+1}^T \Delta W_k = \alpha_h \tilde{\bar{h}}_k^T \Delta W_{k-1} \tag{5.32}$$

Next, the convergence of the cost function errors with adaptive estimation error dynamics $\tilde{\bar{h}}_k$ given by Equation 5.32 is demonstrated for an initial admissible control (Sarangapani 2006) policy. The NCS slowly time-varying system dynamics are known to be asymptotically stable if the initial admissible control and disturbance policy can be applied, provided the system matrices are known. However, introducing the estimated value function results in estimation errors for the stochastic cost function J_k, and therefore the stability of the estimated stochastic cost function needs to be studied. Similar to Xu et al. (2012), adaptive estimator errors can be proven to be asymptotically stable, that is, $\hat{J}_k \to J_k^*$ and $\tilde{\bar{h}}_k \to 0$ when $k \to \infty$. Subsequently, the asymptotic stability of adaptive estimation errors will be used for proving the overall closed-loop system stability in Theorem 5.1 by using an initial admission control policy.

According to Equations 5.19 and 5.20, the estimated control policies and worst-case disturbance signals based on this estimated matrix can be represented as

$$\underset{\tau,\gamma}{E}(\hat{u}_{1k}) = -\underset{\tau,\gamma}{E}[\hat{K}_k y_k^c] = \left(\hat{\bar{H}}_k^{uu} - \hat{\bar{H}}_k^{ud}\left(\hat{\bar{H}}_k^{dd}\right)^{-1}\hat{\bar{H}}_k^{du}\right)^{-1}\left(\hat{\bar{H}}_k^{ud}\left(\hat{\bar{H}}_k^{dd}\right)^{-1}\hat{\bar{H}}_k^{dz} - \hat{\bar{H}}_k^{uz}\right)\underset{\tau,\gamma}{E}(y_k^c)$$

$$\underset{\tau,\gamma}{E}(\hat{d}_{1k}) = -\underset{\tau,\gamma}{E}[\hat{L}_k y_k^c] = \left(\hat{\bar{H}}_k^{du} - \hat{\bar{H}}_k^{du}\left(\hat{\bar{H}}_k^{uu}\right)^{-1}\hat{\bar{H}}_k^{ud}\right)^{-1}\left(\hat{\bar{H}}_k^{du}\left(\hat{\bar{H}}_k^{uu}\right)^{-1}\hat{\bar{H}}_k^{uz} - \hat{\bar{H}}_k^{dz}\right)\underset{\tau,\gamma}{E}(y_k^c) \tag{5.33}$$

Next, the stability of the cost function, control estimation, worst-case disturbance signal estimation, and adaptive estimation error dynamics are considered.

5.2.4 Closed-Loop System Stability

In this section, it will be shown that the slowly time-varying matrix $\bar{H}_k$ and related value function estimation errors dynamics are asymptotically stable. Further, the estimated control inputs and disturbance signals for LNCS (5.33) will approach their optimal control inputs and worst-case disturbance signals asymptotically in the mean. The flowchart of stochastic optimal regulator of LNCS quadratic zero-sum games in input–output form with unknown system dynamics is shown in Figure 5.1.

Next, the initial augment measured system outputs are considered to reside in a set when the initial stabilizing control input and disturbance signals u_{0k}, d_{0k} are being utilized. Further sufficient condition for the adaptive estimator tuning gain α_h is

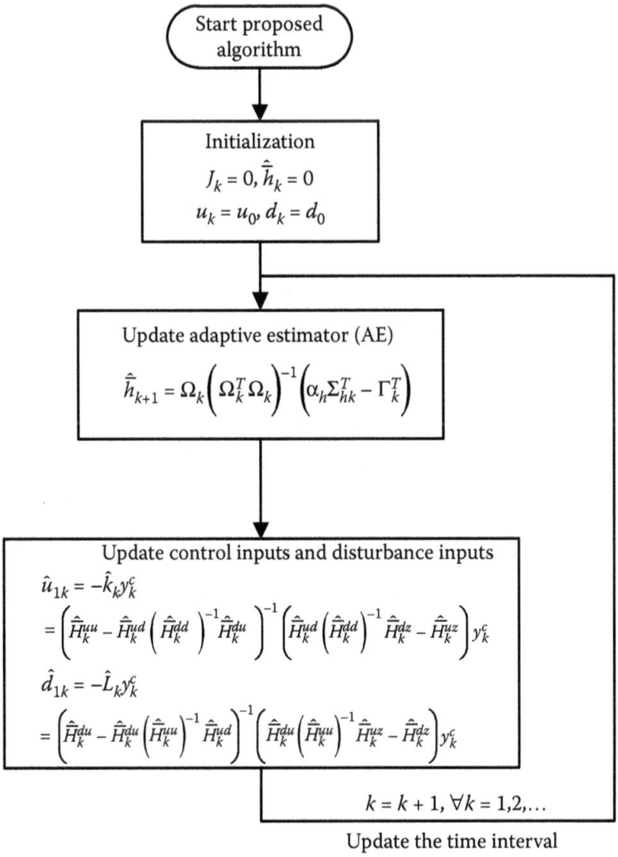

FIGURE 5.1 Flowchart of the stochastic optimal control scheme.

derived to ensure that all future outputs will converge to zero. Then it can be shown that the estimated control inputs and disturbance signals will approach the optimal control inputs and worst-case disturbances asymptotically.

Before convergence proof, the following result is needed to establish bounds on the closed-loop dynamics when the optimal control inputs and worst-case disturbances are applied to the LNCS system (5.9) in input–output form with communication imperfections (i.e., network-induced delays and packet losses). Table 5.1 shows the proposed approach.

Lemma 5.4

Consider a linear quadratic zero-sum game under communication imperfections (5.12), and then there exists a set of optimal strategies such that the following inequality is satisfied

$$\left\| \underset{\tau,\gamma}{E}[Q_k y_k^c + F_k u_k^* + W_k d_k^*] \right\|^2 \le k_a \underset{\tau,\gamma}{E} \left\| y_k^c \right\|^2 \tag{5.34}$$

where $0 < k_a < 1/3$ is a constant.

Proof: See Xu et al. (2014).

Theorem 5.1: Convergence of the Optimal Control Policies and Worst-Case Disturbance Signals

Given the initial conditions for the augment measured system output y_0^c, cost function and adaptive estimator parameter vectors $\hat{\bar{h}}_0$ be bounded in the set Ω, let u_{0k}, d_{0k} be any initial admissible control inputs and disturbance signals for the linear discrete-time quadratic zero-sum game in input–output form with communication imperfections, which can maintain the initial system condition to be bounded in the set Ω while satisfying the bounds given by Equation 5.34 for $0 < k^* < 1/3$. Let the adaptive estimation parameter vector be tuned and estimation of feedback control inputs and worst-case disturbance signals be provided by Equations 5.28 and 5.33, respectively. Then, there exists positive constants α_h given by $0 < \alpha_h < (1/\sqrt{2})$ such that the augment measured system outputs y_k^c and stochastic cost function parameter estimator errors $\tilde{\bar{h}}_k$ are all asymptotically stable in the mean. In other words, as $k \to \infty$, $\underset{\tau,\gamma}{E}(y_k^c) \to 0$, $\tilde{\bar{h}}_k \to 0$, $\underset{\tau,\gamma}{E}(\hat{J}_k) \to \underset{\tau,\gamma}{E}(J_k)$ and $\underset{\tau,\gamma}{E}(\hat{u}_k) \to \underset{\tau,\gamma}{E}(u_k^*)$, $\underset{\tau,\gamma}{E}(\hat{d}_k) \to \underset{\tau,\gamma}{E}(d_k^*)$.

Proof: Refer to Appendix 5C.

Remark 5.4

In traditional ADP (Werbos 1999), policy and value iteration methods are employed during a fixed sampling interval, and system states and inputs are recalculated and

TABLE 5.1
Stochastic Optimal Design for LNCS with Disturbance

Control Input and Disturbance

$$\underset{\tau,\gamma}{E}(\hat{u}_{1k}) = -\underset{\tau,\gamma}{E}(\hat{K}_k \mathbf{y}_k^c) = \left(\hat{\bar{H}}_k^{uu} - \hat{\bar{H}}_k^{ud}\left(\hat{\bar{H}}_k^{dd}\right)^{-1}\hat{\bar{H}}_k^{du}\right)^{-1}\left(\hat{\bar{H}}_k^{ud}\left(\hat{\bar{H}}_k^{dd}\right)^{-1}\hat{\bar{H}}_k^{dz} - \hat{\bar{H}}_k^{uz}\right)\underset{\tau,\gamma}{E}(\mathbf{y}_k^c)$$

$$\underset{\tau,\gamma}{E}(\hat{d}_{1k}) = -\underset{\tau,\gamma}{E}(\hat{L}_k \mathbf{y}_k^c) = \left(\hat{\bar{H}}_k^{du} - \hat{\bar{H}}_k^{du}\left(\hat{\bar{H}}_k^{uu}\right)^{-1}\hat{\bar{H}}_k^{ud}\right)^{-1}\left(\hat{\bar{H}}_k^{du}\left(\hat{\bar{H}}_k^{uu}\right)^{-1}\hat{\bar{H}}_k^{uz} - \hat{\bar{H}}_k^{dz}\right)\underset{\tau,\gamma}{E}(\mathbf{y}_k^c)$$

where $\hat{\bar{H}}_k = \text{vec}^{-1}\left(\hat{\bar{h}}_k\right)$

Bellman Equation Residual Error

$$\underset{\tau,\gamma}{E}\left(\Sigma_{hk}\right) = \underset{\tau,\gamma}{E}\left(\Gamma_k\right) + \hat{\bar{h}}_{k+1}^T \Omega_k$$

Parameter Update

$$\hat{\bar{h}}_{k+1} = \Omega_k(\Omega_k^T \Omega_k)^{-1} \underset{\tau,\gamma}{E}\left(\alpha_h \Sigma_{hk}^T - \Gamma_k^T\right)$$

where α_h is the tuning parameter.

stored for learning the optimal strategy. For example, during time $[kT_s, (k+1)T_s]$, the system states J_{k+1}^i, u_k^i, and d_k^i will be recalculated and stored for learning optimal strategy J_k^*, u_k^*, and d_k^* when iteration index changes from 1 to ∞, that is, $i = 1,2,\ldots,\infty$. Consequently, traditional ADP value and policy iterations can consume a significant amount of time, which may not be practically viable in a real-time environment. However, this stochastic optimal design does not require value and policy iterations while the cost function and control and estimated worst-case disturbances are updated once every sampling interval and therefore will be referred to as time-based ADP. Only the measured real-time data is used to tune the cost function, estimated optimal control inputs and worst-case disturbances, that is, when $k \to \infty$, $\underset{\tau,\gamma}{E}(\hat{J}_k) \to \underset{\tau,\gamma}{E}(J_k^*)$, $\underset{\tau,\gamma}{E}(\hat{u}_k) \to \underset{\tau,\gamma}{E}(u_k^*)$, $\underset{\tau,\gamma}{E}(\hat{d}_k) \to \underset{\tau,\gamma}{E}(d_k^*)$.

5.2.5 Simulation Results

The performances of stochastic optimal strategies for LNCS quadratic zero-sum games in input–output form are evaluated. At the same time, the standard optimal strategy of NCS quadratic zero-sum games with known dynamics is also simulated for comparison.

EXAMPLE 5.1

The continuous-time version of an output feedback batch reactor system dynamics is given by (Dacic and Nesic 2007)

$$\dot{x}(t) = \begin{bmatrix} 1.38 & -0.2077 & 6.715 & -5.676 \\ -0.5814 & -4.29 & 0 & 0.675 \\ 1.067 & 4.273 & -6.654 & 5.893 \\ 0.048 & 4.273 & 1.343 & -2.104 \end{bmatrix} x(t) + \begin{bmatrix} 0 & 0 \\ 5.679 & 0 \\ 1.136 & -3.146 \\ 1.136 & 0 \end{bmatrix} u(t)$$

$$+ \begin{bmatrix} 10 & 0 & 10 & 0 \\ 0 & 5 & 0 & 5 \end{bmatrix}^T d(t) \tag{5.35}$$

$$y(t) = \begin{bmatrix} 0 & 0.3015 & 0.3015 & 0 \\ 0.603 & 0.3015 & 0.603 & 0.3015 \end{bmatrix} x(t)$$

where $x \in \Re^{4\times1}$, $y \in \Re^{2\times1}$, $u \in \Re^{2\times1}$, and $d \in \Re^{2\times1}$. It is important to note that this example has developed over the years as a benchmark example for an NCS, see, for example, Kreisselmeier (1986). Meanwhile, the observability index for this system (5.11) is $N = 2$.

The parameters of this NCS quadratic output feedback zero-sum games are selected as follows (Xu et al. 2012):

1. The sampling time: $T_s = 0.8$ s.
2. The bound of delay is two, that is, $b = 2$.
3. The mean value of random delays: $E(\tau_{sc}) = 0.5$ s, $E(\tau) = 1.1$ s.
4. Packet losses follow the Bernoulli distribution with $p = 0.3$. These values can be changed.
5. To obtain ergodic behavior, Monte Carlo simulation runs with 1000 were utilized.

Incorporating the random delays $\tau(t)$ and packet losses $\gamma(t)$ to the batch reactor system (5.35), the original time-invariant system was represented as a slowly time-varying LNCS given by Equation 5.9. First, by using the ADP value iteration (VI) method and modifying the strategy (Al-Tamim et al. 2007), the control inputs and worst-case disturbances are designed. The ADP VI scheme normally does not require any system dynamics and information of communication imperfections. However, ADP VI-based control cannot maintain the batch reactor system stable in the presence of communication imperfections (i.e., network-induced delays and packet losses) as shown in Figure 5.2.

Then, the proposed adaptive stochastic optimal strategy is implemented for the LNCS quadratic zero-sum games with unknown system dynamics in the presence of communication imperfections (i.e., network-induced delays and packet losses). The augment measured system output is generated as $y_k^c = [y_k \;\; y_{k-1} \;\; u_{k-1} \;\; u_{k-2} \;\; d_{k-1} \;\; d_{k-2}]^T \in \Re^{12} \; \forall k$ or $w_k = [y_k^c \;\; u_k \;\; d_k]^T \in \Re^{16}$. The initial stabilizing policy and disturbance signal for the algorithm were selected as

$$u_0(y_k^c) = \begin{bmatrix} 0.12 & -0.02 & 0.08 & -0.06 & -0.12 & -0.05 & -0.024 & -0.029 & 0.46 & -0.214 & 0.236 & -0.064 \\ -0.045 & 0.015 & -0.03 & 0.035 & 0.06 & 0.008 & 0.027 & 0.006 & -0.13 & 0.108 & -0.073 & 0.051 \end{bmatrix} y_k^c$$

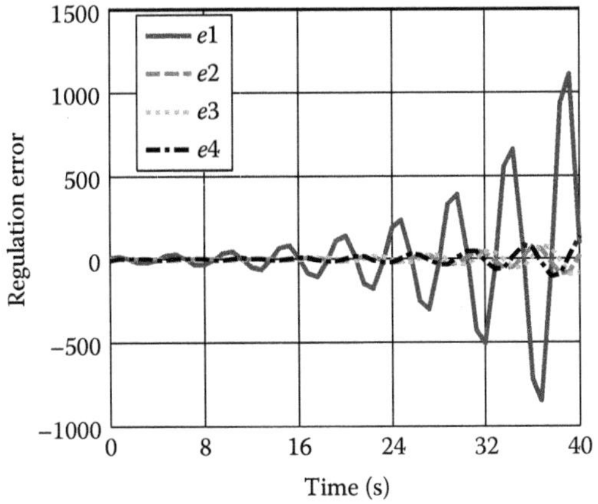

FIGURE 5.2 Performance of the ADP value iteration-based scheme.

and

$$d_0(y_k^c)=\begin{bmatrix} -0.95 & -0.25 & -0.70 & 0.28 & -0.51 & 0.61 \\ 0.41 & -0.12 & 0.25 & -0.297 & -0.77 & -0.135 \end{bmatrix.$$

$$\left. \begin{matrix} -0.234 & 0.274 & -4.591 & 0.055 & -2.07 & 0.02 \\ -0.253 & -0.077 & 1.438 & -1.33 & 0.72 & -0.47 \end{matrix} \right] y_k^c$$

while the regression function for value function was generated as $\{w_1^2, w_1w_2, w_1w_3,\ldots,w_2^2,\ldots,w_{15}^2,\ldots,w_{16}^2\}$ as per Equation 5.21.

In Figures 5.2 through 5.4, the performance of adaptive estimation-based optimal strategy is evaluated. As shown in Figure 5.2, the proposed adaptive estimation-based optimal policy can also force the NCS quadratic zero-sum games state

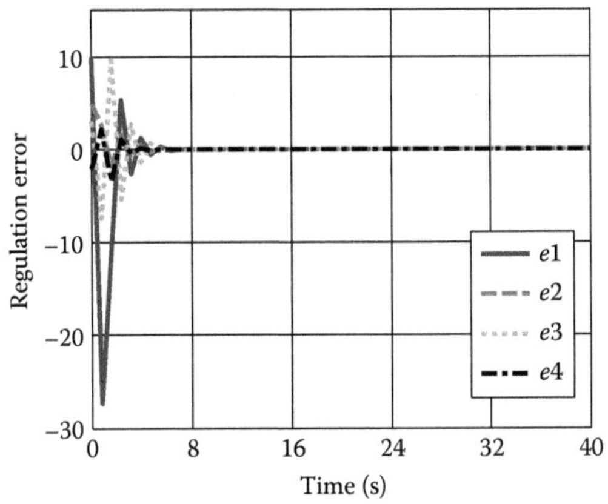

FIGURE 5.3 Performance of the stochastic optimal controller.

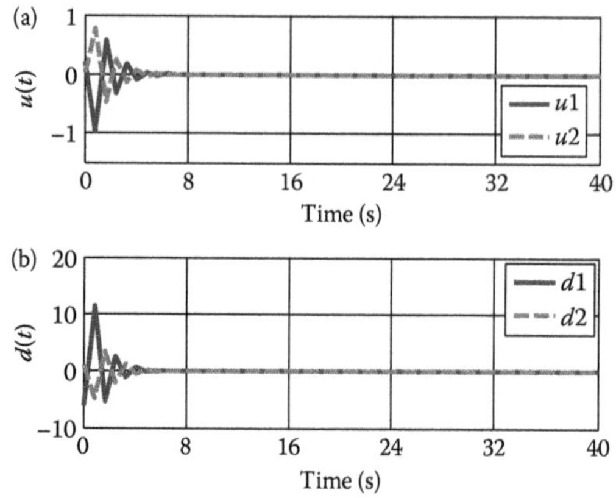

FIGURE 5.4 Performance of the stochastic optimal scheme: (a) control inputs and (b) disturbance input.

regulation errors converge to zero even when the NCS dynamics are unknown which implies that the proposed strategy can make the NCS closed-loop system stable. Owing to an initial online tuning phase needed to learn optimal control and worst-case disturbance signals, there is a slight overshoot at the beginning. In Figure 5.4a and b, the control and worst-case disturbance signals of proposed adaptive estimation-based optimal strategy are shown. Proposed adaptive estimation-based optimal control and worst-case disturbances can make the NCS quadratic zero-sum game converge to zero quickly.

The design parameter for the value function $V(y_k^c, u_k, d_k)$ was selected as $\alpha_h = 10^{-6}$ while initial parameters for the adaptive estimator were set to zeros at the beginning of the simulation. The simulation was run for 40 s, and for the first 22 s, exploration noise with mean zero and variance 0.08 was added to the system in order to ensure that the PE condition holds (Lemma 5.3).

The estimated value function for NCS quadratic zero-sum games are shown in Figure 5.5. The estimated value function is defined in Equation 5.23 as $\hat{V}(y_k^c, u_k, d_k) = [y_k^{cT}\ u_k^T\ d_k^T]^T \hat{\bar{H}}_k [y_k^{cT}\ u_k^T\ d_k^T]$. If all the augment measured system outputs are equal to zero except y_1, y_2, the estimated value function is shown as Figure 5.5a, while Figure 5.5b illustrates the estimated value function when all the augment measured system outputs are equal to zero except y_3, y_4. It is important to note two key points. First, based on the definition of the estimated value function, if all the augment measured system outputs are equal to zero, the estimated value function can be zero. Otherwise, the estimated value function should be a quadratic positive value. This is why a valley is observed in Figure 5.5a and b. Second, the proposed stochastic optimal strategy is designed to minimize the estimated value function.

Based on the results presented in Figures 5.2 through 5.5, and after a short initial tuning time, the proposed adaptive estimation-based stochastic optimal strategy for NCS quadratic zero-sum games with uncertain dynamics and imprecise information on communication imperfections will have nearly the same performance as that

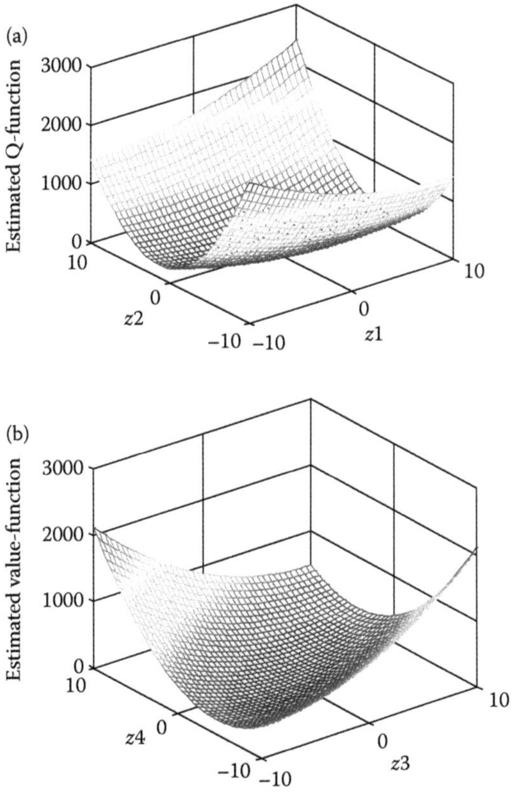

FIGURE 5.5 Estimated value function: (a) in the y_1, y_2 direction and (b) in the y_3, y_4 direction.

of the conventional optimal strategy for NCS quadratic zero-sum games in input–output form when system dynamics and communication imperfections are known.

5.3 CONCLUSIONS

In this chapter, a direct ADP scheme is proposed, which combines the adaptive estimation and the concept of dynamics programming to solve the Bellman equation in real time for the stochastic optimal regulation of LNCS quadratic zero-sum games in input–output form with communication imperfections (i.e., network-induced delays and packet losses). The availability of past output values through perfect measurement ensured that LNCS quadratic zero-sum games dynamics are not needed when an adaptive estimator generates an estimated value function. Initial admissible control and disturbance policies ensured that the adaptive estimator learns the value function $V(y_k^c, u_k, d_k)$ and the matrix $\underset{\tau,\gamma}{E}(H_k)$, cost function and optimal control inputs, and worst-case disturbance signals online. Initial overshoots are observed due to the online learning phase but they are gone quickly with time. All adaptive estimation parameters were tuned online using proposed update laws and Lyapunov theory demonstrated asymptotic stability (AS) in the mean of the closed-loop system.

PROBLEMS

Section 5.3

Problem 5.3.1: Repeat Example 5.1 with the following information.

The parameters of this NCS quadratic output feedback zero-sum games are selected as

1. The sampling time: $T_s = 0.7$ s.
2. The bound of delay is two, that is, $b = 2$.
3. The mean value of random delays: $E(\tau_{sc}) = 0.6$ s, $E(\tau) = 1.3$ s.
4. Packet losses follow the Bernoulli distribution with $p = 0.5$. These values can be changed.
5. Monte Carlo iterations 5000.

Problem 5.3.2: Repeat Problem 5.1.1 with the different optimal weighting matrix, that is, Q, R.

APPENDIX 5A

Proof of Lemma 5.1

Consider Equation 5.9, and observing $t_0 = (k-N)T_s$ as the initial time with $t_f = kT_s$ as the final time instant, we have

$$e^{-At_f} z_{t_f} - e^{-At_0} z_{t_0} = e^{-AkT_s} z_k - e^{-A(k-N)T_s} z_{k-N}$$

$$= \begin{bmatrix} B_{zk} & A_{zk}B_{zk-1} & A_{zk}A_{zk-1}B_{zk-2} \cdots \left(\prod_{i=k-N+2}^{k} A_{zi} \right) B_{zk-N+1} \end{bmatrix} \begin{bmatrix} u_k \\ u_{k-1} \\ \vdots \\ u_{k-N+1} \end{bmatrix} \tag{5A.1}$$

where A_{zk}, B_{zk} are defined in Equation 5.9. Recalling the elements in A_{zk}, B_{zk} (i.e., B_i^k), B_i^k can also be represented as

$$B_i^k = B \cdot \left[\int_{\tau_i^k - iT}^{\tau_{i-1}^k-(i-1)T} e^{A(T-s)} ds \cdot \delta(T + \tau_{i-1}^k - \tau_i^k) \cdot \delta(\tau_i^k - iT) \right] \tag{5A.2}$$

Then, defining $\Theta_C = \left[B_{zk} \ A_{zk}B_{zk-1} \cdots \left(\prod_{i=k-N+2}^{k} A_{zi} \right) B_{zk-N+1} \right]$, Equation 5A.1 can be expressed equivalently as

$$e^{-At_f}z_{t_f} - e^{-At_0}z_{t_0} = \begin{bmatrix} B & \cdots & A^{b-1}B & \cdots & A^{N-2}B & A^{N-1}B \\ I_m & \cdots & 0 & \cdots & 0 & 0 \\ \vdots & \ddots & \vdots & \ddots & \vdots & \vdots \\ 0 & \cdots & I_m & \cdots & 0 & 0 \\ \vdots & \ddots & \vdots & \ddots & \vdots & \vdots \\ 0 & \cdots & 0 & 0 & I_m & 0 \\ 0 & \cdots & 0 & 0 & \cdots & 0 \end{bmatrix} \times \Delta \quad (5A.3)$$

where Δ can be calculated as

$$\Delta = \begin{bmatrix} \displaystyle\int_{\tau_i^k - iT}^{\tau_{i-1}^k - (i-1)T} e^{A(T-s)}\, ds\, u_k \delta(T + \tau_{i-1}^k - \tau_i^k)\delta(\tau_i^k - iT) \\ \vdots \\ \displaystyle\sum_{j=k-b+2}^{k} \sum_{i=1}^{b-1} \int_{\tau_i^j - iT}^{\tau_{i-1}^j - (i-1)T} e^{(j-k+b-1)A(T-s)} ds\, u_j \delta(T + \tau_{i-1}^j - \tau_i^j)\delta(\tau_i^j - iT) \\ \displaystyle\sum_{j=k-b+1}^{k} \sum_{i=1}^{b} \int_{\tau_i^j - iT}^{\tau_{i-1}^j - (i-1)T} e^{(j-k+b)A(T-s)}\, ds\, u_j \delta(T + \tau_{i-1}^j - \tau_i^j)\delta(\tau_i^j - iT) \\ \vdots \\ \displaystyle\sum_{j=k-N}^{k} \sum_{i=1}^{b} \int_{\tau_i^j - iT}^{\tau_{i-1}^j - (i-1)T} e^{(j-k+N)A(T-s)}\, ds\, u_j \delta(T + \tau_{i-1}^j - \tau_i^j)\delta(\tau_i^j - iT) \end{bmatrix} \quad (5A.4)$$

It is important to note that Δ is a vector whose components are functions of the required control inputs. Also, since (A,B) is controllable, $rank([B\ AB \ldots A^{N-1}B]) = n$ and the rank of the controllability matrix for the augmented system (5.9) can be calculated easily by using basic linear algebra equalities (Lay 2011) as

$$rank\left\{ \begin{bmatrix} B & \cdots & A^{b-1}B & \cdots & A^{N-2}B & A^{N-1}B \\ I_m & \cdots & 0 & \cdots & 0 & 0 \\ \vdots & \ddots & \vdots & \ddots & \vdots & \vdots \\ 0 & \cdots & I_m & \cdots & 0 & 0 \\ \vdots & \ddots & \vdots & \ddots & \vdots & \vdots \\ 0 & \cdots & 0 & 0 & I_m & 0 \\ 0 & \cdots & 0 & 0 & \cdots & 0 \end{bmatrix}_{(bm+n)\times Nm} \right\} = bm + n \quad (5A.5)$$

Therefore, according to the definition of controllability (Kreisselmeier 1986), the controllability matrix (5A.5) is of full rank, which indicates that (A_{zk},B_{zk}) is controllable.

Next, for the linear discrete-time system (5.9) with augment states, the observability matrix can be represented as

$$\Theta_O = \begin{bmatrix} C & I_m & \cdots & I_l \\ CA_s & C\gamma_{k-1}B_1^k + I_m & \cdots & C\gamma_{k-b}D_b^k + I_l \\ \vdots & \vdots & \ddots & \vdots \\ CA_s^{N-1} & C\sum_{j=k-N+1}^{k} A_s^j \sum_{i=1}^{b} \gamma_{j-i}B_i^j + I_m & \cdots & CA_s^{N-1}\gamma_{k-b}D_b^k + I_l \end{bmatrix} \tag{5A.6}$$

Since the original system is observable (i.e., (A,C) is observable), $rank([C^T\ (CA)^T \cdots (CA^{N-1})^T]^T) = r$. Also, since I_m, I_l are $m \times m$ and $l \times l$ identity matrices, the rank of the observability matrix, Θ_O, can be obtained by using linear algebra equalities (Kreisselmeier 1986) as

$$rank(\Theta_O) = r + b(m+l) \tag{5A.7}$$

Therefore, according to the definition of observability (Kreisselmeier 1986), observability matrix, Θ_O, is of full rank, which indicates that (A_{zk},C_z) is observable.

APPENDIX 5B

Proof of Lemma 5.2

Since the initial stabilizing control and disturbance inputs are given, linear time-varying discrete-time system can be represented as

$$z_{k+1} = A_{zk}^{new} z_k \tag{5B.1}$$

where

$$A_{zk}^{new} = \begin{bmatrix} A_s - \gamma_k B_0^k K - \gamma_k D_0^k L & \gamma_{k-1}B_1^k & \cdots\cdots & \gamma_{k-b}B_{b-1}^k & \gamma_{k-b}B_b^k & \gamma_{k-1}D_1^k & \cdots\cdots\ 0 & \gamma_{k-b}D_b^k \\ -K & 0 & \cdots\cdots & 0 & 0 & 0 & 0 \cdots 0 & 0 \\ 0 & I_m & 0\ \cdots & 0 & 0 & 0 & 0 \cdots 0 & 0 \\ 0 & 0 & I_m\ \cdots & 0 & 0 & 0 & 0 \cdots 0 & 0 \\ \vdots & \vdots & \ddots & & \vdots & \vdots & 0 \cdots \vdots & \vdots \\ 0 & 0 & \cdots\cdots & I_m & 0 & 0 & 0 \cdots 0 & 0 \\ -L & 0 & \cdots\cdots & 0 & 0 & 0 & 0 \cdots 0 & 0 \\ 0 & 0 & \cdots\cdots & 0 & 0 & I_l & \cdots 0 & 0 \\ 0 & 0 & \cdots\cdots & 0 & 0 & 0 & I_l \cdots 0 & 0 \\ \vdots & & \ddots & \vdots & \vdots & \vdots & \ddots & \vdots \\ 0 & 0 & \cdots\cdots & 0 & 0 & 0 & 0 \cdots I_l & 0 \end{bmatrix}$$

According to the definition of stability for a linear time-varying system (Kreisselmeier 1986), if each eigenvalue of $\underset{\tau,\gamma}{E}(A_{zk}^{new})$ is less than one for any time instant or mean square radius less than one (i.e., $\lambda_i[E(A_{zk}^{new})] < 1, \forall i, \forall k$), then the NCS system (5.9) is stochastically stable. Otherwise, the system is stochastically unstable. Next, using the equivalent linear algebra transformation, $\underset{\tau,\gamma}{E}(A_{zk}^{new})$ can be transferred equivalent as

$$\underset{\tau,\gamma}{E}(A_{zk}^{ET}) = \begin{bmatrix} \underset{\tau,\gamma}{E}(A_s - \gamma_k B_0^k K - \gamma_k D_0^k L) & \underset{\tau,\gamma}{E}(\gamma_{k-b} B_b^k) & \underset{\tau,\gamma}{E}(\gamma_{k-b} D_b^k) & 0 & \cdots & 0 & 0 & 0 & \cdots & 0 & 0 \\ -K & 0 & 0 & 0 & \cdots & 0 & 0 & 0 & \cdots & 0 & 0 \\ -L & 0 & 0 & 0 & \cdots & 0 & 0 & 0 & \cdots & 0 & 0 \\ 0 & 0 & 0 & I_m & \cdots & 0 & 0 & 0 & \cdots & 0 & 0 \\ \vdots & \vdots & \vdots & \vdots & \ddots & \vdots & \vdots & \vdots & \cdots & \vdots & \vdots \\ 0 & 0 & \cdots & 0 & \cdots & I_m & 0 & 0 & \cdots & 0 & 0 \\ 0 & 0 & \cdots & 0 & \cdots & 0 & I_l & 0 & \cdots & 0 & 0 \\ 0 & 0 & \cdots & 0 & \cdots & 0 & 0 & I_l & \cdots & 0 & 0 \\ \vdots & \vdots & \ddots & \vdots & \vdots & \vdots & \vdots & \vdots & \ddots & \vdots & \vdots \\ 0 & 0 & \cdots & 0 & \cdots & \cdots & \cdots & 0 & \cdots & I_l & 0 \\ 0 & 0 & \cdots & 0 & 0 & 0 & 0 & 0 & \cdots & & I_l \end{bmatrix}$$

It is important to know that equivalent linear algebra transformation would not change the value of eigenvalues $\left(\text{i.e., } \lambda\left[\underset{\tau,\gamma}{E}(A_{zk}^{ET})\right] = \lambda\left[\underset{\tau,\gamma}{E}(A_{zk}^{new})\right]\right)$ (Chen, 1999). Since $\underset{\tau,\gamma}{E}(A_{zk}^{ET})$ is a block diagonal matrix, eigenvalues of $\underset{\tau,\gamma}{E}(A_{zk}^{ET})$ are made up by the eigenvalues of the right bottom $\underset{\tau,\gamma}{E}(A_{zk}^{ET})$ block and the left upper $\underset{\tau,\gamma}{E}(A_{zk}^{ET})$ block (Chen 1999). Obviously, if the magnitude of the eigenvalues of two nonzero diagonal matrix blocks is inside a unit sphere, then the NCS system (5.9) is stochastically stable. The magnitude of the eigenvalues of the right bottom matrix block of $\underset{\tau,\gamma}{E}(A_{zk}^{ET})$ is all equal to one. For maintaining the stochastic stability of the system, we need the magnitude of each eigenvalue of $\underset{\tau,\gamma}{E}(A_s - \gamma_k B_0^k K - \gamma_k D_0^k L)$ less than one at any time instant, that is,

$$\lambda_i\left[\underset{\tau,\gamma}{E}(A_s - \gamma_k B_0^k K - \gamma_k D_0^k L)\right] < 1 \quad \forall i, k \tag{5B.2}$$

Since K, L are the initial fixed stabilizing control gain and disturbance inputs gain for linear discrete-time system, we have

$$\lambda_i(A_s - B_s K - D_s L)) = \lambda_i^s < 1 \quad \forall i \tag{5B.3}$$

where $B_s = \int_0^T e^{A(T-s)} ds\, B$ and $D_s = \int_0^T e^{A(T-s)} ds\, D$. Then, since network-induced delay and packet losses are independent and using Equation 5.8, $\underset{\tau,\gamma}{E}(A_s - \gamma_k B_0^k K - \gamma_k D_0^k L)$ can be presented as

$$\underset{\tau,\gamma}{E}(A_s-\gamma_k B_0^k K-\gamma_k D_0^k L)=A_s-\underset{\gamma}{E}(\gamma_k)\underset{\tau}{E}(B_0^k)K-\underset{\gamma}{E}(\gamma_k)\underset{\tau}{E}(D_0^k)L$$

$$=\left[I-\min\left\{\frac{\underset{\gamma}{E}(\gamma_k)\underset{\tau}{E}\left(\int_{\tau_0^k}^T e^{A(T-s)}\,ds\,B\right)}{\int_0^T e^{A(T-s)}\,ds\,B},\frac{\underset{\gamma}{E}(\gamma_k)\underset{\tau}{E}\left(\int_{\tau_0^k}^T e^{A(T-s)}\,ds\right)}{\int_{\tau_0^k}^T e^{A(T-s)}\,ds}\right\}\right]A_s$$

$$+\min\left\{\frac{\underset{\gamma}{E}(\gamma_k)\underset{\tau}{E}\left(\int_{\tau_0^k}^T e^{A(T-s)}\,ds\right)}{\int_0^T e^{A(T-s)}\,ds},\frac{\underset{\gamma}{E}(\gamma_k)\underset{\tau}{E}\left(\int_{\tau_0^k}^T e^{A(T-s)}ds\right)}{\int_{\tau_0^k}^T e^{A(T-s)}\,ds}\right\}A_s$$

$$-\frac{\underset{\gamma}{E}(\gamma_k)\underset{\tau}{E}\left(\int_{\tau_0^k}^T e^{A(T-s)}\,ds\right)}{\int_0^T e^{A(T-s)}ds}B_sK-\frac{\underset{\gamma}{E}(\gamma_k)\underset{\tau}{E}\left(\int_{\tau_0^k}^T e^{A(T-s)}ds\right)}{\int_{\tau_0^k}^T e^{A(T-s)}\,ds}D_sL$$

Next, using Equations 5B.2 and 5B.3, we have

$$\lambda_i\left[\underset{\tau,\gamma}{E}(A_s-\gamma_k B_0^k K-\gamma_k D_0^k L)\right]$$

$$\leq\left[1-\min\left\{\frac{\underset{\gamma}{E}(\gamma_k)\underset{\tau}{E}\left(\int_{\tau_0^k}^T e^{A(T-s)}\,ds\right)}{\int_0^T e^{A(T-s)}\,ds},\frac{\underset{\gamma}{E}(\gamma_k)\underset{\tau}{E}\left(\int_{\tau_0^k}^T e^{A(T-s)}\,ds\right)}{\int_{\tau_0^k}^T e^{A(T-s)}ds}\right\}\right] \tag{5B.4}$$

$$\times\lambda_i(A_s)+\min\left\{\frac{\underset{\gamma}{E}(\gamma_k)\underset{\tau}{E}\left(\int_{\tau_0^k}^T e^{A(T-s)}\,ds\right)}{\int_0^T e^{A(T-s)}ds},\frac{\underset{\gamma}{E}(\gamma_k)\underset{\tau}{E}\left(\int_{\tau_0^k}^T e^{A(T-s)}\,ds\right)}{\int_0^T e^{A(T-s)}\,ds}\right\}\lambda_i^s$$

Since $\lambda_i[\underset{\tau,\gamma}{E}(A_s-\gamma_k B_0^k K-\gamma_k D_0^k L)]<1 \quad \forall i,k$ is needed for stability, the expected values of network-induced delays and packet losses should satisfy the following inequality:

$$\min\left\{\frac{\underset{\gamma}{E}(\gamma_k)\underset{\tau}{E}\left(\int_{\tau_0^k}^T e^{A(T-s)}\,ds\right)}{\int_0^T e^{A(T-s)}\,ds},\frac{\underset{\gamma}{E}(\gamma_k)\underset{\tau}{E}\left(\int_{\tau_0^k}^T e^{A(T-s)}\,ds\right)}{\int_{\tau_0^k}^T e^{A(T-s)}\,ds}\right\}$$

$$>1-\frac{1-\min\left\{\frac{\underset{\gamma}{E}(\gamma_k)\underset{\tau}{E}(\int_{\tau_0^k}^T e^{A(T-s)}\,ds)}{\int_0^T e^{A(T-s)}\,ds},\frac{\underset{\gamma}{E}(\gamma_k)\underset{\tau}{E}\left(\int_{\tau_0^k}^T e^{A(T-s)}\,ds\right)}{\int_0^T e^{A(T-s)}\,ds}\right\}\lambda_i^s}{\lambda_i(A_s)} \tag{5B.5}$$

It is important to note that Equation 5B.5 is the constraint for expected values of network-induced delay and packet losses. When the network imperfections satisfy Equation 5B.5, the LNCS system (5.9) is stochastically stable. Otherwise, even initial stabilizing control and disturbance inputs cannot maintain the stability of LNCS system (5.9) due to network imperfections.

APPENDIX 5C

Proof of Theorem 5.1

Consider the following positive definite Lyapunov function candidate:

$$L = L_D(y_k^c) + L_J\left(\tilde{\bar{h}}_k\right) \tag{5C.1}$$

where $L_D(y_k^c)$ is defined as $L_D(y_k^c) = y_k^{cT} y_k^c$ and $L_J\left(\tilde{\bar{h}}_k\right)$ is defined as

$$L_J\left(\tilde{\bar{h}}_k\right) = \Lambda\left(\tilde{\bar{h}}_k \bar{w}_k - \tilde{\bar{h}}_k \bar{w}_{k-1}\right)^2 = \Lambda\left(\tilde{\bar{h}}_k \Delta W_{k-1}\right)^2 \tag{5C.2}$$

where $\Lambda = 6(F_M^2 + W_M^2)$ is a positive constant with $\|F_k\| \le F_M$ and $\|W_k\| \le W_M$ with F_M, W_M being positive constants. The first difference of Equation 5C.3 can be expressed as $\Delta L = \Delta L_D(y_k^c) + \Delta L_J\left(\tilde{\bar{h}}_k\right)$, and considering that $\Delta L_J\left(\tilde{\bar{h}}_k\right) = \left(\tilde{\bar{h}}_{k+1}\Delta W_k\right)^2 - \left(\tilde{\bar{h}}_k \Delta W_{k-1}\right)^2$ with the adaptive estimator, we have

$$\begin{aligned}\Delta L_J\left(\tilde{\bar{h}}_k\right) &= \Lambda\left[\left(\tilde{\bar{h}}_{k+1}\Delta W_k\right)^2 - \left(\tilde{\bar{h}}_k \Delta W_{k-1}\right)^2\right] \\ &= \Lambda\left[\left(\alpha_h \tilde{\bar{h}}_k \Delta W_{k-1}\right)^2 - \left(\tilde{\bar{h}}_k \Delta W_{k-1}\right)^2\right] = -(1-\alpha_h^2)\Lambda\left(\tilde{\bar{h}}_k \Delta W_{k-1}\right)^2\end{aligned} \tag{5C.3}$$

Next, considering the first part $\Delta L_D(y_k^c) = y_{k+1}^{cT} y_{k+1}^c - y_k^{cT} y_k^c$ and applying the LNCS quadratic zero-sum games in input–output form and the Cauchy–Schwartz inequality reveals

$$\begin{aligned}\Delta L_D(y_k^c) &\le \underset{\tau,\gamma}{E}\left\|Q_k y_k^c + F_k u_k + W_k d_k - F_k \tilde{u}_k - W_k \tilde{d}_k\right\|^2 - \underset{\tau,\gamma}{E}(y_k^{cT} y_k^c) \\ &\le 3\underset{\tau,\gamma}{E}\left\|Q_k y_k^c + F_k u_k + W_k d_k\right\|^2 + 3\underset{\tau,\gamma}{E}\left\|F_k \tilde{u}_k\right\|^2 + 3\underset{\tau,\gamma}{E}\left\|W_k \tilde{d}_k\right\|^2 - \underset{\tau,\gamma}{E}(y_k^{cT} y_k^c)\end{aligned} \tag{5C.4}$$

Using Lemma 5.4 and recalling actual optimal control inputs (5.19) and worst-case disturbance signals (5.20) and estimated optimal control inputs and worst-case disturbance signals (5.33), Equation 5C.4 can be expressed as

$$\begin{aligned}\Delta L_D(y_k^c) &\le \underset{\tau,\gamma}{E}\left\|Q_k y_k^c + F_k u_k + W_k d_k - F_k \tilde{u}_k - W_k \tilde{d}_k\right\|^2 - \underset{\tau,\gamma}{E}(y_k^{cT} y_k^c) \\ &\le -(1-3k_a)\underset{\tau,\gamma}{E}\left\|y_k^c\right\|^2 + 3(F_M^2 + W_M^2)(\tilde{\bar{h}}_k \Delta W_{k-1})^2\end{aligned} \tag{5C.5}$$

At the final step, combining Equations 5C.3 and 5C.5, we have

$$\Delta L \le -(1-3k_a)\underset{\tau,\gamma}{E}\left\|\boldsymbol{y}_k^c\right\|^2 + 3(F_M^2 + W_M^2)\left(\tilde{\bar{h}}_k \Delta W_{k-1}\right)^2 - (1-\alpha_h^2)\Lambda \underset{\tau,\gamma}{E}\left(\tilde{\bar{h}}_k \Delta W_{k-1}\right)^2$$

$$\le -(1-3k_a)\underset{\tau,\gamma}{E}\left\|\boldsymbol{y}_k^c\right\|^2 - \left(\frac{1}{2} - \alpha_h^2\right)\Lambda \Delta W_{\min}^2 \underset{\tau,\gamma}{E}\left\|\tilde{\bar{h}}_k\right\|^2 \tag{5C.6}$$

where $0 < \Delta W_{\min}^2 \le \left\|\Delta W_k\right\|^2$ with $\Delta W_{\min}$ being positive constant, which is ensured by the PE condition. Since $0 < k_a < 1/3$ and $0 < \alpha_h < 1/\sqrt{2}$, ΔL is negative definite with L being positive definite. Also, observe that $\left|\sum_{k=k_0}^{\infty} \Delta L_k\right| = \left|L_\infty - L_0\right| < \infty$ since $\Delta L < 0$ as long as Equation 5C.6 holds. Now, taking the limit as $k \to \infty$, the augment measured system output y_k^c and $\tilde{\bar{h}}_k$ converge to zero asymptotically in the mean. In other words, $k \to \infty, \underset{\tau,\gamma}{E}(\boldsymbol{y}_k^c) \to 0$, $\tilde{\bar{h}}_k \to 0$, then $\underset{\tau,\gamma}{E}(\hat{J}_k) \to \underset{\tau,\gamma}{E}(J_k^*)$. Since optimal control $\underset{\tau,\gamma}{E}(u_k^*) = -\underset{\tau,\gamma}{E}\left((1/2)R_y^{-1}F_k^T(\partial J_{k+1}/\partial \boldsymbol{y}_{k+1}^c)\right)$ and $\underset{\tau,\gamma}{E}(\hat{u}_k) = -(1/2)\underset{\tau,\gamma}{E}\left(R_y^{-1}F_k^T(\partial \hat{J}_{k+1}/\partial \boldsymbol{y}_{k+1}^c)\right)$, then $\underset{\tau,\gamma}{E}(\hat{u}_k) \to \underset{\tau,\gamma}{E}(u_k^*)$ when $\underset{\tau,\gamma}{E}(\hat{J}_k) \to \underset{\tau,\gamma}{E}(J_k)$. Also, since worst-case disturbance signals $\underset{\tau,\gamma}{E}(d_k^*) = \frac{1}{2}\underset{\tau,\gamma}{E}\left(S_y^{-1}W_k^T(\partial J_{k+1}/\partial \boldsymbol{y}_{k+1}^c)\right)$ and $\underset{\tau,\gamma}{E}(\hat{d}_k) = \frac{1}{2}\underset{\tau,\gamma}{E}\left(S_y^{-1}W_k^T(\partial \hat{J}_{k+1}/\partial \boldsymbol{y}_{k+1}^c)\right)$, then $\underset{\tau,\gamma}{E}(\hat{d}_k) \to \underset{\tau,\gamma}{E}(d_k^*)$ when $\underset{\tau,\gamma}{E}(\hat{J}_k) \to \underset{\tau,\gamma}{E}(J_k)$.

Remark 5.5

The assumption on the bounds for the system matrices $\left\|F_k\right\| \le F_M$ and $\left\|W_k\right\| \le W_M$ is not strong since the bounds are not required for the design but needed in the proof.

REFERENCES

Al-Tamimi, A., Lewis, F.L., and Abu-Khalaf, M. 2007. Model-free Q-learning designs for linear discrete-time zero-sum games with application to H-infinite control. *Automatica*, 43, 473–481.

Åstrom, K.J. 1970. *Introduction to Stochastic Control Theory*. Academic, New York.

Basar, T. and Olsder, G.J. 1995. *Dynamic Non-Cooperative Game Theory*. 2nd edition, Academic, New York.

Chen, C.T. 1999. *Linear System Theory and Applications*. 3rd edition, Oxford University Press, Oxford.

Chen, H.F. and Guo, L. 1991. *Identification and Stochastic Adaptive Control*. Birkauser Press, Cambridge, MA.

Dacic, D.B. and Nesic, D. 2007. Quadratic stabilization of linear networked control systems via simultaneous protocol and controller design. *Automatica*, 43, 1145–1155.

Dierks, T. and Jagannathan, S. 2012. Online optimal control of affine nonlinear discrete-time systems with unknown internal dynamics by using time-based policy update. *IEEE Transactions on Neural Networks and Learning Systems*, 23, 1118–1129.

Halevi, Y. and Ray, A. 1988. Integrated communication and control systems: Part I–Analysis. *Journal of Dynamic Systems, Measurement, and Control*, 110, 367–373.

Ioannou, P.A. and Sun, J. 1996. *Robust Adaptive Control*. Prentice-Hall Press, Upper Saddle River, NJ.

Kreisselmeier, G. 1986. Adaptive control of a class of slowly time-varying plants. *Systems & Control Letters*, 8, 97–103.

Krtolica, R., Ozguner, U., Chan, H., Goktas, H., Winkelman, J., and Liubakka, M. 1994. Stability of linear feedback systems with random communication delays. *International Journal of Control*, 59, 925–953.

Lay, D.C. 2011. *Linear Algebra and Its Applications*. 4th edition, Addison Wiley, Boston, MA.

Lewis, F.L. and Vamvoudakis, K.G. 2011. Reinforcement learning for partially observable dynamic process: Adaptive dynamic programming using measured output data. *IEEE Transactions on Systems, Man, and Cybernetics, Part B*, 41, 14–25.

Lian, F., Moyne, J., and Tilbury, D. 2001. Analysis and modeling of networked control systems: MIMO case with multiple time delays. *Proceedings of the American Control Conference*, Arlington, VA, USA, vol. 6, pp. 4306–4312.

Lian, F., Moyne, J., and Tilbury, D. 2003. Modeling and optimal controller design of networked control systems with multiple delays. *International Journal of Control*, 76, 591–606.

Lu, L.L., Xie, L.H., and Cai, W.J. 2004. H2 controller design for networked control systems. *Asian Journal of Control*, 6(1), 88–96.

Nilsson, J., Bernhardsson, B., and Wittenmark, B. 1998. Stochastic analysis and control of real-time systems with random time delays. *Automatica*, 34, 57–64.

Sarangapani, J. 2006. *Neural Network Control of Nonlinear Discrete-Time Systems*. CRC Press, Florida, USA.

Stengel, R.F. 1986. *Stochastic Optimal Control: Theory and Application*. Wiley-Interscience, New York.

Watkins, C. 1989. *Learning from Delayed Rewards*, PhD Thesis, Cambridge University, Cambridge, England.

Walsh, G.C., Beldiman, O., and Bushnell, L. 1999b. Asymptotic behavior of networked control systems. *Proceedings of the IEEE International Conference on Control Applications*, pp. 1448–1453.

Walsh, G.C., Ye, H., and Bushnell, L. 1999a. Stability analysis of networked control systems. *Proceedings of the American Control Conference*, pp. 2876–2880.

Werbos, P.J. 1999. *A Menu of Designs for Reinforcement Learning Over Time*. MIT Press, Cambridge, MA.

Wonham, W.M. 1968. On a matrix Riccati equation of stochastic control. *SIAM Journal on Control*, 6, 681–697.

Wu, J. and Chen, T. 2007. Design of networked control systems with packet dropouts. *IEEE Transactions on Automatic Control*, 52, 1314–1319.

Xu, H., Jagannathan, S., and Lewis, F.L. 2012. Stochastic optimal control of unknown linear networked control system in the presence of random delays and packet losses. *Automatica*, 48, 1017–1030.

Xu, H., Jagannathan, S., and Lewis, F.L. 2014. Stochastic optimal output feedback design for unknown linear discrete-time system zero-sum games under communication constraints. *Asian Journal of Control*, 16(5), 1263–1276.

Yang, C., Guan, Z.H., and Huang, J. 2011. Stochastic switched controller design of networked control systems with a random long delay. *Asian Journal of Control*, 13(2), 255–264.

Zhang, W., Branicky, M.S., and Phillips, S. 2001. Stability of networked control systems. *IEEE Control Systems Magazine*, 21, 84–99.

6 Optimal Control of Uncertain Nonlinear Networked Control Systems via Neurodynamic Programming

Nonlinear networked control systems (NNCS) (Tipsuwan and Chow 2003), which utilize a communication network to connect a nonlinear plant with its controller, have been considered in the literature due to the benefits mentioned in the previous chapters. Following the stability analysis for a LNCS (Zhang et al. 2001), Walsh et al. (2001) analyzed the asymptotic behavior of an NNCS in the presence of known network-induced delays. In Wouw et al. (2012), a discrete-time framework is introduced to analyze the NNCS stability with both delays and packet losses, provided the system dynamics and network imperfections are known beforehand. In addition, optimality is generally preferred (Feng et al. 2002; Hu and Zhu 2003; Dehghani 2005; Tabbara 2008) over stability alone. The optimal policy for LNCS is given in Chapter 3.

The neurodynamic programming (NDP) technique proposed by Werbos (1983) and Bertsekas and Tsitsiklis (1996) intends to obtain the optimal control of uncertain nonlinear systems in a forward-in-time manner in contrast to the traditional backward-in-time optimal control technique, which normally requires complete knowledge of system dynamics. Moreover, by using value or policy iterations, reinforcement learning can be incorporated with NDP to obtain optimal control (Lewis and Vrabie 2009; Wang et al. 2011). Zhang et al. (2009) proposed a policy or value iteration-based NDP scheme to obtain finite-horizon optimal control inputs for a nonlinear system with unknown internal system dynamics. However, in order to compute the optimal solution, the iteration-based NDP methods require a significant number of iterations within a fixed sampling interval (Dierks and Jagannathan 2012), which can be a bottleneck for implementation.

Nevertheless, the existing NDP schemes (Zhang et al. 2009; Wang et al. 2011; Dierks and Jagannathan 2012) are unsuitable for finite-horizon optimal control of NNCS since (a) the techniques (Dierks and Jagannathan 2012) are only developed to solve infinite-horizon optimal control, (b) partial system dynamics in the form of control coefficient matrix is needed, and (c) network imperfections resulting from the communication network are ignored.

Hence, in Xu et al. (2012), a model-free infinite-horizon optimal control of LNCS is derived in the presence of uncertain dynamics and network imperfections by using the ADP-based technique. However, finite-horizon optimal control is more difficult to solve than infinite-horizon-based schemes (Hu and Zhu 2003; Dehghani 2005; Tabbara 2008) for NNCS due to terminal constraints. This chapter taken from Xu and Jagannathan (2013, 2015) considers finite-horizon optimal design for NNCS by incorporating terminal constraints before introducing infinite-horizon optimal control of NNCS. The usual assumption of the certainty-equivalence stochastic value function and optimal control policy is taken during the development. In addition, perfect state measurements are required.

In this chapter, an optimal adaptive control scheme using time-driven NDP is undertaken to obtain finite-horizon stochastic optimal regulation of NNCS in the presence of uncertain system dynamics due to network imperfections. First, to relax the NNCS dynamics, a novel neural network (NN) identifier is proposed to learn the control coefficient matrix online. Then, by using an initial admissible control, a critic NN (Sarangapani 2006) is introduced and tuned in a forward-in-time manner to approximate the stochastic value function by using the HJB equation (Stengel 1986; Chen and Guo 1991), given the terminal constraint. Eventually, an actor NN is introduced to generate optimal control input by minimizing the estimated stochastic value function.

Compared with the traditional stochastic optimal controller design from Chen and Lewis (2007), which requires the full knowledge of system dynamics, our proposed stochastic optimal controller design for NNCS can relax the requirement on system dynamics and network imperfections as well as value or policy iterations by using the novel time-based NDP technique. Moreover, the available control techniques for time-delay systems with known deterministic delays (Luck and Ray 1990; Mahmoud and Ismail 2005) are unsuitable here since network imperfections from NNCS result in random delays and packet losses, which cannot be handled by time-delay control techniques.

The main focus of this chapter taken from (Xu and Jagannathan 2013, 2015) includes: (1) a time-driven stochastic optimal control NDP approach over finite- and infinite-horizon for uncertain NNCS by incorporating the terminal constraints, (2) a novel online identifier to obtain the NNCS dynamics; and (3) demonstration of the closed-loop stability via a combination of Lyapunov and geometric sequence analysis, which guarantees the stability.

This chapter is organized as follows: first, Section 6.1 presents the traditional nonlinear optimal design by using the HJB equation-based solution. Then, Section 6.2 develops the finite-horizon optimal control for NNCS (Xu and Jagannathan 2015). Finally, Section 6.3 extends the results of Section 6.2 to an infinite-horizon case (Xu and Jagannathan 2013).

6.1 TRADITIONAL NONLINEAR OPTIMAL CONTROL DESIGN AND HJB EQUATION-BASED SOLUTION

Consider the affine nonlinear discrete-time system

$$x_{d,k+1} = f_d(x_{d,k}) + g_d(x_{d,k})u_{d,k} + w_k \tag{6.1}$$

where $x_{d,k}$, $u_{d,k}$, and w_k represent the system state, control input, and disturbance inputs while $f_d(x_{d,k})$ and $g_d(x_{d,k})$ denote the internal dynamics and control coefficient matrix, respectively. According to Stengel (1986), optimal control input is derived by minimizing the certainty-equivalent stochastic value function expressed as

$$\begin{cases} V_k(x_{d,k},k) &= E\left[\phi_N(x_{d,N})+\sum_{l=k}^{N-1} r(x_{d,l},u_{d,l})\right] \\ &= E\left[\phi_N(x_{d,N})+\sum_{l=k}^{N-1}(Q_d(x_{d,l})+u_{d,l}^T R_d u_{d,l})\right] \quad \forall k=0,1,\ldots,N- \\ V_N(x_{d,N},N) = E[\phi_N(x_{d,N})] \end{cases} \tag{6.2}$$

where the cost-to-go is denoted as $r(x_{d,l},u_{d,l}) = Q_d(x_{d,l})+u_{d,l}^T R_d u_{d,l}, \forall k=0,1,\ldots,N-1$, NT_s is the final time instant, $Q_d(x) \ge 0$, $\phi_N(x) \ge 0$ and R_d are the symmetric positive definite matrix, and $E(\cdot)$ is the expectation operator (the mean value). Here the terminal constraint $\phi_N(x)$ needs to be satisfied in the finite-horizon optimal control design. Equation 6.2 can also be rewritten as

$$\begin{aligned} V_k(x_{d,k},k) &= E\left[r(x_{d,k},u_{d,k})+\phi_N(x_{d,N})+\sum_{l=k}^{N-1} r(x_{d,l},u_{d,l})\right] \\ &= E\left[(Q_d(x_{d,k})+u_{d,k}^T R_d u_{d,k})+V_{k+1}(x_{d,k+1})\right], \quad k=0,\ldots,N-1 \end{aligned} \tag{6.3}$$

According to the observability condition (Chen and Guo 1991), when $x=0$, $V_k(x)=0$, the value function $V_k(x)$ serves as a Lyapunov function (Sarangapani 2006). Based on the Bellman principle of optimality (Stengel 1986; Lewis and Syrmos 1995), the optimal value function also satisfies the certainty-equivalent discrete-time HJB equation given by

$$\begin{cases} V^*(x_{d,k},k) &= \min_{u_{d,k}}(V_k(x_{d,k},k)) \\ &= \min_{u_{d,k}}\left[E[Q_d(x_{d,k})+u_{d,k}^T R_d u_{d,k}+V^*(x_{d,k+1})]\right], \quad \forall k=0,\ldots,N-1 \\ V^*(x_{d,N},N) = E[\varphi_N(x_{d,N})] \end{cases} \tag{6.4}$$

Differentiating Equation 6.4, the optimal control $u_{d,k}^*$ is obtained as

$$E\left[\frac{\partial(Q_d(x_{d,k})+u_{d,k}^T R_d u_{d,k})}{\partial u_{d,k}}+\frac{\partial x_{d,k+1}^T}{\partial u_{d,k}}\frac{\partial V^*(x_{d,k+1})}{\partial x_{d,k+1}}\right]=0 \tag{6.5}$$

In other words,

$$\begin{cases} u^*(x_{d,k}) = -\dfrac{1}{2} E\left[R_d^{-1} g_d^T(x_{d,k}) \dfrac{\partial V^*(x_{d,k+1}, k+1)}{\partial x_{d,k+1}} \right], \quad k = 0,\ldots,N-1 \\ u^*(x_{d,N-1}) = -\dfrac{1}{2} E\left[R_d^{-1} g_d^T(x_{d,N-1}) \dfrac{\partial \varphi_N(x_{d,N}, N)}{\partial x_{d,N}} \right] \end{cases} \tag{6.6}$$

Substituting Equation 6.6 into Equation 6.4, the discrete-time HJB Equation 6.4 becomes

$$V^*(x_{d,k}, k) = E\left[Q_d(x_{d,k}) + \frac{1}{4} \frac{\partial V^{*T}(x_{d,k+1})}{\partial x_{d,k+1}} g_d(x_{d,k+1}) R_d^{-1} \times g_d(x_{d,k+1}) \frac{\partial V^{*T}(x_{d,k+1})}{\partial x_{d,k+1}} + V^*(x_{d,k+1}) \right] \quad \forall k = 0,\ldots,N-1$$

$$V^*(x_{d,N-1}) = E\left[Q_d(x_{d,N-1}) + \frac{1}{4} \frac{\partial \phi_N^T(x_{d,N})}{\partial x_{d,N}} g_d(x_{d,N-1}) R_d^{-1} \times g_d(x_{d,N-1}) \frac{\partial \phi_N(x_{d,N})}{\partial x_{d,N}} + \phi_N(x_{d,N}) \right] \tag{6.7}$$

It is worthwhile to observe that obtaining a closed-form solution to the discrete-time HJB is difficult since the future system state x_{k+1} and system dynamics are needed at kT_s. To circumvent this issue, normally, value and policy iteration-based schemes are utilized (Zhang et al. 2009; Wang et al. 2011). However, iteration-based methods are unsuitable for real-time control due to the large number of iterations needed within a sampling interval. An inadequate number of iterations will lead to instability (Dierks and Jagannathan 2012). Therefore, the time-based NDP finite-horizon optimal controller design from Xu and Jagannathan (2015) is presented next for the NNCS.

6.2 FINITE-HORIZON OPTIMAL CONTROL FOR NNCS

In this section, novel time-based NDP technique is derived to obtain stochastic optimal regulation of NNCS over finite time horizon with uncertain system dynamics due to network imperfections such as random delays and packet losses. First, the background of NNCS is introduced. Then, an online NN identifier is introduced to obtain the control coefficient matrix. Then, the critic NN is proposed to approximate the stochastic value function within the fixed final time. Eventually, by using an actor NN, identified NNCS dynamics and estimated stochastic value function, the finite-horizon stochastic optimal control of NNCS is derived.

6.2.1 Background

The block diagram representation of general NNCS is shown as Figure 6.1 where the control loop is closed by using a communication network. Figure 6.1 shows a typical NNCS, which is practical and utilized in many real-time applications. In addition, since the communication network is shared (Stallings 2002; Goldsmith 2003), the NNCS in this chapter considers network imperfections including: (1) $\tau_{sc}(t)$: sensor-to-controller delay, (2) $\tau_{ca}(t)$: controller-to-actuator delay, and (3) $\gamma(t)$: indicator of network-induced packet losses.

In the recent NCS (Walsh et al. 2001; Zhang et al. 2001; Hu and Zhu 2003; Wouw et al. 2012) and communication network protocol development literature (Stallings 2002; Goldsmith 2003), the following assumption (Hu and Zhu 2003; Xu et al. 2012) is utilized for the controller design.

Assumption 6.1

(a) Owing to the wide area network, two types of network-induced delays are considered independent, ergodic, and unknown whereas their probability distribution functions are considered known. The sensor-to-controller delay is assumed to be less than a sampling interval. (b) The sum of two delays is considered to be bounded while the initial state of the system is assumed to be deterministic (Hu and Zhu 2003).

Incorporating network-induced delays and packet losses, the original affine nonlinear system can be represented as

$$\dot{x}(t) = f(x(t)) + \gamma(t)g(x(t))u(t - \tau(t)) \tag{6.8}$$

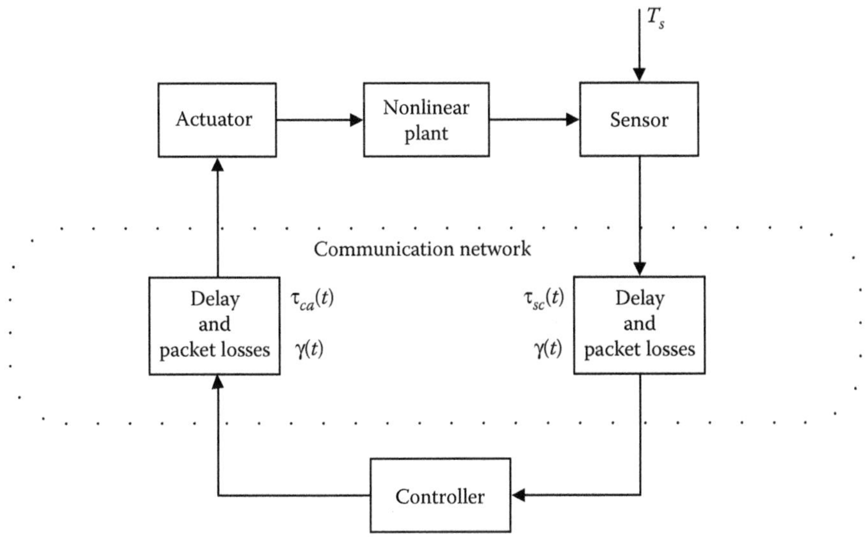

FIGURE 6.1 General NNCS block diagram.

with

$$\gamma(t) = \begin{cases} \boldsymbol{I}^{n\times n} & \text{if control input is received by the actuator at time } t \\ \boldsymbol{0}^{n\times n} & \text{if control input is lost at time } t \end{cases}$$

$\boldsymbol{I}^{n\times n}$ is $n \times n$ identity matrix, $x(t) \in \Re^{n\times n}$, $u(t) \in \Re^{m\times m}$, $f(x) \in \Re^{n\times n}$, and $g(x) \in \Re^{n\times m}$ represents system state, control inputs, nonlinear internal dynamics, and control coefficient matrix, respectively.

Similar to Xu and Jagannathan (2013, 2015), integrating (6.1) over a sampling interval $[kT_s, (k+1)T_s)$ with network-induced delays and packet losses, the NNCS can be represented as

$$x_{k+1} = X_{\tau,\gamma}(x_k, u_{k-1}, \ldots u_{k-\bar{d}}) + P_{\tau,\gamma}(x_k, u_{k-1}, \ldots u_{k-\bar{d}})u_k \tag{6.9}$$

where $\bar{d}T_s$ is the maximum network-induced delay, T_s is the sampling interval, and $\bar{d}$ denotes the number of sampling intervals, $x(kT_s) = x_k$, $u((k-i)T_s) = u_{k-i}$, $\forall i = 0,1,\ldots,\bar{d}$ are discretized system state and previous control inputs, and $X_{\tau,\gamma}(\cdot)$, $P_{\tau,\gamma}(\cdot)$ are defined similar to Xu and Jagannathan (2013).

Next, defining a new augment state variable $z_k = [x_k^T \; u_{k-1}^T \; \ldots \; u_{k-\bar{d}}^T]^T \in \Re^{n+\bar{d}m}$. Equation 6.8 can be expressed equivalently as

$$z_{k+1} = F(z_k) + G(z_k)u_k \tag{6.10}$$

where NNCS internal dynamics and control coefficient matrix $F(\cdot)$ and $G(\cdot)$ are derived similar to Xu and Jagannathan (2013) with $\|G(z_k)\|_F \le G_M$, where $\|\cdot\|_F$ denotes the Frobenius norm (Sarangapani 2006) and G_M is a positive constant. Owing to the presence of network-induced delays and packet losses, Equation 6.10 becomes uncertain and stochastic, thus needing adaptive control methods.

6.2.2 Online NN Identifier Design

The control coefficient matrix is required for the optimal control of an affine nonlinear system (Zhang et al. 2009; Wang et al. 2011). However, the control coefficient matrix is not normally known in advance. To overcome this deficiency, a novel NN identifier is proposed to estimate the control coefficient matrix denoted as $G(z)$. Owing to network imperfections that are presented by random variables, stochastic mathematical treatment will be used throughout the chapter.

Based on Dankert et al. (2006) and the universal function approximation property, the NNCS internal dynamics and control coefficient matrix can be represented on a compact set Ω as

$$\begin{aligned} \underset{\tau,\gamma}{E}\left[F(z_k)\right] &= \underset{\tau,\gamma}{E}[W_F^T \upsilon_F(z_k) + \varepsilon_{F,k}] \quad \forall k = 0,1,\ldots,N \\ \underset{\tau,\gamma}{E}\left[G(z_k)\right] &= \underset{\tau,\gamma}{E}[W_G^T \upsilon_G(z_k) + \varepsilon_{G,k}] \quad \forall k = 0,1,\ldots,N \end{aligned} \tag{6.11}$$

where W_F, W_G denote the target NN weights, $\upsilon_F(\cdot)$, $\upsilon_G(\cdot)$ are activation functions, and $\varepsilon_{F,k}$, $\varepsilon_{G,k}$ represent reconstruction errors, respectively. Substituting Equation 6.11 into Equation 6.10, we get

$$\begin{aligned} \underset{\tau,\gamma}{E}[z_k] &= \underset{\tau,\gamma}{E}[F(z_{k-1})+G(z_{k-1})u_{k-1}] \\ &= \underset{\tau,\gamma}{E}\left[[W_F^T \; W_G^T]\begin{bmatrix} \upsilon_F(z_{k-1}) & 0 \\ 0 & \upsilon_G(z_{k-1}) \end{bmatrix}\begin{bmatrix} 1 \\ u_{k-1} \end{bmatrix} + \varepsilon_{I,k-1}\right] \\ &= \underset{\tau,\gamma}{E}[W_I^T \upsilon_I(z_{k-1})U_{k-1} + \varepsilon_{I,k-1}], \quad \forall k = 0,1,\ldots,N \end{aligned} \tag{6.12}$$

where $W_I = [W_F^T \; W_G^T]^T$, $\upsilon_I(z_{k-1}) = diag[\upsilon_F(z_{k-1})\upsilon_G(z_{k-1})]$ are NN identifier target weight and activation function, respectively, augment control input $U_{k-1} = [1 \quad u_{k-1}^T]^T$ includes historical input values u_{k-1}, $\varepsilon_{I,k-1} = \varepsilon_{F,k-1} + \varepsilon_{G,k-1}u_{k-1}$ represents the NN identifier reconstruction error, and $\underset{\tau,\gamma}{E}(\cdot)$ is the expectation operator. Since the NN activation function and augmented control input from previous time instants are considered bounded with an initial bounded input, the term $\left\| \underset{\tau,\gamma}{E}[\upsilon_I(z_{k-1})U_{k-1}] \right\| \leq \zeta_M$, where ζ_M is a positive constant. Moreover, the NN identifier reconstruction error is considered to be bounded, that is, $\left\| \underset{\tau,\gamma}{E}[\varepsilon_{I,k-1}] \right\| \leq \varepsilon_{I,M}$, where $\varepsilon_{I,M}$ denotes a positive constant. Therefore, given the bounded NN activation functions $\upsilon_F(\cdot)$, $\upsilon_G(\cdot)$, and $\upsilon_I(\cdot)$, the NNCS control coefficient matrix $G(z)$ can be identified once the NN identifier weight matrix, W_I, is obtained. Next, the update law for the NN identifier will be introduced.

The NNCS system state z_k can be approximated by using an NN identifier as

$$\underset{\tau,\gamma}{E}[\hat{z}_k] = \underset{\tau,\gamma}{E}\left[\hat{W}_{I,k}^T \upsilon_I(z_{k-1})U_{k-1}\right], \quad \forall k = 0,1,\ldots,N \tag{6.13}$$

where $\hat{W}_{I,k}$ is an actual identifier NN weight matrix at time instant kT_s, and $\underset{\tau,\gamma}{E}[\bar{\upsilon}_I(z_k)U_k]$ is the basis function of the NN identifier. Based on Equations 6.12 and 6.13, the identification error can be expressed as

$$\underset{\tau,\gamma}{E}(e_{I,k}) = \underset{\tau,\gamma}{E}(z_k - \hat{z}_k) = \underset{\tau,\gamma}{E}(z_k) - \underset{\tau,\gamma}{E}\left[\hat{W}_{I,k}^T \upsilon_I(z_{k-1})U_{k-1}\right] \tag{6.14}$$

Moreover, identification error dynamics can be derived as

$$\underset{\tau,\gamma}{E}(e_{I,k+1}) = \underset{\tau,\gamma}{E}(z_{k+1}) - \underset{\tau,\gamma}{E}\left[\hat{W}_{I,k+1}^T \upsilon_I(z_k)U_k\right], \quad \forall k = 0,1,\ldots,N-1 \tag{6.15}$$

According to Xu and Jagannathan (2013) and using the history from NNCS, an auxiliary identification error vector can be defined as

$$\underset{\tau,\gamma}{E}(\Xi_{I,k}) = \underset{\tau,\gamma}{E}(Z_k - \hat{Z}_k) = \underset{\tau,\gamma}{E}(Z_k) - \underset{\tau,\gamma}{E}\left[\hat{W}_{I,k}^T \bar{\upsilon}_I(z_{k-1})\bar{U}_{k-1}\right] \tag{6.16}$$

where $Z_k = [z_k \; z_{k-1} \ldots z_{k+1-l}]$, $\overline{\upsilon}_I(z_{k-1}) = [\upsilon_I(z_{k-1}) \; \upsilon_I(z_{k-2}) \ldots \upsilon_I(z_{k-l})]$, $\overline{U}_{k-1} = diag$ $[U_{k-1}^T \ldots U_{k-l}^T]^T$, and $\overline{\varepsilon}_{I,k-1} = [\varepsilon_{I,k-1} \ldots \varepsilon_{I,k-l}]$ with $0 < l < k-1$. The l previous identification errors (6.16) are recalculated by using the most recent actual NN identifier weight matrix $\left(\text{i.e.}, \underset{\tau,\gamma}{E}\left(\hat{W}_{I,k}\right)\right)$.

Then, dynamics of auxiliary identification error can be represented as

$$\underset{\tau,\gamma}{E}(\Xi_{I,k+1}) = \underset{\tau,\gamma}{E}(Z_{k+1}) - \underset{\tau,\gamma}{E}\left[\hat{W}_{I,k+1}^T \overline{\upsilon}_I(z_k)\overline{U}_k\right], \quad \forall k = 0,1,\ldots,N-1 \tag{6.17}$$

To force the actual NN identifier weight matrix close to its target within the fixed final time, the stochastic update law for $\underset{\tau,\gamma}{E}(\hat{W}_{I,k})$ can be expressed as

$$\underset{\tau,\gamma}{E}(\hat{W}_{I,k+1}) = \underset{\tau,\gamma}{E}[\overline{U}_k \overline{\upsilon}_I(z_k)(\overline{\upsilon}_I^T(z_k)\overline{U}_k^T \overline{U}_k \overline{\upsilon}_I(z_k))^{-1}(Z_k - \alpha_I \Xi_{I,k})^T],$$
$$\forall k = 0,1,\ldots,N-1 \tag{6.18}$$

where α_I is the tuning parameter satisfying $0 < \alpha_I < 1$. Substituting update law (6.18) into auxiliary error dynamics (6.17), the error dynamics $\underset{\tau,\gamma}{E}(\Xi_{I,k+1})$ can be represented as

$$\underset{\tau,\gamma}{E}(\Xi_{I,k+1}) = \alpha_I \underset{\tau,\gamma}{E}(\Xi_{I,k}), \quad \forall k = 0,1,\ldots,N-1 \tag{6.19}$$

In order to learn the NNCS control coefficient matrix $G(z)$, $\underset{\tau,\gamma}{E}[\overline{\upsilon}_I(z_k)U_k]$ has to be persistently exciting (Sarangapani 2006; Dierks and Jagannathan 2012) long enough. Namely, there exists a positive constant $\zeta_{\min}$ such that $0 < \zeta_{\min} \leq \left\|\underset{\tau,\gamma}{E}[\overline{\upsilon}_I(z_k)U_k]\right\|$ is satisfied for $k = 0,1,\ldots,N$.

Recalling Equation 6.12, the identification error dynamics (6.15) can be represented as

$$\underset{\tau,\gamma}{E}(e_{I,k+1}) = \underset{\tau,\gamma}{E}\left[\tilde{W}_{I,k+1}^T \upsilon_I(z_k)U_k + \varepsilon_{I,k}\right], \quad \forall k = 0,1,\ldots,N-1 \tag{6.20}$$

where $\tilde{W}_{I,k} = W_I - \hat{W}_{I,k}$ is the NN identifier weight estimation error at time kT_s. Using the NN identifier update law and Equation 6.20, the NN weight estimation error dynamics of the NN identifier can be derived as

$$\underset{\tau,\gamma}{E}(\tilde{W}_{I,k+1}^T \upsilon_I(z_k)U_k) = \underset{\tau,\gamma}{E}(\alpha_I(\tilde{W}_{I,k}^T \upsilon_I(z_k)U_{k-1}) + \alpha_I \varepsilon_{I,k-1} - \varepsilon_{I,k}) \tag{6.21}$$

Next, the stability of NN identification error (6.14) and weight estimation errors $\underset{\tau,\gamma}{E}(\tilde{W}_{I,k})$ will be analyzed.

Lemma 6.1: Boundedness in the Mean for the NN Identifier

Given the initial NN identifier weight matrix $W_{I,0}$, which resides in a compact set Ω, let the proposed NN identifier be defined as Equation 6.13, and its update law be given by Equation 6.18. Assuming that $\underset{\tau,\gamma}{E}[\overline{\upsilon}_I(z_k)U_k]$ satisfies the PE condition, there exists a positive tuning parameter α_I with $0<\alpha_I<\left[\zeta_{\min}/\sqrt{2}\zeta_M\right]$ such that identification error (6.14) and NN identifier weight estimation error $\underset{\tau,\gamma}{E}(\tilde{W}_{I,k})$ are all bounded in the mean within the fixed final time $t \in [0, NT_s]$.

Proof: Refer to Xu and Jagannathan (2015).

6.2.3 Stochastic Value Function Setup and Critic NN Design

According to the value function defined in Zhang et al. (2009) and Wang et al. (2011) and given NNCS dynamics (6.10), the certainty-equivalent stochastic value function can be expressed in terms of augment state z_k as

$$\begin{cases} V(z_k,k)=\underset{\tau,\gamma}{E}[\phi_N(z_N)+\sum_{l=k}^{N-1}(Q_z(z_l)+u_l^T R_z u_l)], \quad k=0,\ldots,N-1 \\ V(z_N,N)=\underset{\tau,\gamma}{E}[\phi_N(z_N)] \end{cases} \tag{6.22}$$

with $Q_z(z_k)\geq 0$ and $R_z=(1/\bar{d})R_d$. Compared with stochastic value function under the infinite-horizon case, a terminal constraint (i.e.,$V_N(z_N,N)=\underset{\tau,\gamma}{E}[\phi_N(z_N)]$) needs to be considered while developing a stochastic optimal controller.

Next, according to the universal approximation property of NN (Sarangapani 2006), the stochastic value function (6.22) can be represented by using a critic NN as

$$V(z_k,k)=\underset{\tau,\gamma}{E}(W_V^T\psi(z_k,N-k)+\varepsilon_{V,k}), \quad \forall k=0,1,\ldots,N \tag{6.23}$$

where W_V, $\varepsilon_{V,k}$ denote critic NN target weight matrix and NN reconstruction error, respectively, and $\psi(z_k, N-k)$ represents the time-dependent critic NN activation function. Since the activation function explicitly depends upon time, the finite-horizon design is different and difficult when compared to the infinite-horizon case (Dierks and Jagannathan 2012). The target weight matrix of the critic NN is considered bounded in the mean as $\left\|\underset{\tau,\gamma}{E}(W_V)\right\|\leq W_{VM}$ with W_{VM} being a positive constant, and the reconstruction error is also considered bounded in the mean such that $\left\|\underset{\tau,\gamma}{E}(\varepsilon_{V,k})\right\|\leq\varepsilon_{VM}$ with ε_{VM} being a positive constant. In addition, the gradient of the NN reconstruction error is assumed to be bounded in the mean as $\left\|\underset{\tau,\gamma}{E}(\partial\varepsilon_{V,k}/\partial z_k)\right\|\leq\varepsilon'_{VM}$ with ε'_{VM} being a positive constant (Dierks and Jagannathan 2012).

Next, the approximate stochastic value function (6.23) can be represented as

$$\hat{V}(z_k,k) = \underset{\tau,\gamma}{E}(\hat{W}_{V,k}^T \psi(z_k, N-k)), \quad \forall k = 0,1,\ldots,N \tag{6.24}$$

where $\underset{\tau,\gamma}{E}(\hat{W}_{V,k})$ is the estimated critic NN weight matrix and $\psi(z_k, N-k)$ represents the time-dependent activation function selected from a basis function set whose elements in the set are linearly independent (Dierks and Jagannathan 2012). Also, since the activation function is continuous and smooth, time-independent functions $\psi_{\min}(z_k)$, $\psi_{\max}(z_k)$ can be found such that $\|\psi_{\min}(z_k)\| \le \|\psi(z_k,N-k)\| \le \|\psi_{\max}(z_k)\|$, $k = 0,\ldots,N$. Then, the target stochastic value function is considered bounded as $\|W_V^T \psi_{\min}(z_k)\| \le \|V(z_k,k)\| \le \|W_V^T \psi_{\max}(z_k)\|$ with the stochastic value function estimation error satisfying $\|\tilde{W}_{V,k}^T \psi_{\min}(z_k)\| \le \|\tilde{V}(z_k,k)\| \le \|\tilde{W}_{V,k}^T \psi_{\max}(z_k)\|$.

Recall the HJB Equation 6.3 and substitute Equation 6.22 into Equation 6.3 to get

$$\begin{aligned}\underset{\tau,\gamma}{E}(\varepsilon_{V,k} - \varepsilon_{V,k+1}) &= \underset{\tau,\gamma}{E}[W_V^T(\psi(z_{k+1}, N-k-1) - \psi(z_k, N-k))] \\ &\quad + \underset{\tau,\gamma}{E}(z_k^T Q_z z_k + u_k^T R_z u_k), \quad \forall k = 0,1,\ldots,N-1\end{aligned} \tag{6.25}$$

Namely,

$$\underset{\tau,\gamma}{E}(W_V^T \Delta\psi(z_k, N-k)) + r(z_k,u_k) = \Delta\varepsilon_{V,k}, \quad k = 0,1,\ldots,N-1 \tag{6.26}$$

where $\Delta\psi(z_k, N-k) = \psi(z_{k+1}, N-k-1) - \psi(z_k, N-k)$, $r(z_k,u_k) = \underset{\tau,\gamma}{E}(z_k^T Q_z z_k + u_k^T R_z u_k)$ and $\Delta\varepsilon_{V,k} = \underset{\tau,\gamma}{E}(\varepsilon_{V,k} - \varepsilon_{V,k+1})$ with $\|\Delta\varepsilon_{V,k}\| = \Delta\varepsilon_{VM}, \forall k = 0,\ldots,N-1$. However, Equation 6.26 cannot be held while utilizing the approximated critic NN, $\hat{V}(z_k,k)$, instead of $V(z_k,k)$. Similar to Dierks and Jagannathan (2012) and Xu and Jagannathan (2013), using delayed values for convenience, the residual error dynamics associated with Equation 6.26 are derived as

$$\begin{aligned}\underset{\tau,\gamma}{E}(e_{HJB,k}) &= \underset{\tau,\gamma}{E}(z_k^T Q_z z_k + u_{k-1}^T R_z u_{k-1}) + \hat{V}(z_{k+1}, k+1) - \hat{V}(z_k,k) \\ &= \underset{\tau,\gamma}{E}(\hat{W}_{V,k}^T \Delta\psi(z_k, N-k)) + r(z_k,u_k), \quad k = 0,\ldots,N-1\end{aligned} \tag{6.27}$$

with $\underset{\tau,\gamma}{E}(e_{HJB,k})$ being the residual error of the HJB equation for the finite-horizon scenario. Moreover, since $r(z_k,u_k) = \Delta\varepsilon_{V,k} - \underset{\tau,\gamma}{E}(W_V^T \Delta\psi(z_k, N-k))$, $\forall k = 0,1,\ldots,N$, the residual error dynamics are represented as

$$\begin{aligned}\underset{\tau,\gamma}{E}(e_{HJB,k}) &= \underset{\tau,\gamma}{E}(\hat{W}_{V,k}^T \Delta\psi(z_k, N-k)) - \underset{\tau,\gamma}{E}(W_V^T \Delta\psi(z_k, N-k)) + \Delta\varepsilon_{V,k} \\ &= -\underset{\tau,\gamma}{E}(\tilde{W}_{V,k}^T \Delta\psi(z_k, N-k)) + \Delta\varepsilon_{V,k}, \quad k = 0,\ldots,N-1\end{aligned} \tag{6.28}$$

where $\underset{\tau,\gamma}{E}(\tilde{W}_{V,k}) = \underset{\tau,\gamma}{E}(W_V) - \underset{\tau,\gamma}{E}(\hat{W}_{V,k})$ denotes the critic NN weight estimation error.

Next, in order to take into account the terminal constraint, the estimation error $\underset{\tau,\gamma}{E}(e_{FC,k})$ is defined as

$$\underset{\tau,\gamma}{E}(e_{FC,k}) = \underset{\tau,\gamma}{E}[\phi_N(z_N)] - \underset{\tau,\gamma}{E}(\hat{W}_{V,k}^T\psi(\hat{z}_{N,k},0)), \quad \forall k = 0,1,\ldots,N \tag{6.29}$$

where $\hat{z}_{N,k}$ is the estimated final NNCS system state vector at time kT_s by using the NN identifier (i.e., $\hat{F}(\cdot),\hat{G}(\cdot)$). Recalling Equation 6.23, Equation 6.29 can be represented in terms of the critic NN weight estimation error as

$$\begin{aligned}\underset{\tau,\gamma}{E}(e_{FC,k}) &= \underset{\tau,\gamma}{E}(W_V^T\psi(z_N,0)+\varepsilon_{V,0}) - \underset{\tau,\gamma}{E}(\hat{W}_{V,k}^T\psi(\hat{z}_{N,k},0)) \\ &= \underset{\tau,\gamma}{E}(\tilde{W}_{V,k}^T\psi(\hat{z}_{N,k},0)) + \underset{\tau,\gamma}{E}(W_V^T\tilde{\psi}(z_N,\hat{z}_{N,k},0)) + \underset{\tau,\gamma}{E}(\varepsilon_{V,0}), \quad \forall k = 0,1,\ldots,N\end{aligned} \tag{6.30}$$

with $\tilde{\psi}(z_N,\hat{z}_{N,k},0) = \psi(z_N,0) - \psi(\hat{z}_{N,k},0)$. Since the critic NN activation function $\psi(\cdot)$ is bounded, that is, $\|\psi(\cdot)\| \le \psi_M$ (Dierks and Jagannathan 2012; Xu and Jagannathan 2013), we have $\|\tilde{\psi}(z_N,\hat{z}_{N,k},0)\| \le 2\psi_M$ with ψ_M being a positive constant.

Combining both the HJB residual and terminal constraint estimation errors and using the gradient descent scheme, the stochastic update law of the critic NN weight can be given by

$$\begin{aligned}\underset{\tau,\gamma}{E}(\hat{W}_{V,k+1}) &= \underset{\tau,\gamma}{E}(\hat{W}_{V,k}) + \alpha_V \underset{\tau,\gamma}{E}\left(\frac{\psi(\hat{z}_{N,k})e_{FC,k}^T}{\psi^T(\hat{z}_{N,k})\psi(\hat{z}_{N,k})+1}\right) \\ &\quad - \alpha_V \underset{\tau,\gamma}{E}\left(\frac{\Delta\psi(z_k,N-k)e_{HJB,k}^T}{\Delta\psi^T(z_k,N-k)\Delta\psi(z_k,N-k)+1}\right), \quad k = 0,\ldots,N-1\end{aligned} \tag{6.31}$$

Remark 6.1

When the NNCS system state becomes zero, $z_k = 0$, $\forall k = 0,1,\ldots,N$, both the stochastic value function (6.23) and critic NN approximation (6.24) become zero. Therefore, the critic NN will stop updating once the system state vector converges to zero. According to Dierks and Jagannathan (2012) and Xu and Jagannathan (2013), this can be considered as a PE requirement for the input to the critic NN. In other words, the system state has to be persistently exciting long enough in order for the critic NN to learn the stochastic value function within the finite time $t \in [0, NT_s]$. Similar to many recent NN works (Dierks and Jagannathan 2012; Xu and Jagannathan 2012), the PE requirement can be satisfied by introducing exploration noise such that $0 < \psi_{\min} < \|\psi_{\min}(z)\| \le \|\psi(z,k)\|$ and $0 < \Delta\psi_{\min} < \|\Delta\psi_{\min}(z_k)\| \le \|\Delta\psi(z_k,N-k)\|$ with $\psi_{\min}$ and $\Delta\psi_{\min}$ being positive constants.

Recalling the definition of the critic NN weight estimation error $\underset{\tau,\gamma}{E}(\tilde{W}_{V,k})$, the stochastic dynamics of $\underset{\tau,\gamma}{E}(\tilde{W}_{V,k})$ can be expressed as

$$\underset{\tau,\gamma}{E}(\tilde{W}_{V,k+1}) = \underset{\tau,\gamma}{E}(\tilde{W}_{V,k}) - \alpha_V \underset{\tau,\gamma}{E}\left(\frac{\psi(\hat{z}_{N,k},0)\psi^T(\hat{z}_{N,k},0)\tilde{W}_{V,k}}{\psi^T(\hat{z}_{N,k},0)\psi(\hat{z}_{N,k},0)+1}\right)$$

$$-\alpha_V \underset{\tau,\gamma}{E}\left(\frac{\psi(z_{N,k},0)\tilde{\psi}^T(z_N,z_{N,k},0)W_V}{\psi^T(z_{\hat{N},k},0)\psi(z_{N,k},0)+1}\right) - \alpha_V \underset{\tau,\gamma}{E}\left(\frac{\psi(\hat{z}_{N,k},0)\varepsilon_{V,0}}{\psi^T(\hat{z}_{N,k},0)\psi(\hat{z}_{N,k},0)+1}\right)$$

$$-\alpha_V \underset{\tau,\gamma}{E}\left(\frac{\Delta\psi(z_k,N-k)\Delta\psi^T(z_k,N-k)\tilde{W}_{V,k}}{\Delta\psi^T(z_k,N-k)\Delta\psi(z_k,N-k)+1}\right)$$

$$-\alpha_V \underset{\tau,\gamma}{E}\left(\frac{\Delta\psi(z_k,N-k)\Delta\varepsilon_{V,k}}{\Delta\psi^T(z_k,N-k)\Delta\psi(z_k,N-k)+1}\right), \quad k = 0,\ldots,N-1 \tag{6.32}$$

Next, the boundedness of the critic NN weight estimation error $\underset{\tau,\gamma}{E}(\tilde{W}_{V,k})$ derived by Equation 6.29 is demonstrated.

Theorem 6.1: Boundedness in the Mean of the Critic NN Weight Estimation Error

Given an initial NNCS admissible control policy $u_0(z_k)$, let the critic NN weight update law be designed as Equation 6.31. Then, there exists a positive constant α_V satisfying $0 < \alpha_V < ((2-\chi)/(\chi+5))$ with $0 < \chi = ((\psi_{\min}^2 + \Delta\psi_{\min}^2 + 2)/(\psi_{\min}^2 + 1)(\Delta\psi_{\min}^2 + 1)) < 2$ such that the critic NN weight estimation error (6.32) is *bounded* in the mean within the fixed final time. Further, the bound depends upon the final time, NT_s, and the initial bounded critic NN weight estimation error B_{WV},0.

Proof: Refer to Xu and Jagannathan (2015).

6.2.4 Actor NN Estimation of Optimal Control Policy

According to the universal approximation property of NN, the ideal finite-horizon stochastic optimal control input can be expressed by using the actor NN as

$$\underset{\tau,\gamma}{E}\left[u^*(z_k)\right] = \underset{\tau,\gamma}{E}[W_u^T\vartheta(z_k,k)+\varepsilon_{u,k}], \quad \forall k = 0,1,\ldots,N \tag{6.33}$$

with $\underset{\tau,\gamma}{E}(W_u)$ and $\underset{\tau,\gamma}{E}(\varepsilon_{u,k})$ denoting the target weight and reconstruction error of the actor NN, respectively, and $\vartheta(z_k,k)$ represents the smooth time-varying actor NN activation function. Moreover, two time-independent function $\vartheta_{\min}(z_k)$, $\vartheta_{\max}(z_k)$ can be found such that $\|\vartheta_{\min}(z_k)\| \le \|\vartheta(z_k,k)\| \le \|\vartheta_{\max}(z_k)\|$, $\forall k = 0,1,\ldots,N$. Also, the ideal actor NN weight matrix, activation function, and reconstruction are all considered to be bounded such that $\left\| \underset{\tau,\gamma}{E}(W_u) \right\| \le W_{uM}, \left\| \underset{\tau,\gamma}{E}(\vartheta(z_k,k)) \right\| \le \vartheta_M$ and $\left\| \underset{\tau,\gamma}{E}(\varepsilon_{u,k}) \right\| \le \varepsilon_{uM}$ with W_{Um}, ϑ_M, and ε_{Um} being positive constants.

Next, similar to Dierks and Jagannathan (2012) and Xu and Jagannathan (2013), the actor NN estimation of Equation 6.33 can be represented as

$$\underset{\tau,\gamma}{E}\left[\hat{u}(z_k)\right] = \underset{\tau,\gamma}{E}[\hat{W}_{u,k}^T \vartheta(z_k,k)], \quad \forall k = 0,1,\ldots,N \tag{6.34}$$

where $\underset{\tau,\gamma}{E}(\hat{W}_{u,k})$ represents the estimated weights for the actor NN. Moreover, the estimation error of the actor NN can be defined as the difference between the actual control inputs (6.34) applied to the NNCS and control policy, which minimizes the estimated stochastic value function (6.24) with identified control coefficient matrix $\hat{G}(z_k)$ during the interval $t \in [0,kT_s]$ as

$$\underset{\tau,\gamma}{E}(e_{u,k}) = \underset{\tau,\gamma}{E}\left[\hat{W}_{u,k}^T \vartheta(z_k,k) + \frac{1}{2} R_z^{-1} \hat{G}^T(z_k) \frac{\partial \psi^T(z_{k+1}, N-k-1)}{\partial z_{k+1}} \hat{W}_{V,k} \right] \tag{6.35}$$

Select the stochastic update law for the actor NN actual weight matrix as

$$\underset{\tau,\gamma}{E}(\hat{W}_{u,k+1}) = \underset{\tau,\gamma}{E}(\hat{W}_{u,k}) - \alpha_u \underset{\tau,\gamma}{E}\left[\frac{\vartheta(z_k,k)}{\vartheta^T(z_k,k)\vartheta(z_k,k)+1} e_{u,k}^T \right], \quad \forall k = 0,\ldots,N-1 \tag{6.36}$$

with the tuning parameter α_u satisfying $0 < \alpha_u < 1$. Further, the ideal actor NN output (6.33) should be equal to the control policy that minimizes the ideal stochastic value function (6.23), which is given by

$$\underset{\tau,\gamma}{E}[W_u^T \vartheta(z_k,k) + \varepsilon_{u,k}] = -\frac{1}{2} \underset{\tau,\gamma}{E}\left[R_z^{-1} G^T(z_k) \frac{\partial(\psi^T(z_{k+1}, N-k-1) W_V + \varepsilon_{V,k}^T)}{\partial z_{k+1}} \right] \tag{6.37}$$

In other words,

$$\underset{\tau,\gamma}{E}\left[W_u^T \vartheta(z_k) + \varepsilon_{u,k} + \frac{1}{2} R_z^{-1} G^T(z_k) \left(\frac{\partial \vartheta^T(z_{k+1}) W_V}{\partial z_{k+1}} + \frac{\partial \varepsilon_{V,k}^T}{\partial z_{k+1}} \right) \right] = 0 \tag{6.38}$$

Substituting Equation 6.38 into Equation 6.35, the actor NN estimation error can be represented equivalently as

$$\begin{aligned} \underset{\tau,\gamma}{E}(e_{u,k}) = &- \underset{\tau,\gamma}{E}[\tilde{W}_{u,k}^T \vartheta(z_k,k)] - \frac{1}{2} \underset{\tau,\gamma}{E}\left[R_z^{-1} \hat{G}^T(z_k) \frac{\partial \psi^T(z_{k+1}, N-k)}{\partial z_{k+1}} \tilde{W}_{V,k} \right] \\ &- \frac{1}{2} \underset{\tau,\gamma}{E}\left[R_z^{-1} \tilde{G}^T(z_k) \frac{\partial \psi^T(z_{k+1}, N-k)}{\partial z_{k+1}} W_V \right] - \frac{1}{2} \underset{\tau,\gamma}{E}\left[R_z^{-1} G^T(z_k) \frac{\partial \varepsilon_{V,k}^T}{\partial z_{k+1}} - \varepsilon_{u,k} \right] \end{aligned} \tag{6.39}$$

where $E(\tilde{W}_{u,k}) = E(W_u) - E(\hat{W}_{u,k})$ is the actor NN weight estimation error with $\tilde{G}(z_k) = G(z_k) - \hat{G}(z_k)$.

Next, substituting Equation 6.39 into Equation 6.36, the actor NN stochastic weight estimation error dynamics can be expressed as

$$\begin{aligned}
\underset{\tau,\gamma}{E}(\tilde{W}_{u,k+1}) &= \underset{\tau,\gamma}{E}(\tilde{W}_{u,k}) + \alpha_u \underset{\tau,\gamma}{E}\left[\frac{\vartheta(z_k,k)}{\vartheta^T(z_k,k)\vartheta(z_k,k)+1} e_{u,k}^T\right] \\
&= \underset{\tau,\gamma}{E}(\tilde{W}_{u,k}) - \alpha_u \underset{\tau,\gamma}{E}\left[\frac{\vartheta(z_k,k)}{\vartheta^T(z_k,k)\vartheta(z_k,k)+1} \tilde{W}_{u,k}\right] \\
&\quad - \frac{\alpha_u}{2}\underset{\tau,\gamma}{E}\left[\frac{\vartheta(z_k,k)}{\vartheta^T(z_k,k)\vartheta(z_k,k)+1} R_z^{-1}\hat{G}^T(z_k)\frac{\partial \psi^T(z_{k+1},N-k-1)}{\partial z_{k+1}}\tilde{W}_{V,k}\right] \\
&\quad - \frac{\alpha_u}{2}\underset{\tau,\gamma}{E}\left[\frac{\vartheta(z_k,k)}{\vartheta^T(z_k,k)\vartheta(z_k,k)+1} R_z^{-1}\tilde{G}^T(z_k)\frac{\partial \psi^T(z_{k+1},N-k-1)}{\partial z_{k+1}}W_V\right] \\
&\quad - \frac{\alpha_u}{2}\underset{\tau,\gamma}{E}\left[\frac{\vartheta(z_k,k)}{\vartheta^T(z_k,k)\vartheta(z_k,k)+1} R_z^{-1}G^T(z_k)\frac{\partial \varepsilon_{V,k}^T}{\partial z_{k+1}} - \varepsilon_{u,k}\right], \quad k = 0,\ldots,N-1
\end{aligned} \tag{6.40}$$

In this approach, owing to the novel NN identifier, the need for the NNCS control coefficient matrix, $G(z_k)$, is relaxed, which itself is a contribution when compared to Zhang et al. (2009), Wang et al. (2011), and Dierks and Jagannathan (2012). Next, the closed-loop stability of NNCS with the proposed novel time-based NDP algorithm will be demonstrated.

6.2.5 Closed-Loop Stability

In this section, we will prove that the closed-loop NNCS system is *bounded in the mean* within the fixed final time with the bounds dependent upon the initial conditions and final time. Moreover, when the final time instant goes to infinity, $k \to \infty$, the estimated control input approaches the infinite-horizon optimal control input. Before demonstrating the main theorem on closed-loop stability, the flowchart of the proposed novel time-based NDP finite-horizon optimal control design is presented in Figure 6.2.

Similar to Dierks and Jagannathan (2012) and Xu and Jagannathan (2013), the initial NNCS system state is considered to reside in a compact set Ω due to the initial admissible control input $u_0(z_k)$. Further, the actor NN activation function, the critic NN activation function, and its gradient are all considered bounded in Ω as $\left\|\underset{\tau,\gamma}{E}(\psi(z_k,k))\right\| \le \psi_M$, $\left\|\underset{\tau,\gamma}{E}[(\partial\psi(z_k))/\partial z_k]\right\| \le \psi'_M$, and $\left\|\underset{\tau,\gamma}{E}(\vartheta(z_k,k))\right\| \le \vartheta_M$. In addition, the PE condition will be satisfied by introducing exploration noise (Dierks and Jagannathan 2012; Xu and Jagannathan 2013). The three NN tuning parameters α_I,

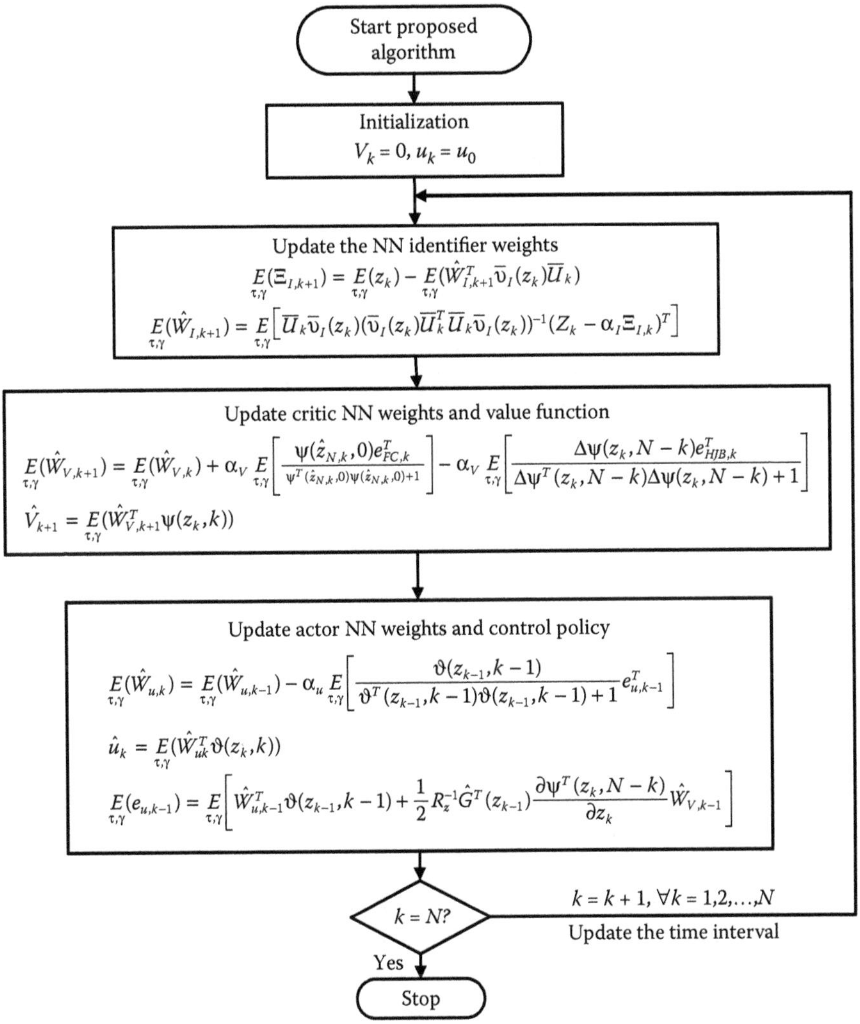

FIGURE 6.2 Flowchart of finite-horizon stochastic optimal control.

α_V, and α_u will be derived to guarantee that the future system state vector remains in Ω. In order to proceed, the following lemma is needed before introducing the theorem. Table 6.1 presents the stochastic optimal regulator.

Lemma 6.2

Let an optimal control policy be utilized for the controllable NNCS (6.10) such that Equation 6.10 is asymptotically stable in the mean (Xu and Jagannathan 2013). Then, the closed-loop NNCS dynamics, $\underset{\tau,\gamma}{E}[F(z_k)+G(z_k)u^*(z_k)]$, satisfy

$$\left\| \underset{\tau,\gamma}{E}[F(z_k)+G(z_k)u^*(z_k)] \right\| \le l_o \left\| \underset{\tau,\gamma}{E}(z_k) \right\|, \quad \forall = 0,1,\ldots,N \tag{6.41}$$

where $u^*(z_k)$ is the optimal control input with $0 < l_o < 1$ being a positive constant.

Proof: Proof follows similar to Dierks and Jagannathan (2012) and Xu and Jagannathan (2013).

Theorem 6.2: Convergence of Stochastic Optimal Control Input

Let $u_0(z_k)$ be any initial admissible control policy for the NNCS (6.10) such that Equation 6.41 holds with $0 < l_o < 1/2$. Given the NN weight update laws for the identifier, critic, and actor NN as Equations 6.18, 6.31, and 6.36, respectively, there exists three positive tuning parameters α_I, α_V, and α_u satisfying

$$0 < \alpha_I < \min\left\{ \frac{1}{2\zeta_M}, \frac{\zeta_{\min}}{\sqrt{2}\zeta_M} \right\}, \quad 0 < \alpha_V < \frac{2-\chi}{\chi+5}$$

with

$$0 < \chi = \frac{\psi_{\min}^2 + \Delta\psi_{\min}^2 + 2}{(\psi_{\min}^2+1)(\Delta\psi_{\min}^2+1)} < 2$$

defined in Theorem 6.1 and $0 < \alpha_u < 1$ such that the NNCS system state $\underset{\tau,\gamma}{E}(z_k)$, identification error $\underset{\tau,\gamma}{E}(e_{I,k})$, NN identifier weight estimation error $\underset{\tau,\gamma}{E}(\tilde{W}_{I,k})$, and critic and actor NN weight estimation errors $\underset{\tau,\gamma}{E}(\tilde{W}_{V,k})$, $\underset{\tau,\gamma}{E}(\tilde{W}_{u,k})$ are all *bounded* in the mean within the fixed final time $t \in [0, NT_s]$. In addition, the bounds are dependent upon the final time instant, NT_s, bounded initial state B_z,0, identification error B_{eI},0, and weight estimation error for the NN identifier, critic, and actor NN, B_{WI},0, B_{WV},0, B_{Wu},0. Moreover, $\underset{\tau,\gamma}{E}\left[\left\| u_k^* - \hat{u}_k \right\|\right] \le B_u$, where B_u is small bound.

Proof: Refer to Xu and Jagannathan (2015).

6.2.6 Simulation Results

The performance of the proposed finite-horizon stochastic optimal regulation control of NNCS in the presence of unknown system dynamics and network imperfections has been evaluated.

EXAMPLE 6.1

The continuous-time version of original two-link robot system is shown as (Lewis et al. 1999)

$$\dot{x} = f(x) + g(x)u \tag{6.42}$$

TABLE 6.1
Finite-Horizon Stochastic Optimal Control

Finite-Horizon Control Input (Actor NN)

$$\underset{\tau,\gamma}{E}[\hat{u}(z_k)] = \underset{\tau,\gamma}{E}[\hat{W}_{u,k}^T \vartheta(z_k,k)], \quad \forall k = 0,1,\ldots,N$$

NN Identifier

Update Law

$$\underset{\tau,\gamma}{E}(\hat{W}_{I,k+1}) = \underset{\tau,\gamma}{E}[\bar{U}_k \bar{\upsilon}_I(z_k)(\bar{\upsilon}_I^T(z_k)\bar{U}_k^T \bar{U}_k \bar{\upsilon}_I(z_k))^{-1}(Z_k - \alpha_I \Xi_{I,k})^T]$$

Identification Error

$$\underset{\tau,\gamma}{E}(\Xi_{I,k+1}) = \underset{\tau,\gamma}{E}(Z_{k+1}) - \underset{\tau,\gamma}{E}[\hat{W}_{I,k+1}^T \bar{\upsilon}_I(z_k)\bar{U}_k], \quad \forall k = 0,1,\ldots,N-1$$

HJB Equation Residual Error and Terminal Constraint Estimation Error

HJB Equation Residual Error

$$\underset{\tau,\gamma}{E}(e_{HJB,k}) = \underset{\tau,\gamma}{E}(\hat{W}_{V,k}^T \Delta\psi(z_k, N-k)) - \underset{\tau,\gamma}{E}(W_V^T \Delta\psi(z_k, N-k)) + \Delta\varepsilon_{V,k}$$

Terminal Constraint Estimation Error

$$\underset{\tau,\gamma}{E}(e_{FC,k}) = \underset{\tau,\gamma}{E}[\phi_N(z_N)] - \underset{\tau,\gamma}{E}(\hat{W}_{V,k}^T \psi(\hat{z}_{N,k},0)), \quad \forall k = 0,1,\ldots,N$$

Critic NN Parameter Update

$$\underset{\tau,\gamma}{E}(\hat{W}_{V,k+1}) = \underset{\tau,\gamma}{E}(\hat{W}_{V,k}) + \alpha_V \underset{\tau,\gamma}{E}\left(\frac{\psi(\hat{z}_{N,k})e_{FC,k}^T}{\psi^T(\hat{z}_{N,k})\psi(\hat{z}_{N,k})+1}\right)$$
$$-\alpha_V \underset{\tau,\gamma}{E}\left(\frac{\Delta\psi(z_k,N-k)e_{HJB,k}^T}{\Delta\psi^T(z_k,N-k)\Delta\psi(z_k,N-k)+1}\right), \quad k = 0,\ldots,N-1$$

where α_V is the tuning parameter.

where internal dynamics $f(x)$ and control coefficient matrix $g(x)$ are given as

$$f(x) = \begin{bmatrix} x_3 \\ x_4 \\ \dfrac{\begin{pmatrix} -(2x_3x_4 + x_4^2 - x_3^2 - x_3^2\cos x_2)\sin x_2 \\ +20\cos x_1 - 10\cos(x_1+x_2)\cos x_2 \end{pmatrix}}{\cos^2 x_2 - 2} \\ \dfrac{\begin{pmatrix} (2x_3x_4 + x_4^2 + 2x_3x_4\cos x_2 + x_4^2\cos x_2 + 3x_3^2 \\ +2x_3^2\cos x_2 + 20(\cos(x_1+x_2) - \cos x_1)\times \\ (1+\cos x_2) - 10\cos x_2\cos(x_1+x_2) \end{pmatrix}}{\cos^2 x_2 - 2} \end{bmatrix}, \tag{6.43}$$

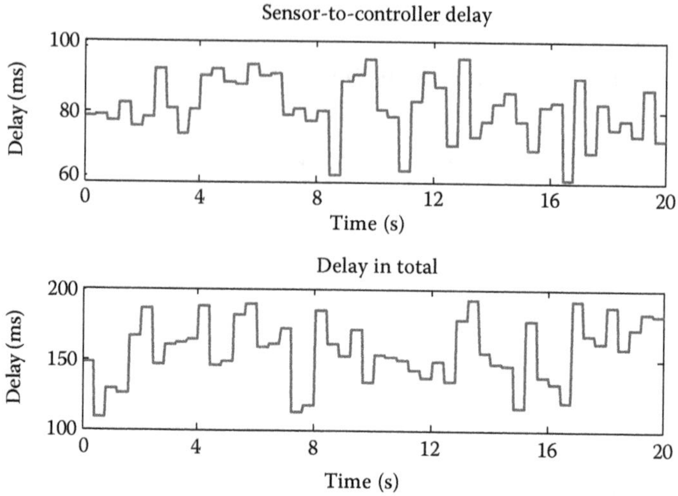

FIGURE 6.3 Distribution of network-induced delay in an NNCS.

$$g(x) = \begin{bmatrix} 0 & 0 \\ 0 & 0 \\ \dfrac{1}{2-\cos^2 x_2} & \dfrac{-1-\cos x_2}{2-\cos^2 x_2} \\ \dfrac{-1-\cos x_2}{2-\cos^2 x_2} & \dfrac{3+2\cos x_2}{2-\cos^2 x_2} \end{bmatrix}$$

The network parameters are selected as follows (Xu and Jagannathan 2013): (1) sampling time is given by $T_s = 10$ ms; (2) the upper bound of network-induced delay is given as $\bar{d}T_s = 20\,\text{ms}$; (3) the network-induced delays: $E(\tau_{sc}) = 8$ ms and $E(\tau) = 15$ ms; (4) network-induced packet losses follow the Bernoulli distribution with $\bar{\gamma} = 0.3$; (5) the final time is set as $t_f = NT_s = 20$ s with simulation time steps $N = 2000$; and (6) to obtain the ergodic performance, Monte Carlo simulation runs were utilized with 1000 iterations. The distribution of network-induced delays and packet losses is shown in Figures 6.3 and 6.4.

6.2.6.1 State Regulation Error and Controller Performance

Note that the problem attempted in this chapter is optimal regulation, which implies that NNCS states should converge to the origin in an optimal manner. After incorporating the network imperfections into NNCS, the augment state vector is presented as $z_k = [x_k \; u_{k-1} \; u_{k-2}]^T \in R^8$ and admissible control and the initial state are selected as

$$u_o(z_k) = \begin{bmatrix} -100 & 0 & -100 & 0 & 0 & 0 & 0 & 0 \\ 0 & -100 & 0 & -100 & 0 & 0 & 0 & 0 \end{bmatrix} z_k$$

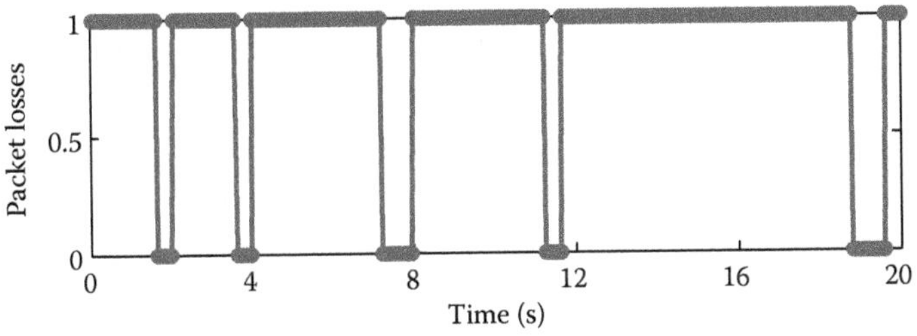

FIGURE 6.4 Distribution of network-induced packet losses in NNCS ("1" denotes packet is received and "0" denotes packet is lost).

and

$$x_0 = \begin{bmatrix} -\frac{\pi}{6} & \frac{\pi}{6} & 0 & 0 \end{bmatrix}^T$$

respectively. Moreover, the NN identifier activation function is given as $\tanh\{(z_{k,1})^2, z_{k,1} z_{k,2}, \ldots, (z_{k,8})^2, \ldots, (z_{k,1})^6, (z_{k,1})^5(z_{k,2}), \ldots, (z_{k,8})^6\}$, the state-dependent part of the activation function for the critic NN is defined as sigmoid of sixth-order polynomial (i.e., $\text{sigmoid}\{(z_{k,1})^2, z_{k,1} z_{k,2}, \ldots, (z_{k,8})^2, \ldots, (z_{k,1})^6, (z_{k,1})^5(z_{k,2}), \ldots, (z_{k,8})^6\}$) and the time-dependent part of critic NN activation function is selected as saturation polynomial time function (i.e., $sat\{(N-k)^{31}, (N-k)^{30}, \ldots, 1; \ldots; 1, (N-k)^{31}, \ldots, N-k\}$), and the activation function of the actor NN is selected as the gradient of critic NN activation function. The saturation operator for the time function is added in order to ensure the magnitude of time function stays within a reasonable range such that the NN weights are computable. Moreover, all three NNs have two layers where the first layer is the input layer and second layer is the hidden layer. For NN identifier, the hidden layer has 39 neurons whereas the critic NN and actor NN have 32 hidden neurons. Feed-forward NNs structure is selected for all the NNs.

The tuning parameters of the NN identifier, critic NN, and actor NN are defined as $\alpha_I = 0.03$, $\alpha_V = 0.01$, and $\alpha_u = 0.5$ where the initial weights of the NN identifier and critic NN in the hidden layer are selected as zero whereas the actor NN weight matrix in the hidden layer is set to reflect the initial admissible control at the beginning of simulation, whereas the weights of three NNs' input layer are all chosen as ones. The results are shown in Figures 6.5 through 6.9.

First, the state regulation error and stochastic optimal control inputs are studied. As shown in Figures 6.5 and 6.6, the proposed stochastic optimal controller can force the NNCS state regulation errors converge close to zero within the fixed final time even in the presence of uncertain NNCS dynamics and network imperfections. Moreover, the stochastic control signal is also bounded in the mean. The initial admission control selection affects the transient performance similar to the NN tuning parameters and these have to be carefully selected for suitable transient performance.

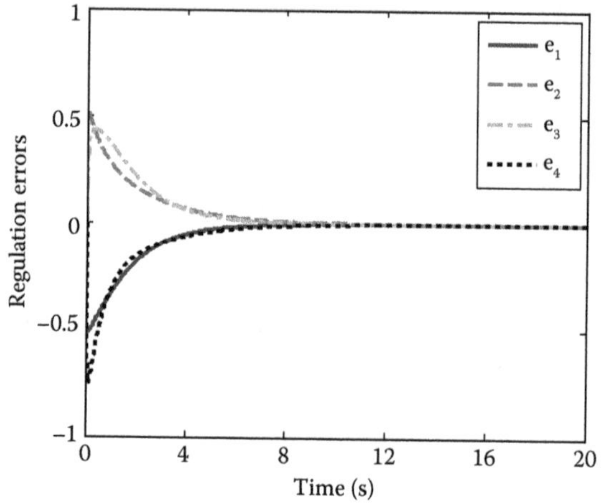

FIGURE 6.5 State regulation errors with the proposed controller.

Next, the effect of network imperfections is evaluated. According to Lewis et al. (1999) and Sarangapani (2006), an NN-based feedback linearization control input can maintain the stability of the two-link robot system (6.42). However, after introducing the network-induced delays and packet losses shown in Figures 6.3 and 6.4, this feedback linearization controller cannot retain the stability of the NNCS as shown in Figure 6.7. This in turn confirms that a controller should be carefully designed after incorporating the effects of network imperfections.

Now, the evolution of the NN weights is studied. In Figure 6.8, the actual weights of the critic and actor NN are shown. Within the fixed final time (i.e., $t \in [0{,}20s]$), the actual weights of the critic and actor NN converge and remain *bounded* in the

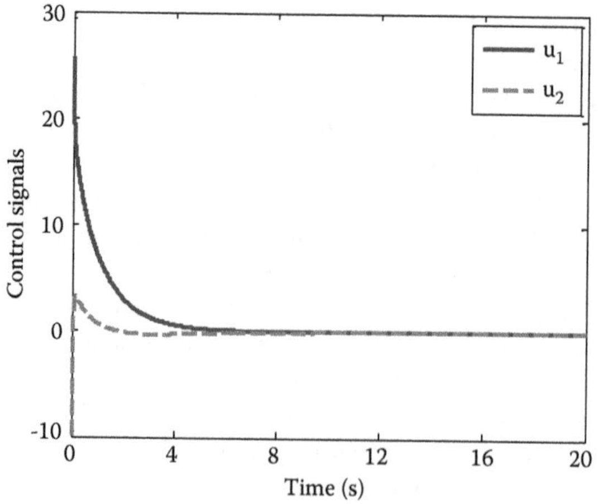

FIGURE 6.6 Stochastic optimal control inputs.

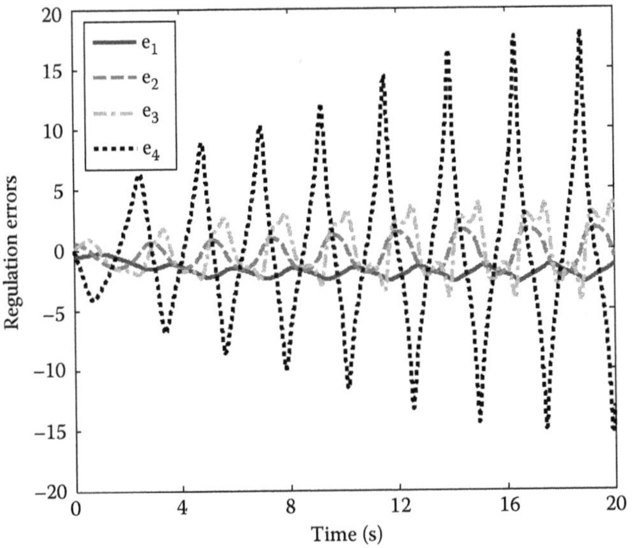

FIGURE 6.7 State regulation errors by using NN feedback linearizing controller with network imperfections.

mean consistent with Theorem 6.2. Further, as shown in Figure 6.9, the identification error converges close to zero, which indicates that the NN identifier learns the system dynamics properly.

6.2.6.2 HJB Equation and Terminal Constraint Estimation Errors

In this part, the HJB equation and terminal constraint estimation errors have been analyzed. It is well known that the proposed control input will approach finite-horizon optimal control (Lewis and Syrmos 1995; Chen and Lewis 2007) input only when the control input satisfies both HJB and terminal constraint errors. If the HJB equation error is near zero, then the solution of the HJB equation is optimal and the control input that uses the value function becomes optimal. In Figure 6.10, within the fixed final time, $t \in [0,20s]$, not only the HJB equation error but also the terminal constraint estimation errors converge close to zero. This indicates that the proposed stochastic optimal control inputs approach the finite-horizon optimal control inputs.

6.2.6.3 Cost Function Comparison

Subsequently, the cost function of the proposed finite-horizon stochastic optimal controller is studied. For comparison, with known system dynamics and network imperfections, both a conventional NN-based feedback linearization control (Sarangapani 2006) and an ideal offline finite-horizon optimal control (Chen and Lewis 2007) of NNCS have been included. In Figure 6.11, the cost function comparison result is shown for the three controllers. Compared with conventional NN-based NNCS feedback linearization control, the proposed optimal control design can deliver a

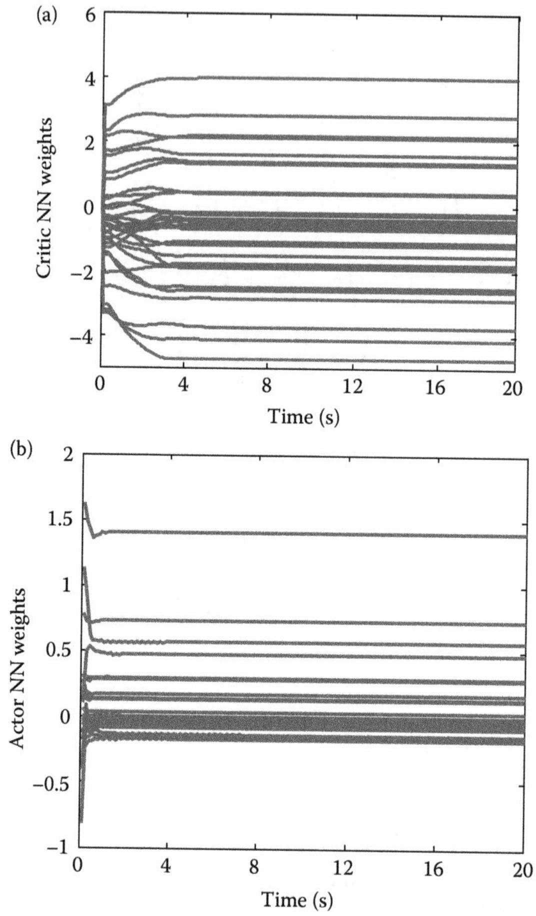

FIGURE 6.8 Estimated NN weights of (a) critic NN and (b) actor NN.

much better performance since optimality is neglected in feedback linearization control. Moreover, in contrast to traditional offline NNCS finite-horizon optimal control (Chen and Lewis 2007), the cost function of the proposed scheme is slightly higher due to system uncertainties and NN approximation whereas the proposed design is more practical since the prior knowledge on network imperfections and system dynamics are not needed unlike in the case of traditional offline optimal controller design.

However, the computational complexity of an optimal controller is higher than a traditional controller. Despite the increase in computational cost for the optimal controller, these advanced controllers can still be realized cheaply in practice due to a drastic increase in processor speed. Therefore, the advanced controllers, such as the one proposed, can be utilized on NNCS for generating an improvement in performance over traditional controllers.

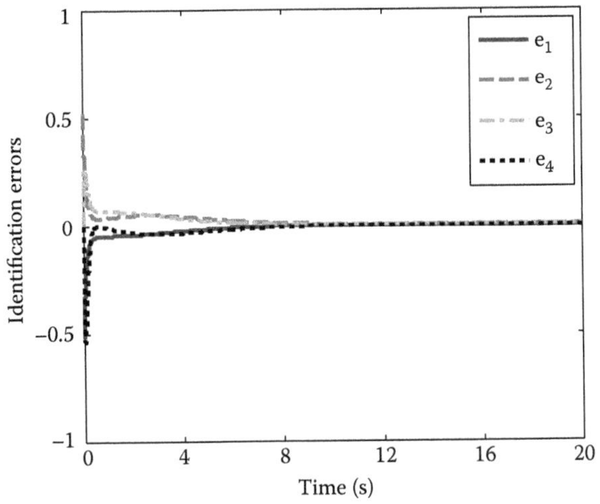

FIGURE 6.9 Identification errors.

In the end, the simulation results given in Figures 6.3 through 6.11 confirm that the proposed time-based NDP scheme renders acceptable performance in the presence of uncertain NNCS dynamics due to network imperfections.

6.3 EXTENSIONS TO INFINITE HORIZON

It is important to note that the infinite-horizon optimal design for NNCS can be derived when the terminal constraint is ignored and time goes to infinity, that is, $N \to \infty$. The details of infinite-horizon optimal control design for NNCS are given from Xu and Jagannathan (2013).

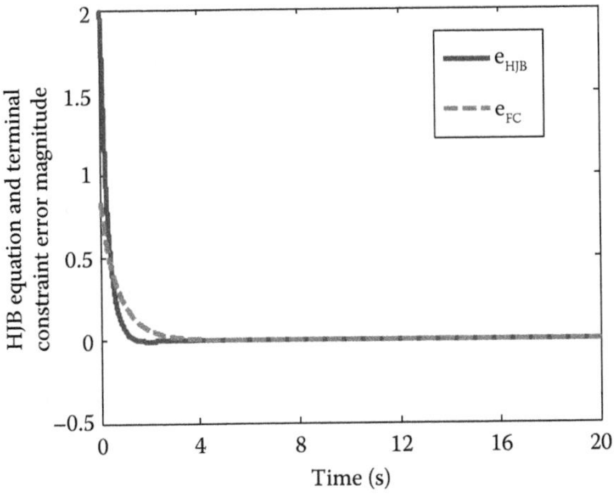

FIGURE 6.10 Estimated HJB equation and terminal constraint errors.

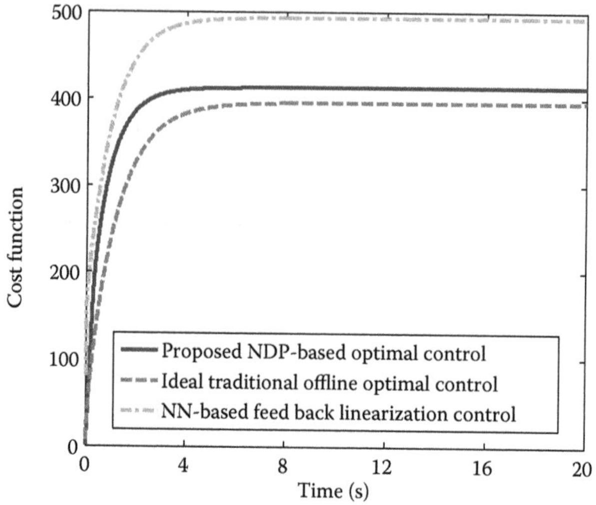

FIGURE 6.11 Comparison of cost functions.

6.3.1 Optimal Stochastic Value Function Approximation and Control Policy Design

By using universal approximation property of the NN, the stochastic value function and control policy can be represented with the critic and actor NN (Xu and Jagannathan 2013) as

$$V(y_k^o) = \underset{\tau,\gamma}{E}(W_V^T \vartheta(y_k^o) + \varepsilon_{Vk}) \tag{6.44}$$

and

$$\underset{\tau,\gamma}{E}[u^*(y_k^o)] = \underset{\tau,\gamma}{E}[W_u^T \varphi(y_k^o) + \varepsilon_{uk}] \tag{6.45}$$

respectively, where $\underset{\tau,\gamma}{E}(W_V)$ and $\underset{\tau,\gamma}{E}(W_u)$ represent the constant target NN weights, ε_{Vk}, ε_{uk} are the reconstruction errors for the critic and actor NN, respectively, and $\vartheta(\cdot)$ and $\varphi(\cdot)$ are the vector activation functions for the two NNs, respectively. The upper bounds in the mean for the two target NN weights are defined as $\left\|\underset{\tau,\gamma}{E}(W_V)\right\| \le W_{VM}$ and $\left\|\underset{\tau,\gamma}{E}(W_u)\right\| \le W_{uM}$ where W_{VM}, W_{uM} are positive constants (Sarangapani 2006), and the approximation errors are also considered bounded in the mean as $\left\|\underset{\tau,\gamma}{E}(\varepsilon_{Vk})\right\| \le \varepsilon_{VM}$ and $\left\|\underset{\tau,\gamma}{E}(\varepsilon_{uk})\right\| \le \varepsilon_{uM}$ where ε_{VM}, ε_{uM} are also positive constants (Sarangapani 2006), respectively. Additionally, the gradient of the approximation error is assumed to be bounded in the mean as $\left\| \underset{\tau,\gamma}{E}(\partial \varepsilon_{Vk} / \partial y_{k+1}^{o}) \right\| \le \varepsilon'_{VM}$ with ε'_{VM} being a positive constant (Sarangapani 2006; Dierks and Jagannathan 2012).

The critic and actor NN approximation of Equations 6.44 and 6.45 can be expressed as (Sarangapani 2006; Dierks and Jagannathan 2012)

$$\hat{V}(y_k^o) = \underset{\tau,\gamma}{E}\left[\hat{W}_{Vk}^T \vartheta(y_k^o)\right] \tag{6.46}$$

and

$$\underset{\tau,\gamma}{E}\left[\hat{u}(y_k^o)\right] = \underset{\tau,\gamma}{E}\left[\hat{W}_{uk}^T \varphi(y_k^o)\right] \tag{6.47}$$

where $\underset{\tau,\gamma}{E}(\hat{W}_{Vk})$ and $\underset{\tau,\gamma}{E}(\hat{W}_{uk})$ represent the estimated values of target weights W_V and W_u, respectively. In this work, the activation functions $\vartheta(\cdot)$, $\varphi(\cdot)$ are selected to be a basis function set and linearly independent (Kumar 1985; Sarangapani 2006). Since it is required that $V\left(\underset{\tau,\gamma}{E}(y_k^o) = 0\right) = 0$ and $u\left(\underset{\tau,\gamma}{E}(y_k^o) = 0\right) = 0$, the basis functions $\vartheta(\cdot)$, $\varphi(\cdot)$ are chosen such that $\vartheta\left(\underset{\tau,\gamma}{E}(y_k^o) = 0\right) = 0$, $\varphi\left(\underset{\tau,\gamma}{E}(y_k^o) = 0\right) = 0$, respectively.

Substituting Equation 6.47 into the HJB equation, it can be rewritten as

$$\underset{\tau,\gamma}{E}(W_V^T(\vartheta(y_{k+1}^o) - \vartheta(y_k^o))) + \underset{\tau,\gamma}{E}(y_k^{oT} Q_y y_k + u_k^T R_y u_k) = \underset{\tau,\gamma}{E}(\varepsilon_{Vk} - \varepsilon_{Vk+1})$$

In other words,

$$\underset{\tau,\gamma}{E}\left[W_V^T \Delta\vartheta(y_{k+1}^o)\right] + \underset{\tau,\gamma}{E}\left[r(y_k^o, u_k)\right] = \Delta\varepsilon_{Vk} \tag{6.48}$$

where $r(y_k^o, u_k) = y_k^{oT} Q_y y_k^o + u_k^T R_y u_k$, $\underset{\tau,\gamma}{E}[\Delta\vartheta(y_k^o)] = \underset{\tau,\gamma}{E}[\vartheta(y_{k+1}^o) - \vartheta(y_k^o)]$, and $\Delta\varepsilon_{vk} = \underset{\tau,\gamma}{E}(\varepsilon_{vk} - \varepsilon_{vk+1})$. However, when implementing the estimated value function (6.46), Equation 6.48 does not hold. Therefore, using delayed values for convenience, the residual error or cost-to-go error with Equation 6.48 can be expressed as

$$\begin{aligned} \underset{\tau,\gamma}{E}(e_{Vk}) &= \underset{\tau,\gamma}{E}(y_k^{oT} Q_y y_k^o + u_k^T R_y u_k) + \hat{V}(y_{k+1}^o) - \hat{V}(y_k^o) \\ &= \underset{\tau,\gamma}{E}[r(y_k^o, u_k)] + \underset{\tau,\gamma}{E}[\hat{W}_{Vk}^T \Delta\vartheta(y_k^o)] \end{aligned} \tag{6.49}$$

Based on gradient descent algorithm, the update law of the critic NN weights is given by

$$\underset{\tau,\gamma}{E}(\hat{W}_{Vk+1}) = \underset{\tau,\gamma}{E}(\hat{W}_{Vk}) - \alpha_v \underset{\tau,\gamma}{E}\left(\frac{\Delta\vartheta(y_k^o)}{\Delta\vartheta^T(y_k^o)\Delta\vartheta(y_k^o)+1} e_{Vk}^T\right)$$
$$= \underset{\tau,\gamma}{E}(\hat{W}_{Vk}) - \alpha_v \underset{\tau,\gamma}{E}\left[\frac{\Delta\vartheta(y_k^o)}{\Delta\vartheta^T(y_k^o)\Delta\vartheta(y_k^o)+1}[r^T(y_k^o,u_k) + \Delta\vartheta^T(y_k^o)\hat{W}_{Vk}]\right] \tag{6.50}$$

As a final step in the critic NN design, define the weight estimation error as $\underset{\tau,\gamma}{E}(\tilde{W}_{Vk}) = \underset{\tau,\gamma}{E}(W_V) - \underset{\tau,\gamma}{E}(\hat{W}_{Vk})$. Since $r^T(y_k^o,u_k) = -\underset{\tau,\gamma}{E}[\Delta\vartheta^T(y_{k+1}^o)W_V] + \Delta\varepsilon_{Vk}^T$ in Equation 6.48, the dynamics of the critic NN weight estimation error can be rewritten as

$$\underset{\tau,\gamma}{E}(\tilde{W}_{Vk+1}) = \underset{\tau,\gamma}{E}(\tilde{W}_{Vk}) - \alpha_V \underset{\tau,\gamma}{E}\left(\frac{\Delta\vartheta(y_k^o)\Delta\vartheta^T(y_k^o)}{\Delta\vartheta^T(y_k^o)\Delta\vartheta(y_k^o)+1}\tilde{W}_{Vk}\right)$$
$$+\alpha_V \underset{\tau,\gamma}{E}\left(\frac{\Delta\vartheta(y_k^o)\Delta\varepsilon_{Vk}}{\Delta\vartheta^T(y_k^o)\Delta\vartheta(y_k^o)+1}\right) \tag{6.51}$$

Next, the boundedness in the mean of the critic NN estimation error dynamics $\underset{\tau,\gamma}{E}(\tilde{W}_{Vk})$ given by Equation 6.51 is demonstrated in the following theorem.

Theorem 6.3: Boundedness in the Mean of the Critic NN Estimation Errors

Let $u_0(y_k^o)$ be any admissible control policy for NNCS, and let the critic NN weights update law be given by Equation 6.50. Then there exists a positive constant α_V satisfying $0 < \alpha_V < 1/2$ and computable positive constant B_{W_V}, such that the critic NN weight estimation error (6.51) is *bounded* in the mean with bounds given by $\left\|\underset{\tau,\gamma}{E}(\tilde{W}_{Vk})\right\| \le B_{W_V}$.

Proof: Refer to Xu and Jagannathan (2013).

Now we need to find the control policy via the actor NN (6.47), which minimizes the approximated value function (6.46). First, the actor NN estimation errors are defined to be the difference between the actual optimal control input (6.47) that is being applied to NNCS and the control input that minimizes the estimated value function (6.46) with identified NNCS dynamics $\hat{G}(y_k^o)$. This error can be expressed as

$$\underset{\tau,\gamma}{E}(e_{uk}) = \underset{\tau,\gamma}{E}\left[\hat{W}_{uk}^T\varphi(y_k^o) + \frac{1}{2}R_y^{-1}\hat{G}^T(y_k^o)\frac{\partial\vartheta^T(y_{k+1}^o)}{\partial y_{k+1}^o}\hat{W}_{Vk}\right] \tag{6.52}$$

The update law for actor NN weights is defined as

$$\underset{\tau,\gamma}{E}(\hat{W}_{uk+1}) = \underset{\tau,\gamma}{E}(\hat{W}_{uk}) - \alpha_u \underset{\tau,\gamma}{E}\left[\frac{\varphi(y_k^o)}{\varphi^T(y_k^o)\varphi(y_k^o)+1}e_{uk}^T\right] \tag{6.53}$$

where $0 < \alpha_u < 1$ is a positive constant. By selecting the control policy u_k to minimize the desired value function (6.44), it follows that

$$\underset{\tau,\gamma}{E}[W_u^T\varphi(y_k^o)+\varepsilon_{uk}] = -\frac{1}{2}\underset{\tau,\gamma}{E}\left[R_y^{-1}\hat{G}^T(y_k^o)\left(\frac{\partial\vartheta^T(y_{k+1}^o)}{\partial y_{k+1}^o}W_V + \frac{\partial\varepsilon_{Vk}^T}{\partial y_{k+1}^o}\right)\right]$$

In other words,

$$\underset{\tau,\gamma}{E}\left[W_u^T\varphi(y_k^o)+\varepsilon_{uk}+\frac{1}{2}R_y^{-1}\hat{G}^T(y_k^o)\left(\frac{\partial\vartheta^T(y_{k+1}^o)}{\partial y_{k+1}^o}W_V + \frac{\partial\varepsilon_{Vk}^T}{\partial y_{k+1}^o}\right)\right] = 0 \tag{6.54}$$

Substituting Equation 6.54 into Equation 6.52, the actor NN estimation error dynamics can be rewritten as

$$\begin{aligned}\underset{\tau,\gamma}{E}(e_{uk}) = &-\underset{\tau,\gamma}{E}[\tilde{W}_{uk}^T\varphi(y_k^o)] - \frac{1}{2}\underset{\tau,\gamma}{E}\left[R_y^{-1}\hat{G}^T(y_k^o)\frac{\partial\vartheta^T(y_{k+1}^o)}{\partial y_{k+1}^o}\tilde{W}_{Vk}\right]\\ &+\frac{1}{2}\underset{\tau,\gamma}{E}\left[R_y^{-1}\tilde{G}^T(y_k^o)\frac{\partial\varepsilon_{Vk}^T}{\partial y_{k+1}^o} - \varepsilon_{ek}\right]\end{aligned} \tag{6.55}$$

where $\underset{\tau,\gamma}{E}(\tilde{W}_{uk}) = \underset{\tau,\gamma}{E}(W_u) - \underset{\tau,\gamma}{E}(\hat{W}_{uk})$, $\tilde{G}(y_k^o) = G(y_k^o) - \hat{G}(y_k^o)$, $\varepsilon_{ek} = \underset{\tau,\gamma}{E}\big[\varepsilon_{uk} + (1/2)R_y^{-1}G^T(y_k^o)(\partial\varepsilon_{Vk}^T/\partial y_{k+1}^o)\big]$ satisfying $\left\|\underset{\tau,\gamma}{E}(\varepsilon_{ek})\right\| \le \varepsilon_{eM}$ with ε_{eM} being a positive constant, and $\left\|\underset{\tau,\gamma}{E}[\partial\varepsilon_{Vk}^T/\partial y_{k+1}^o]\right\| \le \varepsilon'_{VM}$.

The actor NN weight estimation error dynamics can be represented as

$$\begin{aligned}\underset{\tau,\gamma}{E}(\tilde{W}_{uk+1}) &= \underset{\tau,\gamma}{E}(\tilde{W}_{uk}) + \alpha_u\underset{\tau,\gamma}{E}\left[\frac{\varphi(y_k^o)}{\varphi^T(y_k^o)\varphi(y_k^o)+1}e_{uk}^T\right]\\ &= \underset{\tau,\gamma}{E}(\tilde{W}_{uk}) - \alpha_u\underset{\tau,\gamma}{E}\left[\frac{\varphi(y_k^o)}{\varphi^T(y_k^o)\varphi(y_k^o)+1}\Big[\tilde{W}_{uk}^T\varphi(y_k^o)\right.\\ &\left.\left.+\frac{1}{2}R_y^{-1}\hat{G}^T(y_k^o)\frac{\partial\vartheta^T(y_{k+1}^o)}{\partial y_{k+1}^o}\tilde{W}_{Vk} - \frac{1}{2}R_y^{-1}\tilde{G}^T(y_k^o)\frac{\partial\varepsilon_{Vk}^T}{\partial y_{k+1}^o} + \varepsilon_{ek}\right]\right]\end{aligned} \tag{6.56}$$

Remark 6.2

In this chapter, the proposed NN-based identifier relaxes the need for partial NNCS dynamics $G(y_k^o)$. Compared to Dierks and Jagannathan (2012), the knowledge of the input transformation matrix $G(y_k^o)$ and internal dynamics $F(y_k^o)$ are considered unknown here.

Next, the stability in the mean of NN-based identification error dynamics, NN identifier weight estimation errors, critic NN estimation, and actor NN estimation error dynamics are considered. Table 6.2 presents the proposed optimal control design.

Theorem 6.4: Convergence of the Optimal Control Signal

Let $u_0(y_k^o)$ be any initial stabilizing control policy for the NNCS, which satisfy the bounds in the mean in Xu and Jagannathan (2013) and $0 < k^* < 1/2$. Let the NN weight tuning for the identifier, critic, and actor NN be provided as Xu and Jagannathan (2013), Equation 6.50, and Equation 6.53, respectively. Then, there are several positive constants α_C, α_u, α_V satisfying $0 < \alpha_C < \min\{1,((\Psi_{\min})/2\sqrt{2}\Psi_M)\}$, $(1/4) < \alpha_V < (3+\sqrt{3})/12$, and $1/6 < \alpha_u < 1/3$, respectively, and positive constants b_y, b_V, b_{WC}, b_{ey}, and b_u such that the system output vector $\underset{\tau,\gamma}{E}(y_k^o)$, NN identification error $\underset{\tau,\gamma}{E}(e_{yk})$, weight estimation errors $\underset{\tau,\gamma}{E}(\tilde{W}_{Ck})$, and critic and actor NN weight estimation errors $\underset{\tau,\gamma}{E}(\tilde{W}_{Vk})$ and $\underset{\tau,\gamma}{E}(\tilde{W}_{uk})$, respectively, are all *bounded* in the mean for all $k \geq k_0 + T$ with bounds given by $\left\|\underset{\tau,\gamma}{E}(y_k^o)\right\| \leq b_y$, $\left\|\underset{\tau,\gamma}{E}(e_{yk})\right\| \leq b_{ey}$, $\left\|\underset{\tau,\gamma}{E}(\tilde{W}_{Ck})\right\| \leq b_{WC}$, $\left\|\underset{\tau,\gamma}{E}(\tilde{W}_{Vk})\right\| \leq b_V$, and $\left\|\underset{\tau,\gamma}{E}(\tilde{W}_{uk})\right\| \leq b_u$. Further, $\left\|\underset{\tau,\gamma}{E}\left[\hat{u}(y_k^o) - u^*(y_k^o)\right]\right\| \leq \delta_u$ for a small positive constant δ_u.

Proof: Refer to Xu and Jagannathan (2013).

6.3.2 Simulation Results

In this section, the stochastic optimal control of the NNCS with uncertain dynamics in the presence of unknown random delays and packet losses is evaluated over the infinite horizon.

EXAMPLE 6.2

The continuous-time version of the original nonlinear affine system is given by

$$\dot{x} = f(x) + g(x)u, \quad y = Cx \tag{6.57}$$

where

$$f(x) = \begin{bmatrix} -x_1 + x_2 \\ -0.5x_1 - 0.5x_2(1 - (\cos(2x_1) + 2)^2) \end{bmatrix}, \quad g(x) = \begin{bmatrix} 0 \\ \cos(2x_1) + 2 \end{bmatrix}$$

and

$$C = \begin{bmatrix} 0 & 1 \\ 2 & 0 \end{bmatrix}$$

TABLE 6.2
Infinite-Horizon Stochastic Optimal Control

Infinite-Horizon Control Input (Actor NN)

$$\underset{\tau,\gamma}{E}[\hat{u}(y_k^o)] = \underset{\tau,\gamma}{E}[\hat{W}_{uk}^T \varphi(y_k^o)]$$

NN Identifier

Update Law

$$\underset{\tau,\gamma}{E}(\hat{W}_{Ck+1}) = \underset{\tau,\gamma}{E}\left(\bar{U}_k \Delta\psi_{Ck}(\Delta\psi_{Ck}^T \bar{U}_k^T \bar{U}_k \Delta\psi_{Ck})^{-1}(Y_{k+1}^o - \alpha_C \Sigma_{yk})^T\right)$$

Identification Error

$$\underset{\tau,\gamma}{E}(\Sigma_{yk+1}) = \underset{\tau,\gamma}{E}(Y_{k+1}^o) - \underset{\tau,\gamma}{E}(\hat{W}_{Ck+1}^T \Delta\psi_{Ck} \bar{U}_k)$$

HJB Equation Residual Error

$$\underset{\tau,\gamma}{E}(e_{Vk}) = \underset{\tau,\gamma}{E}(y_k^{oT} Q_y y_k^o + u_k^T R_y u_k) + \hat{V}(y_{k+1}^o) - \hat{V}(y_k^o)$$
$$= \underset{\tau,\gamma}{E}[r(y_k^o, u_k)] + \underset{\tau,\gamma}{E}[\hat{W}_{Vk}^T \Delta\vartheta(y_k^o)]$$

Critic NN Parameter Update

$$\underset{\tau,\gamma}{E}(\hat{W}_{Vk+1}) = \underset{\tau,\gamma}{E}(\hat{W}_{Vk}) - \alpha_v \underset{\tau,\gamma}{E}\left(\frac{\Delta\vartheta(y_k^o)}{\Delta\vartheta^T(y_k^o)\Delta\vartheta(y_k^o)+1} e_{Vk}^T\right)$$
$$= \underset{\tau,\gamma}{E}(\hat{W}_{Vk}) - \alpha_v \underset{\tau,\gamma}{E}\left[\frac{\Delta\vartheta(y_k^o)}{\Delta\vartheta^T(y_k^o)\Delta\vartheta(y_k^o)+1}[r^T(y_k^o, u_k) + \Delta\vartheta^T(y_k^o)\hat{W}_{Vk}]\right]$$

where α_v is the tuning parameter.

The network parameters of the NNCS are selected as (1) the sampling time: $T_s = 100$ ms; (2) the bound of delay is set as two, that is, $\bar{d} = 2$; (3) the mean random delay values are given by $E(\tau_{sc}) = 80$ ms $E(\tau) = 150$ ms; (4) packet losses follow the Bernoulli distribution with $p = 0.3$; and (5) to obtain ergodic behavior, Monte Carlo simulation runs were conducted using 1000 iterations.

First, the effect of random delays and packet losses for the NNCS is studied. The initial state is taken as $x_0 = [5 \quad -3]^T$. The initial static control $u_k = [-2 \quad -5]x_k$, which maintains the original nonlinear affine system (6.57) stable, is shown in Figure 6.12a. By contrast, this controller cannot maintain the system stable for NNCS in the presence of random delays and packet losses as Figure 6.12b.

Next, the proposed stochastic optimal control is implemented for the NNCS with unknown system dynamics in the presence of random delays and packet losses. The augment state y_k^o is generated as $y_k^o = [y_k \ u_{k-1} \ u_{k-2}]^T \in \Re^{4\times 1}, \ \forall k$, and the initial stabilizing policy for the proposed algorithm was selected as

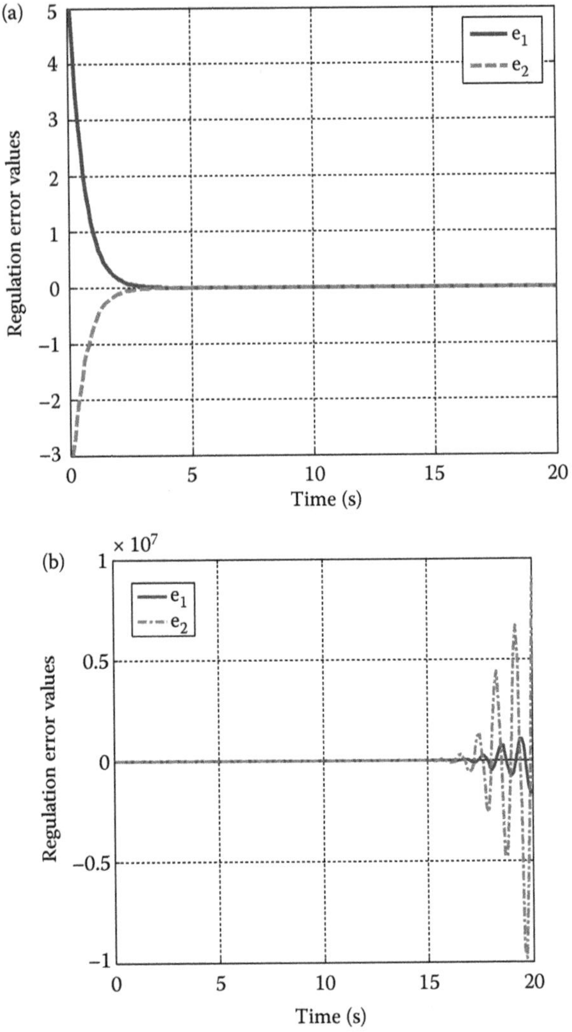

FIGURE 6.12 Performance of a static feedback controller: (a) random delays and packet losses are not present; (b) with random delays and packet losses.

$u_o(y_k^o)=[-2\ \ -5\ \ -1\ \ -2]y_k^o$ generated by using the standard pole placement method, while the activation functions for the NN-based identifier were generated as $\tanh\{(y_1^o)^2, y_1^o y_2^o, \ldots, (y_4^o)^2, (y_1^o)^4, (y_1^o)^3 y_2^o, \ldots, (y_4^o)^6\}$, the critic NN activation function was selected as sigmoid of the sixth-order polynomial $\{(y_1^o)^2, y_1^o y_2^o, \ldots, (y_4^o)^2, (y_1^o)^4, (y_1^o)^3 y_2^o, \ldots, (y_4^o)^6\}$ and the actor NN activation function was generated from the gradient of critic NN activation function.

The design parameters for the NN-based identifier, critic NN, and actor NN were selected as $\alpha_C = 0.002$, $\alpha_V = 10^{-4}$, and $\alpha_u = 0.005$ while the NN-based identifier and critic NN weights are set to zero at the beginning of the simulation. The initial

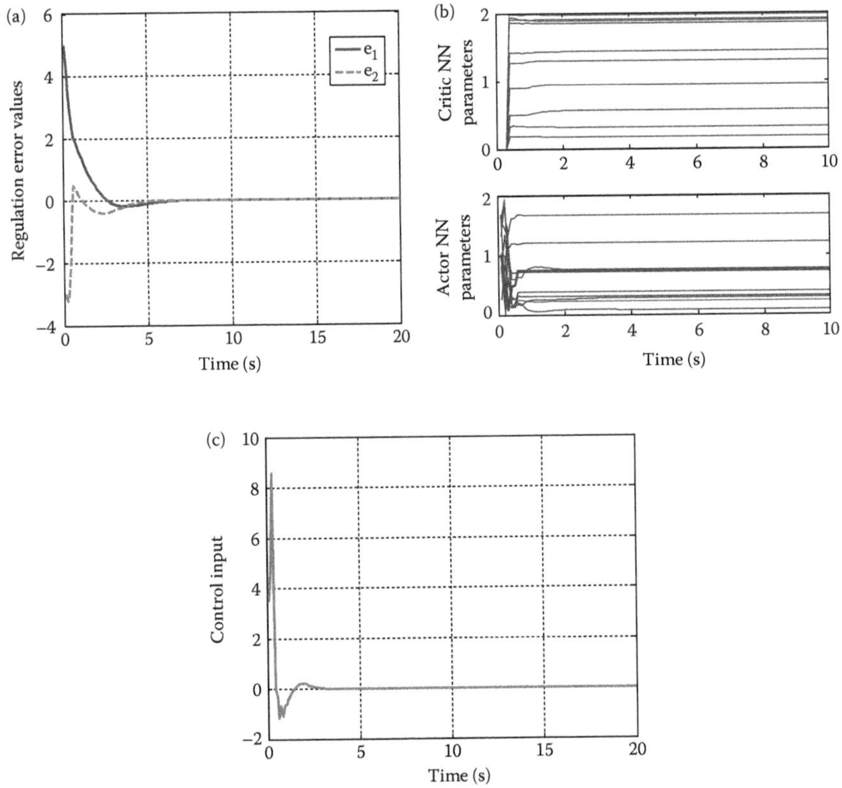

FIGURE 6.13 Performance of stochastic optimal controller for NNCS: (a) state regulation errors; (b) critic NN and actor NN parameters; and (c) control input.

weights of the actor NN are chosen to reflect the initial stabilizing control. The simulation was run for 20 s (200 time steps); for the first 10 s (100 time steps), exploration noise with mean zero and variance 0.06 was added to the system in order to ensure the PE condition (Xu and Jagannathan 2013).

The performance of the proposed stochastic optimal controller is evaluated from several aspects: (1) as shown in Figure 6.13a, the proposed stochastic optimal controller can make the NNCS state regulation errors converge to zero even when the NNCS dynamics are uncertain, which implies that the proposed controller can make the NNCS system stable in the mean; (2) the proposed critic NN and actor NN parameters converge to constant values and remain bounded in the mean consistent with Theorem 6.3 as shown in Figure 6.13b; and (3) the optimal control input for NNCS with uncertain dynamics is shown in Figure 6.13c, which is bounded in the mean.

For comparison, the HDP VI (Al-Tamimi and Lewis 2008; Lewis and Vrabie 2009)-based scheme is also implemented for an NNCS with known dynamics $G(\cdot)$ by incorporating the $g(\cdot)$ (6.57) and the information of delays and packet losses that

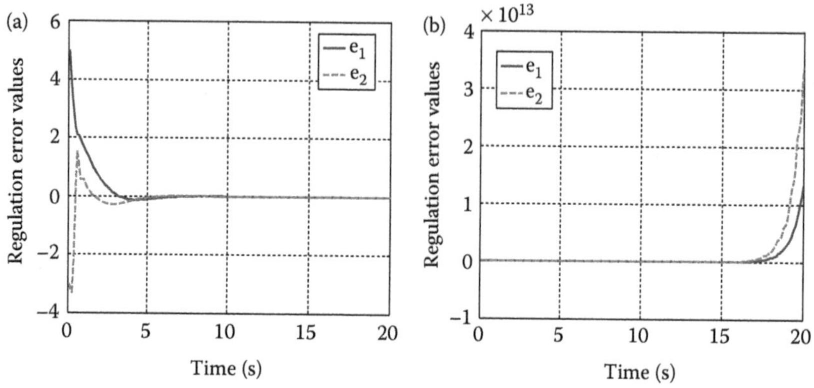

FIGURE 6.14 Performance of HDP value iteration for NNCS: (a) iterations = 100 times/sample and (b) iterations = 10 times/sample.

are normally not known beforehand. The initial admissible control, critic NN, and actor NN activation function are the same as the proposed time-based stochastic optimal control.

As shown in Figure 6.14a, the HDP VI method can make the NNCS state regulation errors converge to zero when the number of iterations is 100 times/sample. By contrast, HDP VI cannot maintain the stability of NNCS in the mean when iterations become 10 times/sample as shown in Figure 6.14b. It implies that the HDP VI scheme needs not only partial knowledge of original nonlinear affine system dynamics, $g(\cdot)$, but also information on delays and packet losses. The number of iterations required for a given nonlinear system is unknown. Owing to these drawbacks, the HDP VI is not preferred for NNCS implementation in real time.

Based on the results presented in Figures 6.12 through 6.14, the proposed stochastic optimal control scheme with uncertain NNCS dynamics and unknown network imperfections can overcome the drawbacks of the HDP-based VI method and will render nearly the same performance as that of an optimal controller for an NNCS when the system dynamics, random delays, and packet losses are known.

6.4 CONCLUSIONS

This chapter presented both finite- and infinite-horizon optimal adaptive control of nonlinear systems in the presence of network imperfections, mainly random delays and packet losses. The incorporation of delays and packet losses results in an augmented stochastic nonlinear time-varying system with uncertain dynamics. The traditional HJB-based solution is not possible since the system dynamics are needed while the proposed NDP-based optimal control solution based on the HJB equation in both finite and infinite horizon relaxed the system dynamics. Time history information of the previous cost-to-go errors help with the convergence without using value and policy iterations.

PROBLEMS

Section 6.2

Problem 6.2.1: Repeat Example 6.1 using the following network information: (1) sampling time is given by $T_s = 15$ ms; (2) the upper bound of network-induced delay is given as $\bar{d}T_s = 30\,\text{ms}$; (3) the network-induced delays: $E(\tau_{sc}) = 6$ ms and $E(\tau) = 16$ ms; (4) network-induced packet losses follow the Bernoulli distribution with $\bar{\gamma} = 0.4$; (5) the final time is set as $t_f = NT_s = 20$ s with simulation time steps $N = 2000$, and (6) Monte Carlo iteration number $N_{MC} = 5000$. Repeat the simulation with several initial conditions.

Problem 6.2.2: Repeat Problem 6.2.1 with Monte Carlo iteration number $N_{MC} = 100$.

Section 6.3

Problem 6.3.1: Repeat Example 6.2 with the following information: (1) the sampling time: $T_s = 90$ ms; (2) the bound of delay is set as two, that is, $\bar{d} = 2$; (3) the mean random delay values are given by $E(\tau_{sc}) = 80$ ms $E(\tau) = 170$ ms; (4) packet losses follow the Bernoulli distribution with $p = 0.4$; and (5) Monte Carlo iteration number $N_{MC} = 5000$.

Problem 6.3.2: Repeat Problem 6.3.1 with different initial condition.

REFERENCES

Al-Tamimi, A. and Lewis, F.L. 2008. Discrete-time nonlinear HJB solution using approximate dynamic programming: Convergence proof. *IEEE Transactions on Systems, Man, and Cybernetics, Part B*, 38, 943–949.

Bertsekas, D.P. and Tsitsiklis, J. 1996. *Neuro-Dynamics Programming.* Athena Scientific, MIT Press, MA.

Chen, H.F. and Guo, L. 1991. *Identification and Stochastic Adaptive Control.* Cambridge Press, Cambridge, MA.

Chen, T. and Lewis, F.L. 2007. Fixed-final-time-constrained optimal control of nonlinear systems using neural network HJB approach. *IEEE Transactions on Neural Networks*, 18, 1725–1737.

Dankert, J., Yang, L., and Si, J. 2006. A performance gradient perspective on approximate dynamic programming and its application to partially observable Markov decision process. *Proceedings of the IEEE International Symposium on Intelligent Control*, Munich, Germany, pp. 458–463.

Dehghani, A.K. 2005. Optimal networked control system design: A dual-rate approach. *Proceedings of the Canadian Conference on ECE*, Saskatoon, Canada, pp. 790–793.

Dierks, T. and Jagannathan, S. 2012. Online optimal control of affine nonlinear discrete-time systems with unknown internal dynamics by using time-based policy update. *IEEE Transactions on Neural Networks and Learning Systems*, 23, 1118–1129.

Feng, L.L., Moyne, J., and Tilbury, D. 2002. Optimal control design and evaluation for a class of networked control systems with distributed constant delays. *Proceedings of the IEEE American Control Conference*, Anchorage, AK, pp. 3009–3014.

Goldsmith, A. 2003. *Wireless Communication.* Cambridge University Press, Cambridge, UK.

Hu, S.S. and Zhu, Q.X. 2003. Stochastic optimal control and analysis of stability of networked control systems with long delay. *Automatica*, 39, 1877–1884.

Kumar, P.R. 1985. A survey of some results in stochastic adaptive control. *SIAM Journal on Control and Optimization*, 23, 329–380.

Lewis, F.L., Jagannathan, S., and Yesilderik, A. 1999. *Neural Network Control of Robot Manipulators and Nonlinear Systems*. Taylor & Francis, Florence, KY.

Lewis, F.L. and Syrmos, V.L. 1995. *Optimal Control*. 2nd edition, Wiley Press, New York.

Lewis, F.L. and Vrabie, D. 2009. Reinforcement learning and adaptive dynamic programming for feedback control. *IEEE Circuits and Systems Magazine*, 9, 32–50.

Luck, R. and Ray, A. 1990. An observer-based compensator for distributed delays. *Automatica*, 26, 903–908.

Mahmoud, M.S. and Ismail, A. 2005. New results on delay-dependent control of time-delay systems. *IEEE Transactions on Automatic Control*, 50, 95–100.

Sarangapani, J. 2006. *Neural Network Control of Nonlinear Discrete-Time Systems*. CRC Press, Florida, USA.

Stallings, W. 2002. *Wireless Communications and Networks*. 1st edition, Prentice-Hall Press, New Jersey, NJ.

Stengel, R.F. 1986, *Stochastic Optimal Control: Theory and Application*. Wiley-Interscience, New York.

Tabbara, M. 2008. A linear quadratic Gaussian framework for optimal networked control system design. *Proceedings of the IEEE American Control Conference*, Seattle, WA, pp. 3804–3809.

Tipsuwan, Y. and Chow, M.Y. 2003. Control methodologies in networked control systems. *Control Engineering Practice*, 11(10), 1099–1111.

Walsh, G.C., Beldiman, O., and Bushnell, L.G. 2001. Asymptotic behavior of nonlinear networked control systems. *IEEE Transactions on Automatic Control*, 46, 1093–1097.

Wang, F.Y., Jin, N., Liu, D., and Wei, Q.L. 2011. Adaptive dynamic programming for finite horizon optimal control of discrete-time nonlinear systems with ϵ-error bound. *IEEE Transactions on Neural Networks*, 22, 24–36.

Werbos, P.J. 1983. A menu of designs for reinforcement learning over time. *Journal of Neural Networks Control*, 3(1), 835–846.

Wouw, N.V.D., Nesic, D., and Heemels, W.P.H. 2012. A discrete-time framework for stability analysis of nonlinear networked control systems. *Automatica*, 48(12), 1144–1153.

Xu, H. and Jagannathan, S. 2013. Stochastic optimal controller design for uncertain nonlinear networked control system via neuro dynamic programming. *IEEE Transactions on Neural Networks and Learning Systems*, 24, 471–484.

Xu, H. and Jagannathan, S. 2015. Neural network based finite horizon stochastic optimal control design for nonlinear networked control systems. *IEEE Transactions on Neural Networks and Learning Systems*, 26, 472–485.

Xu, H., Jagannathan, S., and Lewis, F.L. 2012. Stochastic optimal control of unknown linear networked control system in the presence of random delays and packet losses. *Automatica*, 48, 1017–1030.

Zhang, W., Branicky, M.S., and Phillips, S. 2001. Stability of networked control systems. *IEEE Control Systems Magazine*, 21, 84–99.

Zhang, H., Luo, Y.H., and Liu, D. 2009. Neural network based near optimal control for a class of discrete-time affine nonlinear systems with control constraints. *IEEE Transactions on Neural Networks*, 20, 1490–1503.

7 Optimal Design for Nonlinear Two-Player Zero-Sum Games under Communication Constraints

Nonlinear networked control systems (Hespanha et al. 2007), which brings in a communication network to close the feedback loop between a nonlinear plant and remote controller, has been considered recently due to many benefits. Liu et al. (2007) analyzed and derived the stability region for LNCS. Subsequently, Walsh et al. (2001) considered the asymptotic behavior of NNCS with network-induced delays alone. A discrete-time framework has been proposed by Wouw et al. (2010) to analyze NNCS stability in the presence of both delays and packet losses. Liu et al. (2007), Walsh et al. (2001), and Wouw et al. (2010) requires the knowledge of system dynamics and network imperfections to maintain the stability of the NCS. Moreover, optimality is preferred besides stability (Hu and Zhu 2003; Wang et al. 2007; Jia et al. 2009).

Therefore, Xu et al. (2012) and Xu and Jagannathan (2013) proposed an infinite-horizon optimal control for both LNCS and NNCS under uncertain dynamics and unknown network imperfections. The finite-horizon optimal problem is more difficult due to terminal constraints (Heydari and Balakrishnan 2011) and the time-varying solution to the HJB equation. In addition, existing designs (Hu and Zhu 2003; Wang et al. 2007; Jia et al. 2009) ignored the worst-case disturbance referred to as the NCS two-player zero-sum game while considering optimality (Burgin 1969; Basar and Olsder 1995; Al-Tamimi et al. 2007).

To optimize the performance of the two-player zero-sum game, the min–max optimization H_∞ control problem (Burgin 1969; Basar and Olsder 1995; Al-Tamimi et al. 2007) is introduced, where the controller is viewed as a minimizing player while disturbance is treated as the maximizing player.

The NN implementation can be incorporated with NDP (Werbos 1983; Bertsekas and Tsitsiklis 1996) to obtain optimal policies (Seiffertt et al. 2008) by using value or policy iterations. Wang et al. (2011) proposed a policy or value iteration-based NDP scheme to obtain finite-horizon optimal control signals for a nonlinear system by using offline NN training. Vamvoudakis and Lewis (2010) utilized synchronous policy iteration to attain the infinite-horizon optimal design for nonlinear two-player zero-sum games. However, these iteration-based NDP methods (Vamvoudakis and

Lewis 2010; Heydari and Balakrishnan 2011; Wang et al. 2011) require a significant number of iterations within a fixed sampling interval; otherwise, the iteration-based schemes (Wang et al. 2011) can become unstable. To circumvent this issue, Dierks and Jagannathan (2012) developed a time-based NDP approach to solve the infinite-horizon optimal control of the affine nonlinear discrete-time system in the presence of unknown internal dynamics.

However, these existing NDP schemes (Vamvoudakis and Lewis 2010; Heydari and Balakrishnan 2011; Wang et al. 2011; Dierks and Jagannathan 2012) are not suitable for the finite-horizon NNCS zero-sum game since (a) the worst-case disturbance input has not been considered; (b) only the infinite-horizon optimal control is addressed, (c) the knowledge of system dynamics in terms of control and disturbance coefficient matrices (Vamvoudakis and Lewis 2010) is still required, and (d) network imperfections stemming from an unreliable communication network are ignored (Vamvoudakis and Lewis 2010).

In contrast, an optimal adaptive control scheme using time-based NDP is undertaken in this chapter to solve finite-horizon stochastic optimal strategies for the NNCS two-player zero-sum game in the presence of unknown system dynamics and network imperfections, and bounded disturbances. First, to relax the need for system dynamics, a novel NN identifier is proposed to learn both the control and disturbance coefficient matrices online. Then, a critic NN is introduced and tuned forward-in-time to learn the stochastic time-varying value function of the NNCS two-player zero-sum game within a finite time by using Hamilton–Jacobi–Isaacs (HJI) equation (Basar and Olsder 1995) given the terminal constraint. Eventually, two action NNs are utilized to estimate the stochastic optimal control and disturbance inputs by minimizing and maximizing, respectively, the stochastic value function.

Further, designing control and disturbance policies for the NNCS two-player zero-sum game with network imperfections is quite different from the time-delay system control with known deterministic delays (Ge et al. 2003; Hong et al. 2005) since NNCS network imperfections result in random delays and packet losses that cannot be handled in time-delay systems. Therefore, control and disturbance approaches for time-delay systems (Ge et al. 2003; Hong et al. 2005) are unsuitable for the NNCS two-player zero-sum game.

The main contribution of this chapter includes the development of a time-based finite-horizon stochastic optimal design for the uncertain NNCS two-player zero-sum game using the HJI equation without using policy or value iterations and while incorporating the terminal constraints and worst-case disturbance inputs. In addition, the need for system dynamics is relaxed. Finally, closed-loop stability is demonstrated.

First, Section 7.1 presents the traditional finite-horizon optimal strategies. Subsequently, Section 7.2 provides the background of the NNCS two-player zero-sum game. Then, the novel finite-horizon stochastic optimal design with an online identifier is developed in Section 7.3 where the stability is also analyzed by using Lyapunov theory. Section 7.4 illustrates the effectiveness of the proposed approach while the concluding remarks are given in Section 7.5.

7.1 TRADITIONAL STOCHASTIC OPTIMAL CONTROL DESIGN FOR TWO-PLAYER ZERO-SUM GAME

Consider the nonlinear two-player zero-sum game as

$$x_{DT,k+1} = f_{DT}(x_{DT,k}) + g_{DT}(x_{DT,k})u_{DT,k} + h_{DT}(x_{DT,k})d_{DT,k} \tag{7.1}$$

where $x_{DT,k}$, $u_{DT,k}$, $d_{DT,k}$ represent the system state, control input, and disturbance signals and $f_{DT}(x_{DT,k})$, $g_{DT}(x_{DT,k})$, $h_{DT}(x_{DT,k})$ denote the deterministic internal dynamics, control and disturbance coefficient matrices, respectively, in discrete time. According to the traditional optimal control (Burgin 1969; Basar and Olsder 1995; Al-Tamimi et al. 2007) for the two-player zero-sum game, the finite-horizon optimal strategies can be derived to minimize the certainty-equivalent stochastic value function (Stengel 1986), which is expressed as

$$\begin{cases} V_k(x_{DT,k},k) = E\left[\phi_N(x_{DT,N}) + \sum_{l=k}^{N-1} r(x_{DT,l},u_{DT,l},d_{DT,l})\right], \quad k=0,\ldots,N-1 \\ V_N(x_{DT,N},N) = E[\phi_N(x_{DT,N})] \end{cases} \tag{7.2}$$

where cost-to-go is denoted as $r(x_{DT,k},u_{DT,k},D_{DT,k}) = Q_{DT}(x_{DT,k}) + u_{DT,k}^T R_{DT} u_{DT,k} - d_{DT,k}^T S_{DT} d_{DT,k}$, $\forall k = 0,\ldots,N-1$, NT_s is the final time instant, and $Q_{DT}(x) \geq 0$, $\phi_N(x) \geq 0$ and R_{DT}, S_{DT} are symmetric positive definite matrices. In contrast to the infinite-horizon design (Seiffertt et al. 2008), $\phi_N(x)$ is the terminal constraint that needs to be satisfied in the finite-horizon two-player zero-sum game optimal design. According to the dynamic programming technique (Werbos 1983; Bertsekas and Tsitsiklis 1996), Equation 7.2 can also be represented for $k = 0,\ldots,N-1$ as

$$\begin{aligned} V_k(x_{DT,k},k) &= E\left[r(x_{DT,k},u_{DT,k},d_{DT,k}) + \phi_N(x_{DT,N}) + \sum_{l=k}^{N-1} r(x_{DT,l},u_{DT,l},d_{DT,l})\right] \\ &= E\left[(Q_{DT}(x_{DT,k}) + u_{DT,k}^T R_{DT} u_{DT,k}) + V_{k+1}(x_{DT,k+1},k+1)\right] \end{aligned} \tag{7.3}$$

Similar to Dierks and Jagannathan (2012), when $x = 0$, $V_k(x,k) = 0$, the value function $V_k(x,k)$ serves as the Lyapunov function (Lewis and Syrmos 1995; Sarangapani 2006). Based on the Bellman principle of optimality (Lewis and Syrmos 1995; Sarangapani 2006), the optimal value function also satisfies the discrete-time HJI equation as

$$\begin{cases} V^*(x_{DT,k},k) = \min_{u_{DT,k}} \max_{d_{DT,k}} (V(x_{DT,k},k)) \\ \qquad = \min_{u_{DT,k}} \max_{d_{DT,k}} \left[E[Q_{DT}(x_{DT,k}) + u_{DT,k}^T R_{DT} u_{DT,k} - d_{DT,k}^T S_{DT} d_{DT,k} + V^*(x_{DT,k+1})]\right] \\ V^*(x_{DT,N},N) = E[\phi_N(x_{DT,N})] \end{cases} \tag{7.4}$$

For the finite-horizon nonlinear two-player zero-sum game, controller u_k is viewed as the player minimizing the cost function while the disturbance d_k maximizes the cost. According to Burgin (1969), Basar and Olsder (1995), and Al-Tamimi et al. (2007), this two-player zero-sum game optimal design has a unique solution while the Nash condition holds and is given by

$$\min_u \max_d V(x,0) = \max_d \min_u V(x,0) \tag{7.5}$$

According to Burgin (1969), Basar and Olsder (1995), and Al-Tamimi et al. (2007), the optimal control and disturbance $u^*_{DT,k}, d^*_{DT,k}$ can be derived by differentiating Equation 7.4 as

$$E\left[\frac{\partial(Q_{DT}(x_{DT,k}) + u^T_{DT,k}R_{DT}u_{DT,k} - d^T_{DT,k}S_{DT}d_{DT,k})}{\partial u_{DT,k}} + \frac{\partial x^T_{DT,k+1}}{\partial u_{DT,k}}\frac{\partial V^*(x_{DT,k+1})}{\partial x_{DT,k+1}}\right] = 0 \tag{7.6}$$

and

$$E\left[\frac{\partial(Q_{DT}(x_{DT,k}) + u^T_{DT,k}R_{DT}u_{DT,k} - d^T_{DT,k}S_{DT}d_{DT,k})}{\partial d_{DT,k}} + \frac{\partial x^T_{DT,k+1}}{\partial d_{DT,k}}\frac{\partial V^*(x_{DT,k+1})}{\partial x_{DT,k+1}}\right] = 0$$

In other words, the certainty-equivalent control policies are given by

$$\begin{cases} E[u^*(x_{DT,k})] = -\dfrac{1}{2}E\left[R^{-1}_{DT}g^T_{DT}(x_{DT,k})\dfrac{\partial V^*(x_{DT,k+1}, k+1)}{\partial x_{DT,k+1}}\right], & k = 0,\ldots,N-1 \\ E[u^*(x_{DT,N-1})] = -\dfrac{1}{2}E\left[R^{-1}_{DT}g^T_{DT}(x_{DT,N})\dfrac{\partial \phi_N(x_{DT,N}, N)}{\partial x_{DT,N}}\right] \end{cases} \tag{7.7}$$

and

$$\begin{cases} \underset{\tau,\gamma}{E}\left[d^*(x_{DT,k})\right] = \dfrac{1}{2}E\left[S^{-1}_{DT}h^T_{DT}(x_{DT,k})\dfrac{\partial V^*(x_{DT,k+1}, k+1)}{\partial x_{DT,k+1}}\right], & k = 0,\ldots,N-1 \\ \underset{\tau,\gamma}{E}\left[d^*(x_{DT,N-1})\right] = \dfrac{1}{2}E\left[S^{-1}_{DT}h^T_{DT}(x_{DT,N})\dfrac{\partial \phi_N(x_{DT,N}, N)}{\partial x_{DT,N}}\right] \end{cases}$$

Substituting Equation 7.7 into Equation 7.4, the certainty-equivalent discrete-time HJI Equation 7.4 is given by

$$V^*(x_{DT,k},k) = E\left[Q_{DT}(x_{DT,k}) + \frac{1}{4}\frac{\partial V^{*T}(x_{DT,k+1})}{\partial x_{DT,k+1}} g_{DT}(x_{DT,k}) \times R_{DT}^{-1} g_{DT}^T(x_{DT,k}) \frac{\partial V^{*T}(x_{DT,k+1})}{\partial x_{DT,k+1}} - \frac{1}{4}\frac{\partial V^{*T}(x_{DT,k+1})}{\partial x_{DT,k+1}} h_{DT}(x_{DT,k}) \times S_{DT}^{-1} h_{DT}^T(x_{DT,k}) \frac{\partial V^{*T}(x_{DT,k+1})}{\partial x_{DT,k+1}} + V^*(x_{DT,k+1},k+1)\right], \quad k = 0,\dots,N-1 \tag{7.8}$$

and

$$V^*(x_{DT,N-1},N-1) = E\left[Q_{DT}(x_{DT,N-1}) + \frac{1}{4}\frac{\partial \phi_N^T(x_{DT,N})}{\partial x_{DT,N}} g_{DT}(x_{DT,N-1}) \times R_{DT}^{-1} g_{DT}^T(x_{DT,N-1}) \frac{\partial V^{*T}(x_{DT,N})}{\partial x_{DT,N}} - \frac{1}{4}\frac{\partial V^{*T}(x_{DT,N})}{\partial x_{DT,N}} h_{DT}(x_{DT,N-1}) \times S_{DT}^{-1} h_{DT}^T(x_{DT,N-1}) \frac{\partial V^{*T}(x_{DT,N})}{\partial x_{DT,N}} + \phi_N(x_{DT,N})\right], \quad k = 0,\dots,N-1$$

Remark 7.1

Since future system state x_{k+1} and system dynamics are unknown at time kT_s, it is difficult to solve discrete-time HJI. Moreover, there is no closed-form solution for the DT HJI equation (Burgin 1969; Basar and Olsder 1995; Al-Tamimi et al. 2007) even when the dynamics are known. To overcome these deficiencies, value and policy iteration-based schemes are normally utilized (Heydari and Balakrishnan 2011; Wang et al. 2011). However, iteration-based methods are unsuitable for real-time control due to the need for a large number of iterations within a sampling interval. An inadequate number of iterations could lead to instability (Dierks and Jagannathan 2012). Therefore, the time-based NDP finite-horizon optimal design is presented for the NNCS two-player zero-sum game in Section 7.3.

7.2 NNCS TWO-PLAYER ZERO-SUM GAME

The block diagram of NNCS is shown as Figure 3.1 where the feedback control loop is closed by using a communication network.

Using the Assumption 3.1 and incorporating the network-induced delays and packet losses, the nonlinear two-player zero-sum game in affine form can be represented as

$$\dot{x}(t) = f(x(t)) + \gamma(t)g(x(t))u(t-\tau(t)) + \gamma(t)h(x(t))d(t-\tau(t)) \tag{7.9}$$

with

$$\gamma(t) = \begin{cases} \mathbf{I}^{n\times n} & \text{if control input is received by the actuator at time } t \\ \mathbf{0}^{n\times n} & \text{if control input is lost at time } t \end{cases}$$

$\mathbf{I}^{n\times n}$ is $n \times n$ identity matrix, $x(t) \in \Re^{n\times n}$, $u(t) \in \Re^{m\times m}$, $d(t) \in \Re^{l\times l}$, $f(x) \in \Re^{n\times n}$, $g(x) \in \Re^{n\times m}$, and $h(x) \in \Re^{n\times l}$ represent the system state, control inputs, disturbance inputs, nonlinear internal dynamics, control coefficient, and disturbance coefficient matrices, respectively.

Similar to Xu and Jagannathan (2013), integrating Equation 7.9 over a sampling interval $[kT_s,(k+1)T_s)$ with network-induced delays and packet losses, the NNCS two-player zero-sum game can be represented as

$$x_{k+1} = X_{\tau,\gamma}(x_k,u_{k-1},\ldots,u_{k-\bar{b}},d_{k-1},\ldots,d_{k-\bar{b}}) + P_{\tau,\gamma}(x_k,u_{k-1},\ldots,u_{k-\bar{b}},d_{k-1},\ldots,d_{k-\bar{b}})u_k$$
$$+ K_{\tau,\gamma}(x_k,u_{k-1},\ldots,u_{k-\bar{b}},d_{k-1},\ldots,d_{k-\bar{b}})d_k \tag{7.10}$$

where $\bar{b}T_s$ is the upper bound on the network-induced delay, T_s is the sampling interval, $x(kT_s) = x_k$, $u((k-i)T_s) = u_{k-i}$, $d((k-i)T_s) = d_{k-i}$, $\forall i = 0,1,\ldots,\bar{b}$ are discretized NNCS system state, previous control inputs, and disturbance signals. Different from Xu and Jagannathan (2013), the $X_{\tau,\gamma}(\cdot)$, $P_{\tau,\gamma}(\cdot)$, $K_{\tau,\gamma}(\cdot)$ for the NNCS two-play zero-sum game have been derived as

$$X_{\tau,\gamma}(x_k,u_{k-1},\ldots,u_{k-\bar{b}},d_{k-1},\ldots,d_{k-\bar{b}}) = x_k + \int_{kT_s}^{(k+1)T_s} f(x(t))\,dt$$
$$+\gamma_{k-\bar{b}}\left(\int_{kT_s}^{\tau_d-\bar{b}T_s} g(x(t))\,dt\right)u_{k-\bar{b}} + \cdots + \gamma_{k-1}\left(\int_{\tau_2-2T_s}^{\tau_1-T_s} g(x(t))\,dt\right)u_{k-1}$$
$$+\gamma_{k-\bar{b}}\left(\int_{kT_s}^{\tau_d-\bar{b}T_s} h(x(t))\,dt\right)d_{k-\bar{b}} + \cdots + \gamma_{k-1}\left(\int_{\tau_2-2T_s}^{\tau_1-T_s} h(x(t))\,dt\right)d_{k-1}$$

$$P_{\tau,\gamma}(x_k,u_{k-1},\ldots,u_{k-\bar{b}},d_{k-1},\ldots,d_{k-\bar{b}}) = \gamma_k\left(\int_{\tau_0}^{(k+1)T_s} g(x(t))\,dt\right)$$

and

$$K_{\tau,\gamma}(x_k,u_{k-1},\ldots,u_{k-\bar{b}},d_{k-1},\ldots,d_{k-\bar{b}}) = \gamma_k\left(\int_{\tau_0}^{(k+1)T_s} h(x(t))\,dt\right)$$

To simplifying representation (7.10), a new augment state variable is defined as $z_k = \left[x_k^T \; u_{k-1}^T \ldots u_{k-\bar{b}}^T \; d_{k-1}^T \ldots d_{k-\bar{b}}^T\right]^T \in \Re^{n+\bar{d}(m+l)}$. Then, Equation 7.10 can be expressed equivalently as

$$z_{k+1} = F(z_k) + G(z_k)u_k + H(z_k)d_k \tag{7.11}$$

where $I_m \in \Re^{m\times m}$, $I_l \in \Re^{l\times l}$ are the identity matrices,

$$F(z_k) = \begin{bmatrix} Z_{\tau,\gamma}(z_k) \\ 0 \\ u_{k-1} \\ \vdots \\ u_{k-\bar{b}} \\ d_{k-1} \\ \vdots \\ d_{k-\bar{b}} \end{bmatrix}, \; G(z_k) = \begin{bmatrix} P_{\tau,\gamma}(z_k) \\ I_m \\ 0 \\ 0 \\ \vdots \\ 0 \\ \vdots \\ 0 \end{bmatrix}, \text{ and } H(z_k) = \begin{bmatrix} K_{\tau,\gamma}(z_k) \\ 0 \\ \vdots \\ 0 \\ I_l \\ 0 \\ \vdots \\ 0 \end{bmatrix}$$

Further, note that $F(\cdot)$, $G(\cdot)$, and $H(\cdot)$ represent the NNCS two-player zero-sum game internal dynamics, control, and disturbance coefficient matrices with $\|G(z_k)\|_F \le G_M$ and $\|H(z_k)\|_F \le H_M$ (Dierks and Jagannathan 2012) where $\|\cdot\|_F$ denotes the Frobenius norm (Dierks and Jagannathan 2012) and G_M, H_M are positive constants $\|H(z_k)\|_F \le H_M$ (Dierks and Jagannathan 2012). Moreover, since network-induced delays and packet losses have been incorporated into the NNCS two-player zero-sum game representation, Equation 7.11 becomes uncertain and stochastic, thus needing adaptive control methods. Thus, the objective of this chapter is to develop an optimal adaptive scheme for the NNCS two-player zero-sum game using Equation 7.11.

7.3 FINITE-HORIZON OPTIMAL ADAPTIVE DESIGN

In this section, a novel time-based NDP technique is utilized to obtain the finite-horizon stochastic optimal design for the NNCS two-player zero-sum game in the presence of network imperfections. First, an online identifier is developed to obtain the control and disturbance coefficient matrices. Then, a critic NN is proposed to approximate the time-varying stochastic value function within the finite horizon. Eventually, by using action NNs, identified NNCS two-player zero-sum game dynamics and estimated stochastic value function, the novel finite-horizon stochastic optimal design of the NNCS two-player zero-sum game has been derived. All the details are given below.

7.3.1 Online NN Identifier Design

Recalling Equation 7.7, the control and disturbance coefficient matrices are needed for deriving the optimal design of the nonlinear two-player zero-sum game (Burgin 1969; Basar and Olsder 1995; Al-Tamimi et al. 2007). However, these coefficient matrices are not known in a practical system, which need to be estimated. To circumvent this issue, a novel NN-based identifier is proposed to estimate the NNCS two-player zero-sum game control and disturbance coefficient matrices (i.e., $G(z)$, $H(z)$).

According to Xu and Jagannathan (2013) and the universal NN function approximation property, the NNCS two-player zero-sum game internal dynamics, control, and disturbance coefficient matrices can be represented on a compact set Ω as

$$
\begin{aligned}
\underset{\tau,\gamma}{E}[F(z_k)] &= \underset{\tau,\gamma}{E}[W_F^T \upsilon_F(z_k) + \varepsilon_{F,k}] \quad \forall k = 0,1,\ldots,N \\
\underset{\tau,\gamma}{E}[G(z_k)] &= \underset{\tau,\gamma}{E}[W_G^T \upsilon_G(z_k) + \varepsilon_{G,k}] \quad \forall k = 0,1,\ldots,N \\
\underset{\tau,\gamma}{E}[H(z_k)] &= \underset{\tau,\gamma}{E}[W_H^T \upsilon_H(z_k) + \varepsilon_{H,k}] \quad \forall k = 0,1,\ldots,N
\end{aligned}
\tag{7.12}
$$

where $W_F \in \Re^{p_f \times n}$, $W_G \in \Re^{p_g \times n}$, $W_H \in \Re^{p_h \times n}$ denote the target NN weights, $\upsilon_F(\cdot) \in \Re^{p_f}$, $\upsilon_G(\cdot) \in \Re^{p_g}$, $\upsilon_H(\cdot) \in \Re^{p_h}$ are activation functions, and $\varepsilon_{F,k} \in \Re^{p_f}$, $\varepsilon_{G,k} \in \Re^{p_g}$, $\varepsilon_{H,k} \in \Re^{p_h}$ are reconstruction errors, respectively.

Substituting Equation 7.12 into the NNCS two-player zero-sum game dynamics (7.11), we get

$$
\begin{aligned}
\underset{\tau,\gamma}{E}(z_k) &= \underset{\tau,\gamma}{E}[F(z_{k-1}) + G(z_{k-1})u_{k-1} + H(z_{k-1})d_{k-1}] \\
&= \underset{\tau,\gamma}{E}\left[[W_F^T \; W_G^T \; W_H^T] \begin{bmatrix} \upsilon_F(z_{k-1}) & 0 & 0 \\ 0 & \upsilon_G(z_{k-1}) & 0 \\ 0 & 0 & \upsilon_H(z_{k-1}) \end{bmatrix} \begin{bmatrix} 1 \\ u_{k-1} \\ d_{k-1} \end{bmatrix} + \varepsilon_{I,k-1} \right] \\
&= \underset{\tau,\gamma}{E}[W_I^T \upsilon_I(z_{k-1})\beta_{k-1} + \varepsilon_{I,k-1}], \quad \forall k = 0,1,\ldots,N
\end{aligned}
\tag{7.13}
$$

where $W_I = [W_F^T \; W_G^T \; W_H^T]^T$ and $\upsilon_I(z_{k-1}) = diag[\upsilon_F(z_{k-1}) \; \upsilon_G(z_{k-1}) \; \upsilon_H(z_{k-1})]$ are NN identifier target weight matrices and activation function, respectively, $\beta_{k-1} = [1 \; u_{k-1}^T \; d_{k-1}^T]^T \in \Re^{1+m+l}$, historical control and disturbance inputs u_{k-1}, d_{k-1} is defined as augment input at time $(k-1)T_s$, and $\varepsilon_{I,k}-1 = \varepsilon_{F,k-1} + \varepsilon_{G,k-1}u_{k-1} + \varepsilon_{H,k-1}d_{k-1}$ denotes the NN identifier reconstruction error. Moreover, $\underset{\tau,\gamma}{E}(\cdot)$ is the expectation operator (i.e., the mean value in this case) on network imperfections such as network-induced delays and packet losses.

Since the NN activation function and augmented input from the previous time instants are bounded, the term $\left\| \underset{\tau,\gamma}{E}[\upsilon_I(z_{k-1})\beta_{k-1}] \right\|$ will also be bounded, that is, $\left\| \underset{\tau,\gamma}{E}[\upsilon_I(z_{k-1})\beta_{k-1}] \right\| \le \zeta_M$, where ζ_M is a positive constant. In addition, the NN identifier reconstruction error is considered to be bounded such that $\left\| \underset{\tau,\gamma}{E}[\varepsilon_{I,k-1}] \right\| \le \varepsilon_{I,M}$ (Sarangapani 2006; Dierks and Jagannathan 2012) where $\varepsilon_{I,M}$ is a positive constant. Using Equation 7.12 and given the NN activation functions $\upsilon_F(\cdot)$, $\upsilon_G(\cdot)$, $\upsilon_H(\cdot)$, and $\upsilon_I(\cdot)$, the control and disturbance coefficient matrices $G(z)$, $H(z)$ can be identified while the NN identifier weight, W_I, is being tuned. Next, the update law for the NN identifier has been derived.

The NNCS two-player zero-sum game system state z_k can be approximated by using the NN identifier as

$$\underset{\tau,\gamma}{E}(\hat{z}_k) = \underset{\tau,\gamma}{E}\left[\hat{W}_{I,k}^T \upsilon_I(z_{k-1})\beta_{k-1}\right], \quad \forall k = 0,1,\ldots,N \tag{7.14}$$

where $\hat{W}_{I,k}$ is the estimated weight matrix of the NN identifier at time kT_s, and $\underset{\tau,\gamma}{E}[\bar{\upsilon}_I(z_k)\beta_k]$ is the activation function of the NN identifier.

Next, the identification error can be expressed from Equations 7.13 and 7.14 as

$$\underset{\tau,\gamma}{E}(e_{I,k}) = \underset{\tau,\gamma}{E}(z_k - \hat{z}_k) = \underset{\tau,\gamma}{E}(z_k) - \underset{\tau,\gamma}{E}\left[\hat{W}_{I,k}^T \upsilon_I(z_{k-1})\beta_{k-1}\right] \tag{7.15}$$

In addition, the dynamics of the identification error can be derived as

$$\underset{\tau,\gamma}{E}(e_{I,k+1}) = \underset{\tau,\gamma}{E}(z_{k+1}) - \underset{\tau,\gamma}{E}\left[\hat{W}_{I,k+1}^T \upsilon_I(z_k)\beta_k\right], \quad \forall k = 0,1,\ldots,N-1 \tag{7.16}$$

Next, similar to Dierks and Jagannathan (2012) and using the history of the NNCS two-player zero-sum game, the auxiliary identification error vector can be represented as

$$\underset{\tau,\gamma}{E}(\Xi_{I,k}) = \underset{\tau,\gamma}{E}\left(Z_k - \hat{Z}_k\right) = \underset{\tau,\gamma}{E}(Z_k) - \underset{\tau,\gamma}{E}\left[\hat{W}_{I,k}^T \bar{\upsilon}_I(Z_{k-1})\bar{\beta}_{k-1}\right] \tag{7.17}$$

with $Z_k = [z_k \; z_{k-1} \ldots z_{k+1-i}]$, $\bar{\upsilon}_I(Z_{k-1}) = [\upsilon_I(z_{k-1}) \; \upsilon_I(z_{k-2}) \ldots \upsilon_I(z_{k-i})]$, $\bar{\beta}_{k-1} = diag[\beta_{k-1}^T \ldots \beta_{k-i}^T]^T$, and $\bar{\varepsilon}_{I,k-1} = [\varepsilon_{I,k-1} \ldots \varepsilon_{I,k-i}]$ with $0 < i < k-1$. The i previous identification errors (7.15) are recomputed by using the most recent estimated NN identifier weights (i.e., $\underset{\tau,\gamma}{E}(\hat{W}_{I,k})$).

Next, the auxiliary identification error dynamics can be derived as

$$\underset{\tau,\gamma}{E}(\Xi_{I,k+1}) = \underset{\tau,\gamma}{E}(Z_{k+1}) - \underset{\tau,\gamma}{E}\left[\hat{W}_{I,k+1}^T \bar{\upsilon}_I(Z_k)\bar{\beta}_k\right], \quad \forall k = 0,1,\ldots,N-1 \tag{7.18}$$

For tuning NN identifier target weights within the finite horizon, the update law for $\underset{\tau,\gamma}{E}\left(\hat{W}_{I,k}\right)$ can be expressed as

$$\underset{\tau,\gamma}{E}\left(\hat{W}_{I,k+1}\right) = \underset{\tau,\gamma}{E}[\bar{U}_k \bar{\upsilon}_I(Z_k)(\bar{\upsilon}_I^T(Z_k)\bar{U}_k^T \bar{U}_k \bar{\upsilon}_I(Z_k))^{-1}(Z_k - \alpha_I \Xi_{I,k})^T] \tag{7.19}$$

where the tuning parameter α_I satisfies $0 < \alpha_I < 1$. Utilizing the update law (7.19) into the auxiliary error dynamics (7.17), the auxiliary error dynamics $\underset{\tau,\gamma}{E}(\Xi_{I,k+1})$ can be represented as

$$\underset{\tau,\gamma}{E}(\Xi_{I,k+1}) = \alpha_I \underset{\tau,\gamma}{E}(\Xi_{I,k}), \quad \forall k = 0,1,\ldots,N-1 \tag{7.20}$$

In order to learn the NNCS control and disturbance coefficient matrices $G(z)$, $H(z)$ by using the proposed online NN identifier, $\underset{\tau,\gamma}{E}[\bar{\upsilon}_I(Z_k)\bar{\beta}_k]$ has to be persistently exciting (PE) (Dierks and Jagannathan 2012; Xu and Jagannathan 2012) long enough. In other words, there exists a positive constant $\zeta_{\min}$ such that $0 < \zeta_{\min} \le \left\| \underset{\tau,\gamma}{E}[\bar{\upsilon}_I(Z_k)\bar{\beta}_k] \right\|$ holds for $k = 0,1,\ldots,N$.

Next, identification error dynamics (7.16) can be expressed as

$$\underset{\tau,\gamma}{E}(e_{I,k+1}) = \underset{\tau,\gamma}{E}\left[\tilde{W}_{I,k+1}^T \upsilon_I(z_k)\beta_k + \varepsilon_{I,k}\right], \quad \forall k = 0,1,\ldots,N-1 \tag{7.21}$$

where $\tilde{W}_{I,k} = W_I - \hat{W}_{I,k}$ is the NN identifier weight estimation error at time kT_s. Using the NN identifier update law and Equation 7.21, the identifier weight estimation error dynamics can be derived as

$$\underset{\tau,\gamma}{E}\left(\tilde{W}_{I,k+1}^T \upsilon_I(z_k)\beta_k\right) = \underset{\tau,\gamma}{E}\left(\alpha_I\left(\tilde{W}_{I,k}^T \upsilon_I(z_k)\beta_{k-1}\right) + \alpha_I \varepsilon_{I,k-1} - \varepsilon_{I,k}\right) \tag{7.22}$$

Next, the stability of the NN identification error (7.15) and weight estimation errors $\underset{\tau,\gamma}{E}\left(\tilde{W}_{I,k}\right)$ will be considered.

Theorem 7.1: Boundedness in the Mean for NN Identifier

Given the initial NN identifier weight matrix $W_{I,0}$ resides in a compact set Ω, let the proposed NN identifier be defined as Equation 7.14, its update law be given as Equation 7.19 and let $\underset{\tau,\gamma}{E}[\bar{\upsilon}_I(z_k)\bar{\beta}_k]$ satisfy the PE condition within the finite horizon (i.e., $t \in [0, NT_s]$). Then, there exists a positive tuning parameter α_I satisfying $0 < \alpha_I < \left(\zeta_{\min}/\sqrt{2}\zeta_M\right)$ such that the identification error (7.15) and the NN identifier weight estimation error $\underset{\tau,\gamma}{E}\left(\tilde{W}_{I,k}\right)$ are all bounded in the mean.

Proof: Proof follows similar to Xu and Jagannathan (2013).

7.3.2 Stochastic Value Function

According to the value function defined by Xu and Jagannathan (2013), the certainty-equivalent stochastic value function for the NNCS two-player zero-sum game can be expressed in terms of augment state z_k as

$$\begin{cases} V(z_k,k) = \underset{\tau,\gamma}{E}\left[\phi_N(z_N) + \sum_{l=k}^{N-1}(Q_z(z_l) + u_l^T R_z u_l - d_l^T S_z d_l)\right], & k = 0,\ldots,N-1 \\ V(z_N,N) = \underset{\tau,\gamma}{E}[\phi_N(z_N)] \end{cases} \tag{7.23}$$

where $Q_z(z_k) \geq 0$ and R_z, S_z are positive definite matrices. In contrast to the stochastic value function under the infinite-horizon case, a terminal constraint (i.e., $(V_N(z_N,N) = \underset{\tau,\gamma}{E}[\phi_N(z_N)])$) is incorporated while deriving the finite-horizon stochastic optimal design.

According to the universal approximation property of NN (Sarangapani 2006), the time-varying stochastic value function (7.23) can be represented by using a critic NN as

$$V(z_k,k) = \underset{\tau,\gamma}{E}(W_V^T \varphi(z_k,N-k) + \varepsilon_{V,k}), \quad \forall k = 0,1,\ldots,N \tag{7.24}$$

where $W_V \in \Re^r$, $\varepsilon_{V,k} \in \Re$ represent the critic NN target weight matrix and reconstruction error, respectively, and $\varphi(z_k,N-k) \in \Re$ denotes the time-dependent critic NN activation function. It is important to note that the activation function is explicitly dependent upon time and this makes the finite-horizon problem different and difficult over the infinite-horizon case (Seiffertt et al. 2008; Dierks and Jagannathan 2012). Moreover, the critic NN target weight and reconstruction error are bounded in the mean as $\left\|\underset{\tau,\gamma}{E}(W_V)\right\| \leq W_{VM}, \left\|\underset{\tau,\gamma}{E}(\varepsilon_{V,k})\right\| \leq \varepsilon_{VM}$ with W_{VM}, ε_{VM} being positive bounding constants. In addition, the reconstruction error gradient is assumed to be bounded in the mean as $\left\|\underset{\tau,\gamma}{E}(\partial\varepsilon_{V,k}/\partial z_k)\right\| \leq \varepsilon'_{VM}$ with ε'_{VM} being a positive constant (Dierks and Jagannathan 2012).

Next, the critic NN approximation of time-varying stochastic value function (7.24) can be expressed as

$$\hat{V}(z_k,k) = \underset{\tau,\gamma}{E}\left(\hat{W}_{V,k}^T \varphi(z_k,N-k)\right), \quad \forall k = 0,1,\ldots,N \tag{7.25}$$

where $\underset{\tau,\gamma}{E}\left(\hat{W}_{V,k}\right)$ is the estimated critic NN weight and time-dependent activation function $\varphi(z_k,N-k)$ has been selected from a basis function set whose elements in the set are linearly independent (Dierks and Jagannathan 2012). Also, since the activation function is continuous and smooth, two time independent functions $\varphi_{\min}(z_k)$, $\varphi_{\max}(z_k)$ can be found such that $\|\varphi_{\min}(z_k)\| \leq \|\varphi(z_k,N-k)\| \leq \|\varphi_{\max}(z_k)\|$, $k = 0,\ldots,N$. Moreover, the stochastic value function (7.24) can be bounded as

$$\left| \|\underset{\tau,\gamma}{E}(W_V)\| \|\varphi_{\min}(z_k)\| - \varepsilon_{VM} \right|^2 \leq \|V(z_k,k)\|^2 \leq \left| \|\underset{\tau,\gamma}{E}(W_V)\| \|\varphi_{\min}(z_k)\| + \varepsilon_{VM} \right|^2.$$

In addition, recalling Equations 7.15 and 7.16, the value function estimation error, $\tilde{V}(z_k,k)=\tilde{W}_{V,k}^T\varphi(z_k,N-k)+\varepsilon_{V,k}$, can also be bounded as

$$\left|\| \underset{\tau,\gamma}{E}\left(\tilde{W}_{V,k}\right)\|\|\varphi_{\min}(z_k)\| - \varepsilon_{VM}\right|^2 \le \left\|V(z_k,k)\right\|^2 \le \left|\| \underset{\tau,\gamma}{E}\left(\tilde{W}_{V,k}\right)\|\|\varphi_{\max}(z_k)\| + \varepsilon_{VM}\right|^2$$

Recalling HJI Equation 7.4, the following equation can be derived by substituting Equation 7.23 into Equation 7.4 as

$$\begin{aligned}\underset{\tau,\gamma}{E}(\varepsilon_{V,k}-\varepsilon_{V,k+1}) &= \underset{\tau,\gamma}{E}[W_V^T(\varphi(z_{k+1},N-k-1)-\varphi(z_k,N-k))]\\ &\quad+\underset{\tau,\gamma}{E}(z_k^TQ_zz_k+u_k^TR_zu_k-d_k^TS_zd_k),\quad \forall k=0,1,\ldots,N-1\end{aligned} \tag{7.26}$$

In the other words,

$$\underset{\tau,\gamma}{E}(W_V^T\Delta\varphi(z_k,N-k))+r(z_k,u_k,d_k)=\Delta\varepsilon_{V,k},\quad k=0,\ldots,N-1 \tag{7.27}$$

where $\Delta\varphi(z_k,N-k)=\varphi(z_{k+1},N-k-1)-\varphi(z_k,N-k)$, $r(z_k,u_k,d_k)=\underset{\tau,\gamma}{E}(z_k^TQ_zz_k+u_k^TR_zu_k-d_k^TS_zd_k)$, and $\Delta\varepsilon_{V,k}=\underset{\tau,\gamma}{E}(\varepsilon_{V,k}-\varepsilon_{V,k+1})$ with $\left\|\Delta\varepsilon_{V,k}\right\|=\Delta\varepsilon_{VM}$, $\forall k=0,\ldots,N-1$ (Dierks and Jagannathan 2012). However, when an estimated value of critic NN, $\hat{V}(z_k,k)$, from Equation 7.25 is utilized instead of ideal critic NN output, $V(z_k,k)$, as Equation 7.26 cannot hold. Then, using ideas similar to Dierks and Jagannathan (2012) and Xu and Jagannathan (2013), and incorporating delay values for convenience, the temporal difference (TD) error dynamics associated with Equation 7.26 are introduced as

$$\begin{aligned}\underset{\tau,\gamma}{E}(e_{HJI,k}) &= \underset{\tau,\gamma}{E}(z_k^TQ_zz_k+u_{k-1}^TR_zu_{k-1}-d_{k-1}^TS_zd_{k-1})+\hat{V}(z_{k+1},k+1)-\hat{V}(z_k,k)\\ &= \underset{\tau,\gamma}{E}(\hat{W}_{V,k}^T\Delta\psi(z_k,N-k))+r(z_k,u_k,d_k),\quad k=0,\ldots,N-1\end{aligned} \tag{7.28}$$

where $\underset{\tau,\gamma}{E}(e_{HJI,k})$ denotes the HJI equation TD error for the finite-horizon case (i.e., $t\in[0,NT_s]$). Moreover, since $r(z_k,u_k,d_k)=\Delta\varepsilon_{V,k}-\underset{\tau,\gamma}{E}(W_V^T\Delta\phi(z_k,N-k))$, $\forall k=0,1,\ldots,N$ (7.27), the HJI equation TD error dynamics can be derived as

$$\begin{aligned}\underset{\tau,\gamma}{E}(e_{HJI,k}) &= \underset{\tau,\gamma}{E}\left(\hat{W}_{V,k}^T\Delta\varphi(z_k,N-k)\right)-\underset{\tau,\gamma}{E}(W_V^T\Delta\varphi(z_k,N-k))+\Delta\varepsilon_{V,k}\\ &= -\underset{\tau,\gamma}{E}\left(\tilde{W}_{V,k}^T\Delta\varphi(z_k,N-k)\right)+\Delta\varepsilon_{V,k},\ k=0,\ldots,N-1\end{aligned} \tag{7.29}$$

where $\underset{\tau,\gamma}{E}\left(\tilde{W}_{V,k}\right)=\underset{\tau,\gamma}{E}(W_V)-\underset{\tau,\gamma}{E}\left(\hat{W}_{V,k}\right)$ represents the critic NN weight estimation error.

Next, in order to incorporate the terminal constraint, the estimation error $\underset{\tau,\gamma}{E}(e_{FC,k})$ can be defined as

$$\underset{\tau,\gamma}{E}(e_{FC,k}) = \underset{\tau,\gamma}{E}\left[\phi_N(z_N)\right] - \underset{\tau,\gamma}{E}(\hat{W}_{V,k}^T\varphi(\hat{z}_{N,k},0)), \quad \forall k = 0,1,\ldots,N \tag{7.30}$$

where $\hat{z}_{N,k}$ is the estimated final NNCS two-player zero-sum game system state at time kT_s by using the NN identifier (i.e., $\hat{F}(\cdot),\hat{G}(\cdot),\hat{H}(\cdot)$). Since $\underset{\tau,\gamma}{E}[\varphi_N(z_N)] = V(z_N,0) = \underset{\tau,\gamma}{E}(W_V^T\varphi(z_N,0)+\varepsilon_{V,0})$, the terminal constraint estimation error (i.e., $\underset{\tau,\gamma}{E}(e_{FC,k})$) will be expressed in terms of critic NN weight estimation error as

$$\begin{aligned}\underset{\tau,\gamma}{E}(e_{FC,k}) &= \underset{\tau,\gamma}{E}(W_V^T\varphi(z_N,0)+\varepsilon_{V,0}) - \underset{\tau,\gamma}{E}\left(\hat{W}_{V,k}^T\varphi(\hat{z}_{N,k},0)\right) \\ &= \underset{\tau,\gamma}{E}\left(\tilde{W}_{V,k}^T\varphi(\hat{z}_{N,k},0)\right) + \underset{\tau,\gamma}{E}(W_V^T\tilde{\varphi}(\hat{z}_N,z_{N,k},0)) + \underset{\tau,\gamma}{E}(\varepsilon_{V,0})\end{aligned} \tag{7.31}$$

where $\tilde{\varphi}(z_N,\hat{z}_{N,k},0) = \varphi(z_N,0) - \varphi(\hat{z}_{N,k},0)$. Since the critic NN activation function $\varphi(\cdot)$ is bounded such that $\|\varphi(\cdot)\| \le \varphi_M$ (Dierks and Jagannathan 2012; Xu and Jagannathan 2013), $\tilde{\varphi}(z_N,\hat{z}_{N,k},0)$ will also be bounded as $\|\tilde{\varphi}(z_N,\hat{z}_{N,k},0)\| \le 2\varphi_M$ with φ_M being a positive constant.

Considering the HJI TD error and terminal constraint estimation error jointly and using the gradient descent technique, the update law for the critic NN weight can be derived as

$$\begin{aligned}\underset{\tau,\gamma}{E}\left(\hat{W}_{V,k+1}\right) &= \underset{\tau,\gamma}{E}\left(\hat{W}_{V,k}\right) + \alpha_V \underset{\tau,\gamma}{E}\left(\frac{\varphi(\hat{z}_{N,k},0)e_{FC,k}^T}{\varphi^T(\hat{z}_{N,k},0)\varphi(\hat{z}_{N,k},0)+1}\right) \\ &\quad - \alpha_V \underset{\tau,\gamma}{E}\left(\frac{\Delta\varphi(z_k,N-k)e_{HJI,k}^T}{\Delta\varphi^T(z_k,N-k)\Delta\varphi(z_k,N-k)+1}\right), \quad k = 0,\ldots,N-1\end{aligned} \tag{7.32}$$

Remark 7.2

While the NNCS augmented state becomes zero, $z = 0$, both the ideal stochastic value function (7.24) and the estimated critic NN (7.25) become zero. Therefore, the critic NN will stop updating once the NNCS system state converges to zero. Similar to other literature (Dierks and Jagannathan 2012; Xu and Jagannathan 2013), this could be given as a PE requirement for the input (i.e., u_k, d_k) to the critic NN. In other words, the system state has to be persistently exciting long enough in order for the critic NN to learn the stochastic value function within the finite time (i.e., $t \in [0, NT_s]$). Moreover, similar to Dierks and Jagannathan (2012), the PE requirement can be satisfied by introducing exploration noise such that $0 < \varphi_{\min} < \|\varphi_{\min}(z)\| \le \|\varphi(z,k)\|$ and $0 < \Delta\varphi_{\min} < \|\Delta\varphi_{\min}(z_k)\| \le \|\Delta\varphi(z_k,N-k)\|$ with $\varphi_{\min}$ and $\Delta\varphi_{\min}$ being positive constants (Xu and Jagannathan 2013).

By using the definition of critic NN weight estimation error $\underset{\tau,\gamma}{E}\left(\tilde{W}_{V,k}\right)$ and update law (7.32), the dynamics of $\underset{\tau,\gamma}{E}\left(\tilde{W}_{V,k}\right)$ can be represented as

$$\begin{aligned}\underset{\tau,\gamma}{E}\left(\tilde{W}_{V,k+1}\right) &= \underset{\tau,\gamma}{E}\left(\tilde{W}_{V,k}\right) - \alpha_V \underset{\tau,\gamma}{E}\left(\frac{\varphi(\hat{z}_{N,k},0)\varphi^T(\hat{z}_{N,k},0)\tilde{W}_{V,k}}{\varphi^T(\hat{z}_{N,k},0)\varphi(\hat{z}_{N,k},0)+1}\right)\\ &\quad - \alpha_V \underset{\tau,\gamma}{E}\left(\frac{\varphi(\hat{z}_{N,k},0)\tilde{\varphi}^T(z_N,\hat{z}_{N,k},0)W_V}{\varphi^T(\hat{z}_{N,k},0)\varphi(\hat{z}_{N,k},0)+1}\right) - \alpha_V \underset{\tau,\gamma}{E}\left(\frac{\varphi(\hat{z}_{N,k},0)\varepsilon_{V,0}}{\varphi^T(\hat{z}_{N,k},0)\varphi(\hat{z}_{N,k},0)+1}\right)\\ &\quad - \alpha_V \underset{\tau,\gamma}{E}\left(\frac{\Delta\varphi(z_k,N-k)\Delta\varphi^T(z_k,N-k)\tilde{W}_{V,k}}{\Delta\varphi^T(z_k,N-k)\Delta\varphi(z_k,N-k)+1}\right)\\ &\quad - \alpha_V \underset{\tau,\gamma}{E}\left(\frac{\Delta\varphi(z_k,N-k)\Delta\varepsilon_{V,k}}{\Delta\varphi^T(z_k,N-k)\Delta\varphi(z_k,N-k)+1}\right), \quad k=0,\ldots,N-1\end{aligned} \tag{7.33}$$

Next, the stability of the critic NN weight estimation error $\underset{\tau,\gamma}{E}\left(\tilde{W}_{V,k}\right)$ (7.33) is analyzed and demonstrated.

Theorem 7.2: Boundedness in the Mean of the Critic NN Weight Estimation Error

Given the initial NNCS admissible control and disturbance policy $u_0(z_k)$, $d_0(z_k)$, let critic NN weight update law be developed as Equation 7.32. Then, there exists a positive constant α_V satisfying $0<\alpha_V<(2-\chi)/(\chi+5)$ with $0<\chi=\left((\phi_{\min}^2+\Delta\phi_{\min}^2+2)/(\phi_{\min}^2+1)(\Delta\phi_{\min}^2+1)\right)<2$ such that the critic NN weight estimation error (7.33) is *bounded* in the mean within the finite horizon. Moreover, the bound is contingent on the values of final time (i.e., NT_s) and the initial bounded critic NN weight estimation error $B_{WV,0}$.
Proof: Refer to Appendix 7A.

Consider that the critic NN weight estimation error is initiated as a bound positive constant $B_{WV,0}$ with $\left\|\underset{\tau,\gamma}{E}\left(\tilde{W}_{V,0}\right)\right\| \le B_{WV,0}$, $\forall k=0,1,2,\ldots,N$. Using standard Lyapunov analysis (Sarangapani 2006), geometric sequence theory (Brockett et al. 1983; Dankert et al. 2006), and Equation 7A.2, the Lyapunov function can be expressed within the interval $t \in [0, NT_s]$ as

$$\begin{aligned}L_{V,k} &= \Delta L_{V,k-1} + \Delta L_{V,k-2} + \cdots + \Delta L_{V,0} + L_{V,0}\\ &= \sum_{j=0}^{k-1}(\Delta L_{V,j}) + L_{V,0}, \quad \forall k=0,1,\ldots,N\end{aligned} \tag{7.34}$$

Using Equation 7A.2, Equation 7.34 can be represented as

$$L_{V,k} = \sum_{j=0}^{k-1}\left[-2\alpha_V(2-\chi-(\chi+5)\alpha_V)\left\| \underset{\tau,\gamma}{E}\left(\tilde{W}_{V,k}\right)\right\|^2 + \varepsilon_{TV}^2\right] + L_{V,0}$$

$$\leq [1-2\alpha_V(2-\chi-(\chi+5)\alpha_V)]^k \left\| \underset{\tau,\gamma}{E}\left(\tilde{W}_{V,k}\right)\right\|^2$$

$$+\sum_{j=0}^{k-1}[1-2\alpha_V(2-\chi-(\chi+5)\alpha_V]^j \varepsilon_{TV}^2$$

$$\leq \eta^k \left\| \underset{\tau,\gamma}{E}\left(\tilde{W}_{V,0}\right)\right\|^2 + \frac{1-\eta^k}{1-\eta}\varepsilon_{TV}^2, \quad \forall k = 0,1,\ldots,N \tag{7.35}$$

where $\eta = 1 - 2\alpha_V(2 - \chi - (\chi + 5)\alpha_V)$. Recalling $L_{V,k} = \left\| \underset{\tau,\gamma}{E}\left(\tilde{W}_{V,k}\right)\right\|^2$, we have

$$\left\| \underset{\tau,\gamma}{E}\left(\tilde{W}_{V,k}\right)\right\| \leq \sqrt{\eta^k \left\| \underset{\tau,\gamma}{E}\left(\tilde{W}_{V,k}\right)\right\|^2 + \frac{1-\eta^k}{1-\eta}\varepsilon_{TV}^2}$$

$$\leq \sqrt{\eta^k B_{V,0}^2 + \frac{1-\eta^k}{1-\eta}\varepsilon_{TV}^2} \equiv B_{V,k}, \quad \forall k = 0,1,\ldots,N \tag{7.36}$$

where $0 < \alpha_V < (2 - \chi)/(\chi + 5)$ and $0 < \eta < 1$. Therefore, the term η^k will decrease when the final time kT_s increases. In addition, ultimate bound $B_{V,k}$ will also decrease with increasing kT_s. Namely, the bounded critic NN weight estimation errors will decrease over time. Additionally, when time goes to infinity, $k \to \infty$, the ultimate bound of the critic NN weight estimation error will be equal to the bound of the critic NN weight estimation error in the infinite-horizon NNCS optimal design (Xu and Jagannathan 2013).

7.3.3 Approximation of Optimal Control and Disturbance

Recalling the universal approximation property of NN (Sarangapani 2006), the ideal finite-horizon NNCS two-player zero-sum game stochastic optimal control and disturbance inputs can be expressed by using two action NNs as

$$\begin{aligned} \underset{\tau,\gamma}{E}\left[u^*(z_k)\right] &= \underset{\tau,\gamma}{E}[W_u^T \vartheta(z_k,k) + \varepsilon_{u,k}], \quad \forall k = 0,1,\ldots,N \\ \underset{\tau,\gamma}{E}\left[d^*(z_k)\right] &= \underset{\tau,\gamma}{E}[W_d^T \psi(z_k,k) + \varepsilon_{d,k}], \quad \forall k = 0,1,\ldots,N \end{aligned} \tag{7.37}$$

where $W_u \in \Re^{m\times s}$, $\varepsilon_{u,k} \in \Re^{m\times m}$ represent control action NN target weights and reconstruction error for control input, respectively, $W_d \in \Re^{l\times b}$, $\varepsilon_{d,k} \in \Re^{l\times l}$ denote disturbance action NN target weights and reconstruction error, and $\vartheta(z_k,k) \in \Re^{s\times m}$, $\psi(z_k,k) \in \Re^{l\times b}$ represent smooth and continuous time-varying activation function for control and disturbance action NNs, respectively. Moreover, time-independent functions $\vartheta_{\min}(z_k)$, $\vartheta_{\max}(z_k)$, $\psi_{\min}(z_k)$, and $\psi_{\max}(z_k)$ can be found such that $\|\vartheta_{\min}(z_k)\| \le \|\vartheta(z_k,k)\| \le \|\vartheta_{\max}(z_k)\|$ and $\|\psi_{\min}(z_k)\| \le \|\psi(z_k,k)\| \le \|\psi_{\max}(z_k)\|$, $k=0,\ldots,N$. Additionally, the ideal two action NNs weight matrices, activation functions, and reconstruction errors are all considered bounded (i.e., $\left\|\underset{\tau,\gamma}{E}(W_u)\right\| \le W_{uM}$, $\left\|\underset{\tau,\gamma}{E}(W_d)\right\| \le W_{dM}$, $\left\|\underset{\tau,\gamma}{E}(\vartheta(z_k,k))\right\| \le \vartheta_M$, $\left\|\underset{\tau,\gamma}{E}(\psi(z_k,k))\right\| \le \psi_M$, and $\left\|\underset{\tau,\gamma}{E}(\varepsilon_{u,k})\right\| \le \varepsilon_{uM}$, $\left\|\underset{\tau,\gamma}{E}(\varepsilon_{d,k})\right\| \le \varepsilon_{dM}$ with W_{uM}, W_{dM}, ϑ_M, ψ_M, and ε_{uM}, ε_{dM} are all positive constants (Dierks and Jagannathan 2012; Xu and Jagannathan 2013).

Next, the estimated action NNs for control and disturbance inputs can be represented as

$$\begin{aligned} \underset{\tau,\gamma}{E}\left[\hat{u}(z_k)\right] &= \underset{\tau,\gamma}{E}\left[\hat{W}_{u,k}^T\vartheta(z_k,k)\right], \quad \forall k=0,1,\ldots,N \\ \underset{\tau,\gamma}{E}\left[\hat{d}(z_k)\right] &= \underset{\tau,\gamma}{E}\left[\hat{W}_{d,k}^T\psi(z_k,k)\right], \quad \forall k=0,1,\ldots,N \end{aligned} \tag{7.38}$$

where $\underset{\tau,\gamma}{E}\left(\hat{W}_{u,k}\right)$, $\underset{\tau,\gamma}{E}\left(\hat{W}_{d,k}\right)$ denote the estimated weights for control and disturbance action NNs, respectively. Further, the action NNs estimation errors will be considered as the difference between the actual control and disturbance (7.38) inputs applied to the NNCS where the control policy is obtained by minimizing tuned stochastic value function whereas the disturbance policy can be obtained by maximizing the tuned stochastic value function (7.25) (i.e., $\min_u \max_d V(z)$) with identified control and disturbance coefficient matrices (i.e., $\hat{G}(z_k),\hat{H}(z_k)$) within the finite horizon, $t \in [0,kT_s]$. The estimation errors are expressed as

$$\underset{\tau,\gamma}{E}(e_{u,k}) = \underset{\tau,\gamma}{E}\left[\hat{W}_{u,k}^T\vartheta(z_k,k) + \frac{1}{2}R_z^{-1}\hat{G}^T(z_k)\frac{\partial\varphi^T(z_{k+1},N-k-1)}{\partial z_{k+1}}\hat{W}_{V,k}\right]$$

$$\underset{\tau,\gamma}{E}(e_{d,k}) = \underset{\tau,\gamma}{E}\left[\hat{W}_{d,k}^T\psi(z_k,k) - \frac{1}{2}S_z^{-1}\hat{H}^T(z_k)\frac{\partial\varphi^T(z_{k+1},N-k-1)}{\partial z_{k+1}}\hat{W}_{V,k}\right] \tag{7.39}$$

Using the gradient descent scheme, update law for the estimated action NNs weight matrices can be represented as

$$\begin{aligned} \underset{\tau,\gamma}{E}\left(\hat{W}_{u,k+1}\right) &= \underset{\tau,\gamma}{E}\left(\hat{W}_{u,k}\right) - \alpha_u \underset{\tau,\gamma}{E}\left[\frac{\vartheta(z_k,k)}{\vartheta^T(z_k,k)\vartheta(z_k,k)+1}e_{u,k}^T\right], \quad k=0,\ldots,N-1 \\ \underset{\tau,\gamma}{E}\left(\hat{W}_{d,k+1}\right) &= \underset{\tau,\gamma}{E}\left(\hat{W}_{d,k}\right) - \alpha_d \underset{\tau,\gamma}{E}\left[\frac{\psi(z_k,k)}{\psi^T(z_k,k)\psi(z_k,k)+1}e_{d,k}^T\right], \quad k=0,\ldots,N-1 \end{aligned} \tag{7.40}$$

where the action NNs tuning parameters α_u, α_d satisfy $0<\alpha_u<1$ and $0<\alpha_d<1$. In addition, the ideal action NNs designs (7.37) should be equal to the control and disturbance policies, respectively, which minimizes and maximizes the target stochastic value function (7.24), respectively, such that $\min_u \max_d V(z)$ holds, which becomes

$$\begin{aligned} \underset{\tau,\gamma}{E}[W_u^T\vartheta(z_k,k)+\varepsilon_{u,k}] &= -\frac{1}{2}\underset{\tau,\gamma}{E}\left[R_z^{-1}G^T(z_k)\frac{\partial\varphi^T(z_{k+1},N-k-1)W_V+\varepsilon_{V,k}^T}{\partial z_{k+1}}\right] \\ \underset{\tau,\gamma}{E}[W_d^T\psi(z_k,k)+\varepsilon_{d,k}] &= \frac{1}{2}\underset{\tau,\gamma}{E}\left[S_z^{-1}H^T(z_k)\frac{\partial\varphi^T(z_{k+1},N-k-1)W_V+\varepsilon_{V,k}^T}{\partial z_{k+1}}\right] \end{aligned} \tag{7.41}$$

Namely,

$$\begin{aligned} &\underset{\tau,\gamma}{E}\left[W_u^T\vartheta(z_k,k)+\varepsilon_{u,k}+\frac{1}{2}R_z^{-1}G^T(z_k)\left(\frac{\partial\varphi^T(z_{k+1},N-k-1)W_V}{\partial z_{k+1}}+\frac{\partial\varepsilon_{V,k}^T}{\partial z_{k+1}}\right)\right]=0 \\ &\underset{\tau,\gamma}{E}\left[W_d^T\psi(z_k,k)+\varepsilon_{d,k}-\frac{1}{2}S_z^{-1}H^T(z_k)\left(\frac{\partial\varphi^T(z_{k+1},N-k-1)W_V}{\partial z_{k+1}}+\frac{\partial\varepsilon_{V,k}^T}{\partial z_{k+1}}\right)\right]=0 \end{aligned} \tag{7.42}$$

Substituting Equation 7.42 into Equation 7.39, the action NNs estimation errors can be expressed equivalently as

$$\begin{aligned} \underset{\tau,\gamma}{E}(e_{u,k}) = &-\underset{\tau,\gamma}{E}\left[\tilde{W}_{u,k}^T\vartheta(z_k,k)\right]-\frac{1}{2}\underset{\tau,\gamma}{E}\left[R_z^{-1}\hat{G}^T(z_k)\frac{\partial\varphi^T(z_{k+1},N-k)}{\partial z_{k+1}}\tilde{W}_{V,k}\right] \\ &-\frac{1}{2}\underset{\tau,\gamma}{E}\left[R_z^{-1}\tilde{G}^T(z_k)\frac{\partial\varphi^T(z_{k+1},N-k)}{\partial z_{k+1}}W_V\right] \\ &-\frac{1}{2}\underset{\tau,\gamma}{E}\left[R_z^{-1}G^T(z_k)\frac{\partial\varepsilon_{V,k}^T}{\partial z_{k+1}}-\varepsilon_{u,k}\right],\quad \forall k=0,1,\ldots,N \end{aligned} \tag{7.43}$$

and

$$\begin{aligned} \underset{\tau,\gamma}{E}(e_{d,k}) = &-\underset{\tau,\gamma}{E}\left[\tilde{W}_{d,k}^T\psi(z_k,k)\right]+\frac{1}{2}\underset{\tau,\gamma}{E}\left[S_z^{-1}\hat{H}^T(z_k)\frac{\partial\phi^T(z_{k+1},N-k)}{\partial z_{k+1}}\tilde{W}_{V,k}\right] \\ &+\frac{1}{2}\underset{\tau,\gamma}{E}\left[S_z^{-1}\tilde{H}^T(z_k)\frac{\partial\phi^T(z_{k+1},N-k)}{\partial z_{k+1}}W_V\right] \\ &+\frac{1}{2}\underset{\tau,\gamma}{E}\left[S_z^{-1}H^T(z_k)\frac{\partial\varepsilon_{V,k}^T}{\partial z_{k+1}}-\varepsilon_{d,k}\right],\quad \forall k=0,1,\ldots,N \end{aligned} \tag{7.44}$$

where $\underset{\tau,\gamma}{E}\left(\tilde{W}_{u,k}\right)=\underset{\tau,\gamma}{E}(W_u)-\underset{\tau,\gamma}{E}\left(\hat{W}_{u,k}\right)$, $\underset{\tau,\gamma}{E}\left(\tilde{W}_{d,k}\right)=\underset{\tau,\gamma}{E}(W_d)-\underset{\tau,\gamma}{E}\left(\hat{W}_{d,k}\right)$represent two action NNs weight estimation errors and $\tilde{G}(z_k)=G(z_k)-\hat{G}(z_k)$, $\tilde{H}(z_k)=H(z_k)-\hat{H}(z_k)$.

Then, substituting Equations 7.43 and 7.44 into Equation 7.40, the dynamics for the action NNs weight estimation errors can be represented as

$$\begin{aligned}
\underset{\tau,\gamma}{E}\left(\tilde{W}_{u,k+1}\right) &= \underset{\tau,\gamma}{E}\left(\tilde{W}_{u,k}\right)+\alpha_u \underset{\tau,\gamma}{E}\left[\frac{\vartheta(z_k,k)}{\vartheta^T(z_k,k)\vartheta(z_k,k)+1}e_{u,k}^T\right] \\
&= \underset{\tau,\gamma}{E}\left(\tilde{W}_{u,k}\right)-\alpha_u \underset{\tau,\gamma}{E}\left[\frac{\vartheta(z_k,k)}{\vartheta^T(z_k,k)\vartheta(z_k,k)+1}\tilde{W}_{u,k}\right] \\
&\quad -\frac{\alpha_u}{2}\underset{\tau,\gamma}{E}\left[\frac{\vartheta(z_k,k)}{\vartheta^T(z_k,k)\vartheta(z_k,k)+1}R_z^{-1}\hat{G}^T(z_k)\frac{\partial\varphi^T(z_{k+1},N-k-1)}{\partial z_{k+1}}\tilde{W}_{V,k}\right] \\
&\quad -\frac{\alpha_u}{2}\underset{\tau,\gamma}{E}\left[\frac{\vartheta(z_k,k)}{\vartheta^T(z_k,k)\vartheta(z_k,k)+1}R_z^{-1}\tilde{G}^T(z_k)\frac{\partial\varphi^T(z_{k+1},N-k-1)}{\partial z_{k+1}}W_V\right] \\
&\quad -\frac{\alpha_u}{2}\underset{\tau,\gamma}{E}\left[\frac{\vartheta(z_k,k)}{\vartheta^T(z_k,k)\vartheta(z_k,k)+1}R_z^{-1}G^T(z_k)\frac{\partial\varepsilon_{V,k}^T}{\partial z_{k+1}}-\varepsilon_{u,k}\right],\quad k=0,\ldots,N-1
\end{aligned} \tag{7.45}$$

and

$$\begin{aligned}
\underset{\tau,\gamma}{E}\left(\tilde{W}_{d,k+1}\right) &= \underset{\tau,\gamma}{E}\left(\tilde{W}_{d,k}\right)+\alpha_d \underset{\tau,\gamma}{E}\left[\frac{\psi(z_k,k)}{\psi^T(z_k,k)\psi(z_k,k)+1}e_{d,k}^T\right] \\
&= \underset{\tau,\gamma}{E}\left(\tilde{W}_{d,k}\right)-\alpha_d \underset{\tau,\gamma}{E}\left[\frac{\psi(z_k,k)}{\psi^T(z_k,k)\psi(z_k,k)+1}\tilde{W}_{d,k}\right] \\
&\quad +\frac{\alpha_d}{2}\underset{\tau,\gamma}{E}\left[\frac{\psi(z_k,k)}{\psi^T(z_k,k)\psi(z_k,k)+1}S_z^{-1}\hat{H}^T(z_k)\frac{\partial\varphi^T(z_{k+1},N-k-1)}{\partial z_{k+1}}\tilde{W}_{V,k}\right] \\
&\quad -\frac{\alpha_d}{2}\underset{\tau,\gamma}{E}\left[\frac{\psi(z_k,k)}{\psi^T(z_k,k)\psi(z_k,k)+1}S_z^{-1}\tilde{H}^T(z_k)\frac{\partial\varphi^T(z_{k+1},N-k-1)}{\partial z_{k+1}}W_V\right] \\
&\quad -\frac{\alpha_d}{2}\underset{\tau,\gamma}{E}\left[\frac{\psi(z_k,k)}{\psi^T(z_k,k)\psi(z_k,k)+1}S_z^{-1}H^T(z_k)\frac{\partial\varepsilon_{V,k}^T}{\partial z_{k+1}}-\varepsilon_{u,k}\right],\quad k=0,\ldots,N-1
\end{aligned} \tag{7.46}$$

It is worthwhile to observe that the requirement on the NNCS two-player zero-sum game control and disturbance coefficient matrices, $G(z_k)$, $H(z_k)$, have been relaxed due to the novel NN identifier, which is an important contribution compared to the recent literature (Vamvoudakis and Lewis 2010; Heydari and Balakrishnan 2011; Wang et al. 2011) where it is required.

Eventually, the closed-loop stability of the NNCS two-player zero-sum game with the proposed novel finite-horizon time-based NDP algorithm will be demonstrated.

7.3.4 Closed-Loop System Stability

In this section, we will prove that the closed-loop NNCS two-player zero-sum game system is *bounded* in the mean with the bounds being a function of initial conditions and final time. Additionally, while the final time is increased to infinity $k \to \infty$, the estimated control and disturbance policies converge close to the infinite-horizon optimal control input and disturbance signals, respectively. Before demonstrating the closed-loop stability theorem, the flowchart of the proposed novel time-based NDP finite-horizon optimal design for NNCS two-player zero-sum game with unknown network imperfections and system dynamics is presented in Figure 7.1.

Similar to the most recent literature (Dierks and Jagannathan 2012; Xu and Jagannathan 2013), the initial NNCS two-player zero-sum system state is considered to reside in a compact set Ω due to the initial admissible policy $u_0(z_k)$ and $d_0(z_k)$. In addition, the action NN activation functions, the critic NN activation function, and its gradient are all bounded in a compact set Ω (i.e., $\left\| \underset{\tau,\gamma}{E}(\varphi(z_k,k)) \right\| \le \varphi_M$, $\left\| \underset{\tau,\gamma}{E}\left[\frac{\partial \varphi(z_k,k)}{\partial z_k} \right] \right\| \le \varphi'_M$, $\left\| \underset{\tau,\gamma}{E}(\vartheta(z_k,k)) \right\| \le \vartheta_M$, and $\left\| \underset{\tau,\gamma}{E}(\psi(z_k,k)) \right\| \le \psi_M$ (Dierks and Jagannathan 2012; Xu and Jagannathan 2013). Moreover, the PE condition will be held by adding exploration noise (Dierks and Jagannathan 2012; Xu and Jagannathan 2013), and NN tuning parameters α_I, α_V and α_u, α_d will be chosen properly to ensure that all future system state remains in the compact set. In order to proceed, the following lemma is needed before demonstrating the main theorem. Table 7.1 presents the optimal design.

Lemma 7.1

Let a set of optimal control and disturbance policies be utilized to the NNCS two-player zero-sum game (7.11) such that Equation 7.11 is asymptotically stable in the mean (Xu and Jagannathan 2013). Then, the closed-loop NNCS two-player zero-sum game dynamics, $\underset{\tau,\gamma}{E}[F(z_k)+G(z_k)u^*(z_k)+H(z_k)d^*(z_k)]$, satisfy the following inequality for $k = 0,\ldots,N$.

$$\left\| \underset{\tau,\gamma}{E}[F(z_k)+G(z_k)u^*(z_k)+H(z_k)d^*(z_k)] \right\|^2 \le l_o \left\| \underset{\tau,\gamma}{E}(z_k) \right\|^2 \tag{7.47}$$

where $u^*(z_k)$, $d^*(z_k)$ are the optimal control input and disturbance signal where $0 < l_o < 1/2$ is a positive constant.

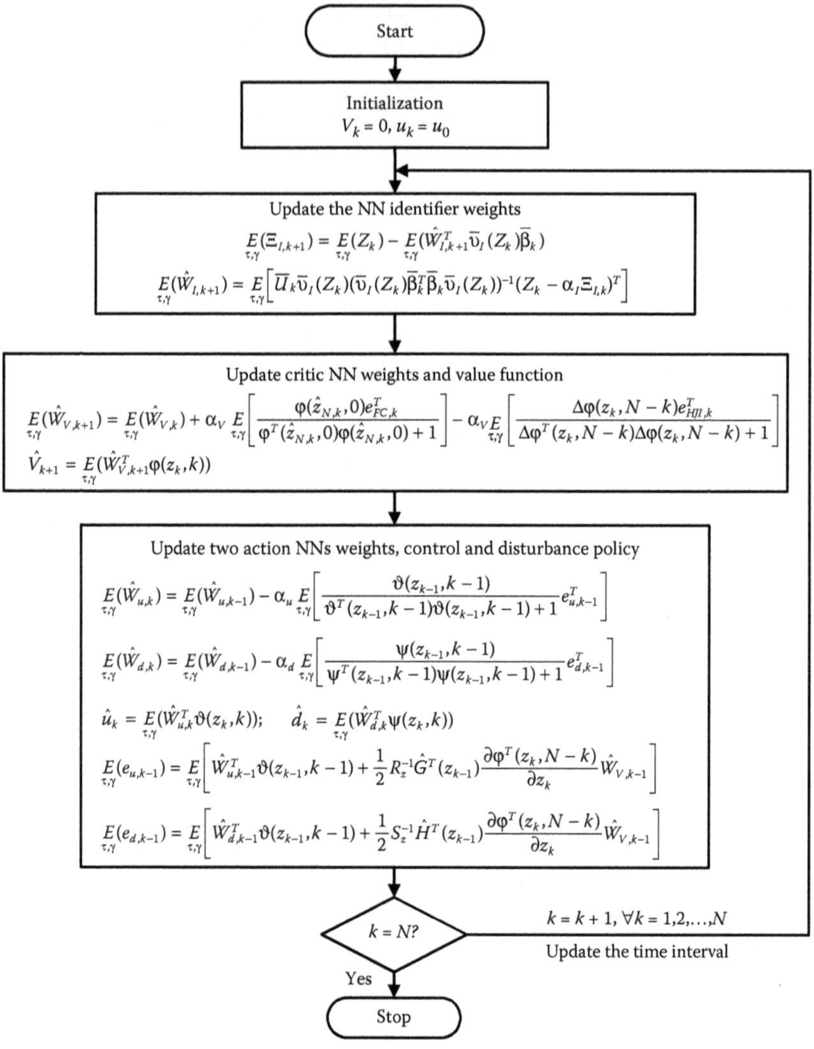

FIGURE 7.1 Flowchart of the proposed finite-horizon optimal design.

Theorem 7.3: Convergence of Stochastic Optimal Control Input and Disturbance

Given $u_0(z_k)$ and $d_0(z_k)$, denote an initial admissible control and disturbance policies for the NNCS two-player zero-sum game (7.11) such that Equation 7.47 holds with $0 < l_o < 1/2$. Consider the NN weight update laws for the identifier, critic, and two action NNs as Equations 7.19, 7.32, and 7.40, respectively, there exists four positive tuning parameters α_I, α_V, α_u, α_d satisfying $0 < \alpha_I < \min\left\{(1/2\zeta_M), \left(\zeta_{\min}/\sqrt{2}\zeta_M\right)\right\}$, $0 < \alpha_V < (2-\chi)/(\chi+5)$ with $0 < \chi = ((\psi_{min}^2 + \Delta\psi_{min}^2 + 2)/((\psi_{\min}^2+1)(\Delta\psi_{\min}^2+1))) < 2$, which is defined in Theorems 7.1 and 7.2 and $0 < \alpha_u < 1/3$, $0 < \alpha_d < 1/3$ such that the NNCS two-player zero-sum game system state $\underset{\tau,\gamma}{E}(z_k)$, identification error $\underset{\tau,\gamma}{E}(e_{I,k})$, NN identifier weight estimation error $\underset{\tau,\gamma}{E}\left(\tilde{W}_{I,k}\right)$, critic and two action NNs weight estimation errors $\underset{\tau,\gamma}{E}\left(\tilde{W}_{V,k}\right), \underset{\tau,\gamma}{E}\left(\tilde{W}_{u,k}\right), \underset{\tau,\gamma}{E}\left(\tilde{W}_{d,k}\right)$ are all bounded in the mean during the finite horizon (i.e., $t \in [0, NT_s]$). Further, the bounds, given in Equation 7B.11, are a function of final time, NT_s, bounded initial NNCS two-player zero-sum game system state B_z,0, bounded initial identification error B_I,0 and bounded initial weight estimation error for NN identifier, critic, and two action NNs $B_{WI,0}$, $B_{WV,0}$, $B_{Wu,0}$, $B_{Wd,0}$, respectively.

Proof: Refer to Appendix 7B.

TABLE 7.1
Finite-Horizon Stochastic Optimal Design

Compute Finite-Horizon Control Signal and Disturbance

$$\underset{\tau,\gamma}{E}[\hat{u}(z_k)] = \underset{\tau,\gamma}{E}\left[\hat{W}_{u,k}^T \vartheta(z_k,k)\right], \quad \forall k = 0,1,\ldots,N$$

$$\underset{\tau,\gamma}{E}[\hat{d}(z_k)] = \underset{\tau,\gamma}{E}\left[\hat{W}_{d,k}^T \psi(z_k,k)\right], \quad \forall k = 0,1,\ldots,N$$

where the update law are given as

$$\underset{\tau,\gamma}{E}\left(\hat{W}_{u,k+1}\right) = \underset{\tau,\gamma}{E}\left(\hat{W}_{u,k}\right) - \alpha_u \underset{\tau,\gamma}{E}\left[\frac{\vartheta(z_k,k)}{\vartheta^T(z_k,k)\vartheta(z_k,k)+1} e_{u,k}^T\right], \quad k = 0,\ldots,N-1$$

$$\underset{\tau,\gamma}{E}\left(\hat{W}_{d,k+1}\right) = \underset{\tau,\gamma}{E}\left(\hat{W}_{d,k}\right) - \alpha_d \underset{\tau,\gamma}{E}\left[\frac{\psi(z_k,k)}{\psi^T(z_k,k)\psi(z_k,k)+1} e_{d,k}^T\right], \quad k = 0,\ldots,N-1$$

Compute Hamilton–Jacobi–Isaacs Equation Residual Error and Terminal Constraint Estimation Error

HJI Equation Residual Error

$$\underset{\tau,\gamma}{E}(e_{HJI,k}) = \underset{\tau,\gamma}{E}\left(\hat{W}_{V,k}^T \Delta\varphi(z_k,N-k)\right) - \underset{\tau,\gamma}{E}(W_V^T \Delta\varphi(z_k,N-k)) + \Delta\varepsilon_{V,k}$$
$$= -\underset{\tau,\gamma}{E}\left(\tilde{W}_{V,k}^T \Delta\varphi(z_k,N-k)\right) + \Delta\varepsilon_{V,k}, k = 0,\ldots,N-1$$

Terminal Constraint Estimation Error

$$\underset{\tau,\gamma}{E}(e_{FC,k}) = \underset{\tau,\gamma}{E}(W_V^T \varphi(z_N,0) + \varepsilon_{V,0}) - \underset{\tau,\gamma}{E}\left(\hat{W}_{V,k}^T \varphi(\hat{z}_{N,k},0)\right)$$
$$= \underset{\tau,\gamma}{E}\left(\tilde{W}_{V,k}^T \varphi(\hat{z}_{N,k},0)\right) + \underset{\tau,\gamma}{E}(W_V^T \tilde{\varphi}(z_N,\hat{z}_{N,k},0)) + \underset{\tau,\gamma}{E}(\varepsilon_{V,0})$$

Parameter Update

$$\underset{\tau,\gamma}{E}\left(\hat{W}_{V,k+1}\right) = \underset{\tau,\gamma}{E}\left(\hat{W}_{V,k}\right) + \alpha_V \underset{\tau,\gamma}{E}\left(\frac{\varphi(\hat{z}_{N,k},0)e_{FC,k}^T}{\varphi^T(\hat{z}_{N,k},0)\varphi(\hat{z}_{N,k},0)+1}\right)$$
$$-\alpha_V \underset{\tau,\gamma}{E}\left(\frac{\Delta\varphi(z_k,N-k)e_{HJI,k}^T}{\Delta\varphi^T(z_k,N-k)\Delta\varphi(z_k,N-k)+1}\right), \quad k = 0,\ldots,N-1$$

where α_V is the tuning parameter.

7.4 SIMULATION RESULTS

In this section, the performance of the proposed stochastic optimal design for NNCS two-player zero-sum game in the presence of unknown system dynamics and network imperfections has been evaluated. Before demonstrating the results, the simulation example is introduced.

EXAMPLE 7.1

The continuous-time version of the original affine nonlinear two-player zero-sum game in affine form (Vamvoudakis and Lewis 2010) is given by

$$\dot{x} = f(x) + g(x)u + h(x)d \tag{7.48}$$

where $x = [x_1^T \; x_2^T]^T \in \mathbb{R}^{2\times 1}$, $u \in \Re$, $d \in \Re$, and internal dynamics $f(x)$, control and disturbance coefficient matrices $g(x)$, $h(x)$ are given as

$$f(x) = \begin{bmatrix} -x_1 + x_2 \\ -0.5x_1 - 0.5x_2(1-(\cos(2x_1)+2)^2) \end{bmatrix},$$

$$g(x) = \begin{bmatrix} 0 \\ \cos(2x_1)+2 \end{bmatrix}, \text{ and } h(x) = \begin{bmatrix} 0 \\ \sin(4x_1)+2 \end{bmatrix}$$

The parameter of the NNCS two-player zero-sum game are selected as (Xu and Jagannathan 2013): (1) The sampling time: $T_s = 100$ ms; (2) the upper bound of the network-induced delay is given as two, that is, $\bar{b} = 2$; (3) the network-induced delays: $E(\tau_{sc}) = 80$ ms and $E(\tau) = 150$ ms; (4) network-induced packet losses follow the Bernoulli distribution with $\bar{\gamma} = 0.3$; (5) the final time is set as $t_f = NT_s = 20$ s with simulation time steps $N = 200$; and (6) to obtain ergodic performance, Monte Carlo simulation were run with $N = 1000$ iterations. The distributions of network-induced delays and packet losses are shown in Figures 7.2 and 7.3. The initial state vector is deterministic.

7.4.1 State Regulation and Control and Disturbance Input Performance

First, the performances of the NNCS two-player zero-sum game state regulation errors and proposed finite-horizon stochastic optimal control and disturbance inputs are analyzed. For incorporating network parameters and the proposed design into the NNCS two-player zero-sum game, the augment state can be defined as $z_k = [x_k^T \; u_{k-1}^T \; u_{k-2}^T \; d_{k-1}^T \; d_{k-2}^T]^T \in \mathbb{R}^{6\times 1}$ and the initial state is chosen as $x_0 = [6 \; -3.5]^T$ and admissible control and disturbance policy are given as $u_o(z_k) = [-2 \; -2.5 \; -1 \; -1 \; 0 \; 1]z_k$ and $d_o(z_k) = [-1 \; -2 \; -1 \; -1 \; 1 \; 0]z_k$, respectively. Similar to Xu and Jagannathan (2013), the activation function for the NN identifier is $\tanh\{(z_{k,1})^2, z_{k,1}z_{k,2}, \ldots, (z_{k,6})^2, \ldots, (z_{k,1})^6, (z_{k,1})^5(z_{k,2}), \ldots, (z_{k,6})^6\}$, the critic NN state-dependent part activation function is chosen as the sigmoid of the sixth-order polynomial (i.e., $sigmoid\{(z_{k,1})^2, z_{k,1}z_{k,2}, \ldots, (z_{k,6})^2, \ldots, (z_{k,1})^6, (z_{k,1})^5(z_{k,2}), \ldots, (z_{k,6})^6\}$) and the time-dependent part of the critic NN activation function is given as the saturation

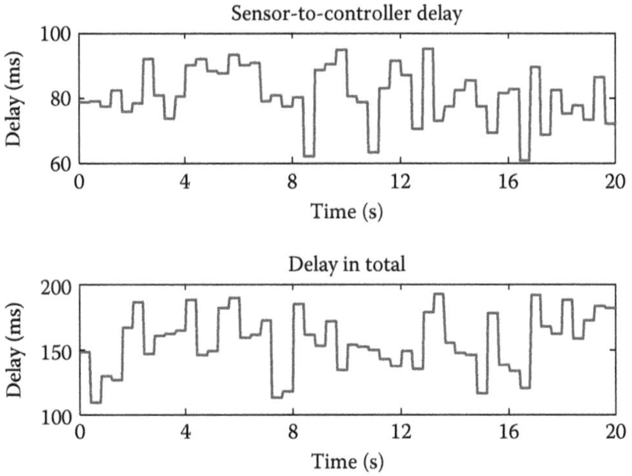

FIGURE 7.2 Distribution of network-induced delay in NNCS.

polynomial time function (i.e., $sat\{(N-k)^{10},(N-k)^{9},\ldots,1;\cdots;1,(N-k)^{10},\ldots,N-k\}$), and activation function of two action NNs are all selected as the gradient of the critic NN activation function. It is important to note that the saturation for the time-dependent part in the critic NN activation function is to guarantee that the magnitude of the time function stays inside a reasonable range such that the NN weights are computable.

Next, the tuning parameters of the NN identifier, the critic NN, and the two action NNs are chosen as $\alpha_I = 0.005$, $\alpha_V = 0.0002$, and $\alpha_u = 0.5$, $\alpha_d = 0.5$, and the weights of the NN identifier and the critic NN are generated as zeros initially and two action NNs weights are defined to reflect the initial admissible control and disturbance policy. In Figures 7.4 and 7.5, the proposed stochastic optimal control and disturbance can force the NNCS two-player zero-sum game state regulation errors approach to zero closely within the finite horizon (i.e., proposed finite-horizon strategies can maintain the NNCS two-player zero-sum game *bounded* in the mean) even in the presence of uncertain NNCS dynamics and network imperfections. Moreover, the proposed stochastic control signal and disturbance signal for the NNCS two-player zero-sum game is also bounded in the mean.

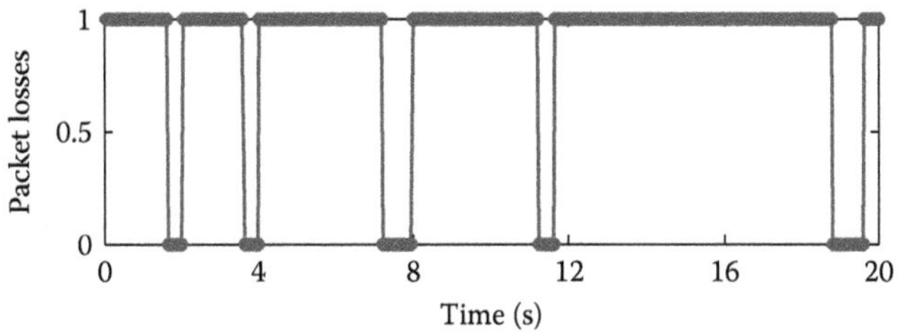

FIGURE 7.3 Distribution of network-induced packet losses in NNCS.

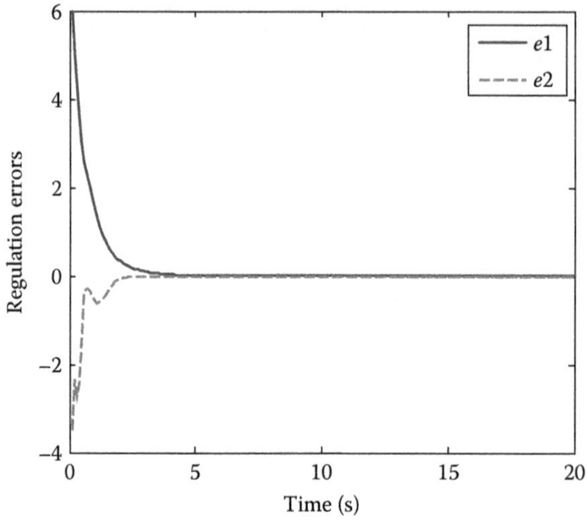

FIGURE 7.4 NNCS two-player zero-sum game state regulation errors with proposed finite-horizon stochastic optimal strategies.

In Figure 7.6, approximated weights for the critic and two action NNs are demonstrated. It is important to note that the estimated weights of the critic and two action NNs converge close to constant values and are *bounded* in the mean within the finite horizon (i.e., $t \in [0, 20s]$), which is consistent as Theorem 7.3.

7.4.2 Hamilton–Jacobi–Isaacs and Terminal Constraint Errors

In this part, the HJI equation error and terminal constraint error have been evaluated in Figure 7.7. As shown in Figure 7.8, during the finite simulation time (i.e.,

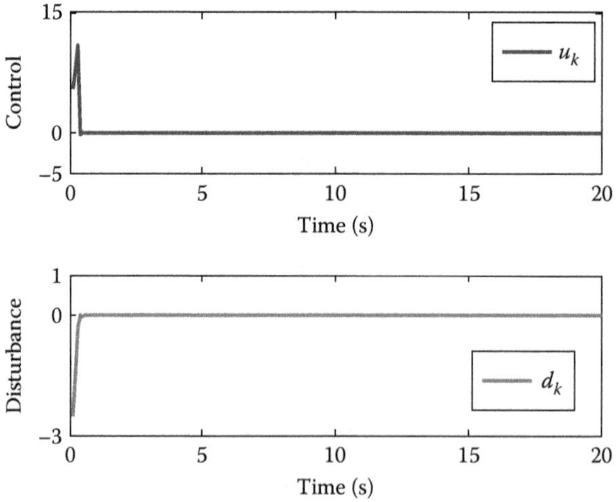

FIGURE 7.5 Proposed finite-horizon stochastic optimal strategies.

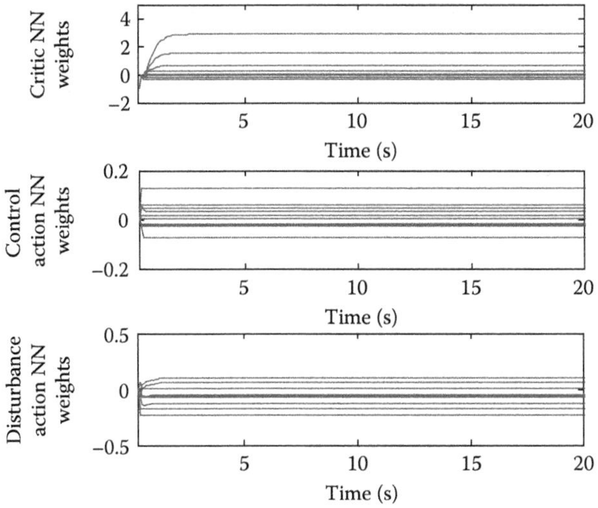

FIGURE 7.6 Estimated critic and two action NNs weights.

$t \in [0,20s]$), both the HJI equation error and terminal error can converge close to zero, which indicates that the proposed stochastic optimal strategies can approach the finite-horizon optimal solution for the NNCS two-player zero-sum game by forcing the HJI equation error converge close to zero while satisfying the terminal constraint. Recalling Theorem 7.3, the convergence performance of HJI and terminal constraint errors are related with tuning parameters, initial conditions, and final time. Therefore, while the final time instant (i.e., NT_s) increases, the upper bound on

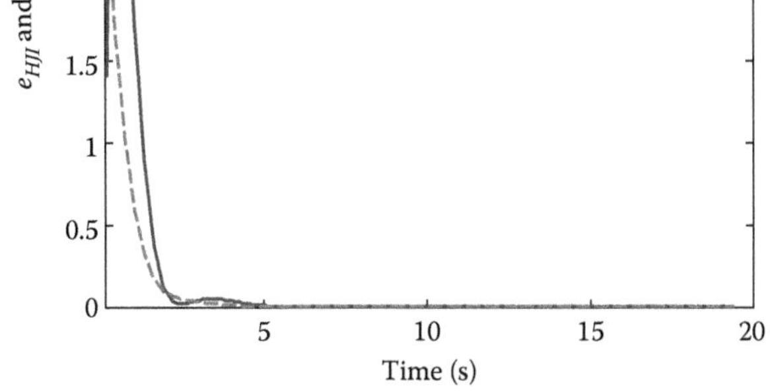

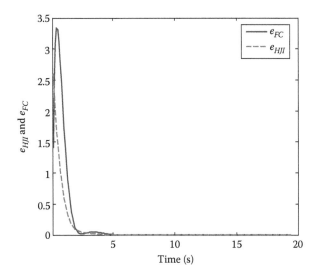

FIGURE 7.7 HJI equation error and terminal constraint error.

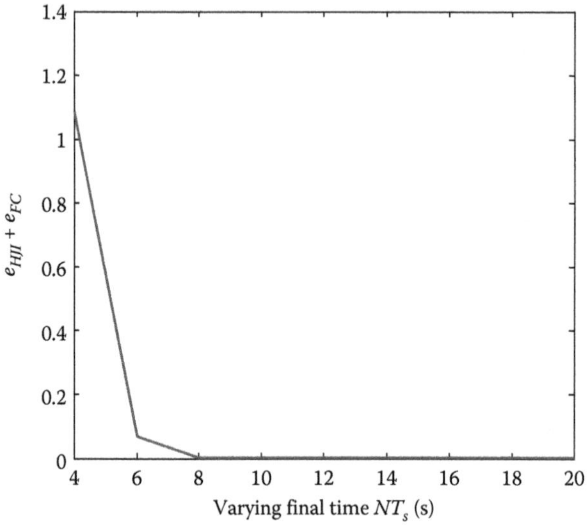

FIGURE 7.8 Effect of increasing final time (i.e., NT_s).

the sum of HJI and terminal constraint errors will decrease as depicted in Figure 7.8. Obviously, when the final time goes to infinity, that is, $N \to \infty$, the sum of HJI and terminal constraint errors will converge to zero and the proposed stochastic optimal design will approach the ideal infinite-horizon optimal design for the NNCS two-player zero-sum game.

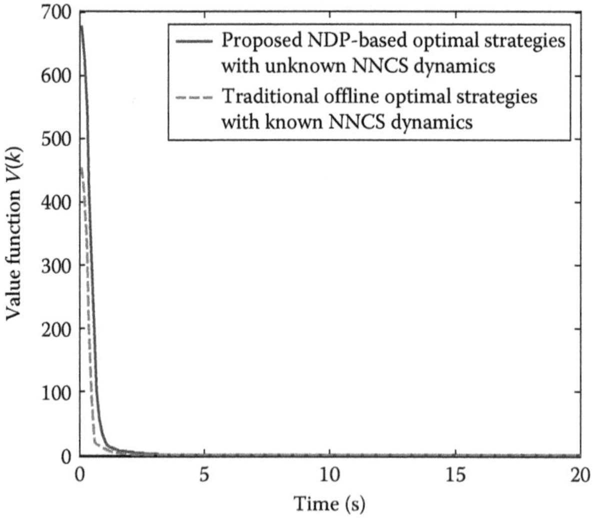

FIGURE 7.9 Comparison of value function between proposed finite-horizon optimal strategies and ideal traditional optimal design with known NNCS two-player zero-sum game dynamics and network imperfections.

7.4.3 Optimal Performance of the Proposed Design

Eventually, the optimal performance (i.e., value function) of the proposed finite-horizon stochastic optimal design for the NNCS two-player zero-sum game is demonstrated. For comparison, a traditional finite-horizon optimal design that solves the HJI equation offline with known system dynamics and network imperfections has been included. In Figure 7.9, the comparison result is illustrated. Although the value function of the proposed scheme is slightly higher than that of the ideal traditional finite-horizon optimal design, the proposed design is more valuable and practical since network imperfections and system dynamics are not required.

According to Figures 7.4 through 7.9, the proposed time-based NDP scheme in the presence of uncertain NNCS two-player zero-sum game dynamics and network imperfections can attain optimal design online without value or policy iteration and render nearly the same performance as the offline traditional optimal design, which requires the knowledge of system dynamics and network imperfections within finite time.

7.5 CONCLUSIONS

In this chapter, a novel time-based finite-horizon NDP scheme has been introduced for the NNCS two-player zero-sum game by using the NN identifier, critic, and two action NNs to solve stochastic optimal control and disturbance policies for the NNCS two-player zero-sum game in the presence of uncertain system dynamics and network imperfections. By using historical inputs and the NN identifier, the requirement on both system internal dynamics, control, and disturbance coefficient matrices were relaxed. Further, critic NN approximates the HJI equation solution online while satisfying the terminal constraint for the NNCS two-player zero-sum game over the finite horizon. An initial admissible control ensures that the system is stable when an NN identifier, critic, and two action NNs are being tuned online. Using Lyapunov theory, NNCS system state, identification error, weight estimation errors of NN identifier, critic, and two action NNs were shown to be *bounded* in the mean with bounds dependent on NNCS initial condition $B_{CL,0}$ and final time NT_s. When the final time NT_s increases, the bounds will decrease and converge to zero as the time goes to infinity.

PROBLEMS

Section 7.4

Problem 7.4.1: Repeat Example 7.1 with the following information.

1. The sampling time: $T_s = 90$ ms.
2. The upper bound of network-induced delay is given as two, that is, $\bar{b} = 2$.
3. The network-induced delays: $E(\tau_{sc}) = 85$ ms and $E(\tau) = 140$ ms.
4. Network-induced packet losses follow the Bernoulli distribution with $\bar{\gamma} = 0.4$.
5. The final time is set as $t_f = NT_s = 20$ s with simulation time steps $N = 200$.
6. Monte Carlo iteration number is $N = 5000$.

Problem 7.4.2: Repeat Problem 7.1 with different initial condition.

APPENDIX 7A

Proof of Theorem 7.2

Similar to Xu and Jagannathan (2013), the Lyapunov function candidate is considered as

$$L_{V,k} = tr\left\{ \underset{\tau,\gamma}{E}\left(\tilde{W}_{V,k}^T \tilde{W}_{V,k}\right)\right\} \quad \forall k = 0,1,\ldots,N \tag{7A.1}$$

Then, using Equation 7.33, the first difference of Equation 7A.1 can be derived for $k = 0,1,\ldots,N$ as

$$\begin{aligned}
\Delta L_{V,k} &= tr\left\{ \underset{\tau,\gamma}{E}\left(\tilde{W}_{V,k+1}^T \tilde{W}_{V,k+1}\right)\right\} - tr\left\{ \underset{\tau,\gamma}{E}\left(\tilde{W}_{V,k}^T \tilde{W}_{V,k}\right)\right\} \\
&\le -2\alpha_V(2-\chi-(\chi+5)\alpha_V)\left\| \underset{\tau,\gamma}{E}\left(\tilde{W}_{V,k}\right)\right\|^2 + \left(1+\frac{5\alpha_V^2}{\varphi_{\min}^2+1}\right)\varepsilon_{VM}^2 \\
&\quad +(4\varphi_M^2+5\alpha_V^2)W_{VM}^2 + \left(1+\frac{5\alpha_V^2}{\Delta\varphi_{\min}^2+1}\right)\Delta\varepsilon_{VM}^2 \\
&\le -2\alpha_V(2-\chi-(\chi+5)\alpha_V)\left\| \underset{\tau,\gamma}{E}\left(\tilde{W}_{V,k}\right)\right\|^2 + \varepsilon_{TV}^2, \quad k = 0,1,\ldots,N
\end{aligned} \tag{7A.2}$$

where $\chi = ((\varphi_{\min}^2 + \Delta\varphi_{\min}^2 + 2)/((\varphi_{\min}^2+1)(\Delta\varphi_{\min}^2+1))) < 2$, $0 < \varphi_{\min} < \|\varphi(z)\|$, $0 < \Delta\varphi_{\min} < \|\Delta\varphi_{\min}(z_k)\| \le \|\Delta\varphi(z_k, N-k)\|$ can be ensured due to the PE condition given in Remark 7.2, and

$$\begin{aligned}
\varepsilon_{TV}^2 &= \left(1+\frac{5\alpha_V^2}{\varphi_{\min}^2+1}\right)\varepsilon_{VM}^2 + (4\varphi_M^2+5\alpha_V^2)W_{VM}^2 + \left(1+\frac{5\alpha_V^2}{\Delta\varphi_{\min}^2+1}\right) \\
&\quad \times \Delta\varepsilon_{VM}^2 \quad \text{for} \quad k = 0,1,\ldots,N
\end{aligned}$$

It is important to note while tuning parameter is selected as $0 < \alpha_V < (2-\chi)/(\chi+5)$, the term $-2\alpha_V(2-\chi-(\chi+5)\alpha_V)\left\| \underset{\tau,\gamma}{E}\left(\tilde{W}_{V,k}\right)\right\|^2$ in Equation 7A.2 is less than zero. Therefore, using Lyapunov theory (Sarangapani 2006) and geometric sequence theory (Brockett et al. 1983; Dankert et al. 2006), the critic NN weight estimation error can be proven *bounded* in the mean with the bound depending upon initial conditions and final time NT_s for the finite-horizon case.

APPENDIX 7B

Proof of Theorem 7.3

Consider Lyapunov function candidate for closed-loop NNCS two-player zero-sum game as

$$L_{CL,k} = L_{z,k} + L_{I,k} + L_{V,k} + L_{u,k} + L_{A,k} + L_{B,k}, \quad \forall k = 0,\dots,N \tag{7B.1}$$

where

$$L_{z,k} = \kappa \left\| \underset{\tau,\gamma}{E}(z_k) \right\|^2, L_{V,k} = tr\left\{ \underset{\tau,\gamma}{E}\left(\tilde{W}_{V,k}^T \Pi \tilde{W}_{V,k} \right) \right\}, L_{I,k} = tr\left\{ \underset{\tau,\gamma}{E}(e_{I,k}^T \Lambda e_{I,k}) \right\},$$

$$+ \; tr\left\{ \underset{\tau,\gamma}{E}\left(\tilde{W}_{I,k}^T \Lambda \tilde{W}_{I,k} \right) \right\}, L_{u,k} = \left\| \underset{\tau,\gamma}{E}\left(\tilde{W}_{u,k} \right) \right\|^2 \text{ and } L_{A,k} = \omega \left\| \underset{\tau,\gamma}{E}\left(\tilde{W}_{I,k} \right) \right\|^4,$$

$$L_{B,k} = \rho \left\| \underset{\tau,\gamma}{E}\left(\tilde{W}_{V,k} \right) \right\|^4, \quad k = 0,\dots,N$$

with

$$\kappa = \frac{\min\left\{ \dfrac{(\alpha_u(1-3\alpha_u)\vartheta_{u\min}^2)}{(\vartheta_{uM}^2+1)}, \dfrac{(\alpha_d(1-3\alpha_d)\psi_{d\min}^2)}{(\psi_{dM}^2+1)} \right\}}{\max\{8G_M^2\vartheta_M^2, 8H_M^2\psi_M^2\}},$$

$$\omega = \frac{(\phi'_{VM})^2\zeta_M^4\lambda_{\max}^2(R_z^{-1})(1+4\alpha_u^2)}{2(\zeta_{\min}^2 - 2\alpha_I^2\zeta_M^2)(2-\zeta_{\min}^2+2\alpha_I^2\zeta_M^2)} + \frac{(\phi'_{VM})^2\zeta_M^4\lambda_{\max}^2(S_z^{-1})(1+4\alpha_d^2)}{2(\zeta_{\min}^2 - 2\alpha_I^2\zeta_M^2)(2-\zeta_{\min}^2+2\alpha_I^2\zeta_M^2)}$$

$$\rho = \frac{(\phi'_{VM})^2[\lambda_{\max}^2(R_z^{-1})(1+4\alpha_u^2)+\lambda_{\max}^2(S_z^{-1})(1+4\alpha_d^2)]}{4\alpha_V(2-\chi-(\chi+5)\alpha_V)[2-2\alpha_V(2-\chi-(\chi+5)\alpha_V)]},$$

$$\Lambda = \left[\frac{W_{VM}^2(\phi'_{VM})^2\zeta_M^2}{(\zeta_{\min}^2 - 2\alpha_I^2\zeta_M^2)} \times [\lambda_{\max}^2(R_z^{-1})(1+2\alpha_u^2)+\lambda_{\max}^2(S_z^{-1})(1+2\alpha_d^2)] \right.$$
$$\left. + \frac{8\omega(1-\zeta_{\min}^2+2\alpha_I^2\zeta_M^2)\Delta\varepsilon_{IM}^2}{(\zeta_{\min}^2 - 2\alpha_I^2\zeta_M^2)} \right] \boldsymbol{I}$$

and

$$\Pi = \left[\frac{(\phi'_{VM})^2[(1+4\alpha_u^2)\lambda_{\max}^2(R_z^{-1})G_M^2] + (\phi'_{VM})^2\lambda_{\max}^2(S_z^{-1})H_M^2}{2\alpha_V(2-\chi-(\chi+5)\alpha_V)} \right.$$
$$\left. + \frac{4\alpha_d^2(\varphi'_{VM})^2\lambda_{\max}^2(S_z^{-1})H_M^2 + 2[1-2\alpha_V(2-\chi-(\chi+5)\alpha_V)]\rho\varepsilon_{TV}^2}{2\alpha_V(2-\chi-(\chi+5)\alpha_V)} \right] \boldsymbol{I}$$

are all positive terms where $\boldsymbol{I}$ is the identity matrix, $\lambda_{\max}(R_z^{-1})$, $\lambda_{\max}(S_z^{-1})$ are the maximum eigenvalue of R_z^{-1}, S_z^{-1}, and ϑ_M, $\vartheta_{\min}$, ψ_M, $\psi_{\min}$, G_M, H_M, φ'_M, φ_M, W_{VM}, υ_{IM}, $\zeta_{\min}$, and ζ_M are defined in Theorems 7.1 and 7.2. Then, the first difference of Equation 7B.1 can be expressed as

$$\Delta L_{CL,k} = \Delta L_{z,k} + \Delta L_{I,k} + \Delta L_{V,k} + \Delta L_{u,k} + \Delta L_{A,k} + \Delta L_{B,k} \tag{7B.2}$$

Incorporating NNCS two-player zero-sum game dynamics (7.11) and using the Cauchy–Schwartz inequality and Lemma 7.1, $\Delta L_{z,k}$ can be expressed as

$$\begin{aligned}\Delta L_{z,k} &= \kappa\left\|\underset{\tau,\gamma}{E}(z_{k+1})\right\|^2 - \kappa\left\|\underset{\tau,\gamma}{E}(z_k)\right\|^2 \le -(1-2l_o)\kappa\left\|\underset{\tau,\gamma}{E}(z_k)\right\|^2 + 8G_M^2\vartheta_M^2\kappa\left\|\underset{\tau,\gamma}{E}\left(\tilde{W}_{u,k}\right)\right\|^2 \\ &\quad +8G_M^2\kappa\varepsilon_{uM}^2 + 8H_M^2\kappa\psi_M^2\left\|\underset{\tau,\gamma}{E}\left(\tilde{W}_{d,k}\right)\right\|^2 + 8H_M^2\kappa\varepsilon_{dM}^2, \quad k=0,\ldots,N\end{aligned} \tag{7B.3}$$

Then, using Equation 7.45, the first difference, $\Delta L_{z,k}$, can be represented as

$$\begin{aligned}\Delta L_u &= tr\left\{\underset{\tau,\gamma}{E}\left(\tilde{W}_{uk+1}^T\tilde{W}_{uk+1}\right)\right\} - tr\left\{\underset{\tau,\gamma}{E}\left(\tilde{W}_{uk}^T\tilde{W}_{uk}\right)\right\} \\ &\le -\frac{2\alpha_u(1-3\alpha_u)\vartheta_{\min}^2}{\vartheta_M^2+1}\left\|\underset{\tau,\gamma}{E}\left(\tilde{W}_{uk}\right)\right\|^2 + \frac{G_M^2(\varphi'_{VM})^2\lambda_{\max}^2(R_z^{-1})(1+4\alpha_u^2)}{2}\left\|\underset{\tau,\gamma}{E}\left(\tilde{W}_{Vk}\right)\right\|^2 \\ &\quad + \frac{W_{VM}^2(\varphi'_{VM})^2\lambda_{\max}^2(R_z^{-1})(1+2\alpha_u^2)\zeta_M^2}{2}\left\|\underset{\tau,\gamma}{E}(\tilde{W}_{I,k})\right\|^2 \\ &\quad + \frac{(\varphi'_{VM})^2\lambda_{\max}^2(R_z^{-1})(1+4\alpha_u^2)}{4}\left\|\underset{\tau,\gamma}{E}(\tilde{W}_{Vk})\right\|^4 \\ &\quad + \frac{(\varphi'_{VM})^2\lambda_{\max}^2(R_z^{-1})(1+4\alpha_u^2)\zeta_M^4}{4}\left\|\underset{\tau,\gamma}{E}(\tilde{W}_{I,k})\right\|^4 + \varepsilon_{uM}, \quad k=0,1,\ldots,N\end{aligned} \tag{7B.4}$$

Similar to Equation 7B.4, recalling Equation 7.46, $\Delta L_{d,k}$ can be derived as

$$\begin{aligned}\Delta L_d &= tr\left\{\underset{\tau,\gamma}{E}\left(\tilde{W}_{dk+1}^T\tilde{W}_{dk+1}\right)\right\} - tr\left\{\underset{\tau,\gamma}{E}\left(\tilde{W}_{dk}^T\tilde{W}_{dk}\right)\right\} \\ &\le -\frac{2\alpha_d(1-3\alpha_d)\psi_{\min}^2}{\psi_M^2+1}\left\|\underset{\tau,\gamma}{E}\left(\tilde{W}_{dk}\right)\right\|^2 + \frac{H_M^2(\varphi'_{VM})^2\lambda_{\max}^2(S_z^{-1})(1+4\alpha_u^2)}{2}\left\|\underset{\tau,\gamma}{E}(\tilde{W}_{Vk})\right\|^2 \\ &\quad + \frac{(\varphi'_{VM})^2\lambda_{\max}^2(S_z^{-1})(1+4\alpha_d^2)\zeta_M^4}{4}\left\|\underset{\tau,\gamma}{E}(\tilde{W}_{I,k})\right\|^4 + \varepsilon_{dM}, \quad k=0,1,\ldots,N\end{aligned} \tag{7B.5}$$

Next, recalling Equation 7.22 and Theorem 7.1, $\Delta L_{A,k}$ can be expressed as

$$\begin{aligned}\Delta L_{A,k} &= \omega \left\| \underset{\tau,\gamma}{E}\left(\tilde{W}_{I,k+1}\right)\right\|^4 - \omega \left\| \underset{\tau,\gamma}{E}\left(\tilde{W}_{I,k}\right)\right\|^4 \\ &\le -\omega(\zeta_{\min}^2 - 2\alpha_I^2\zeta_M^2)(2-\zeta_{\min}^2 + 2\alpha_I^2\zeta_M^2)\left\| \underset{\tau,\gamma}{E}\left(\tilde{W}_{I,k}\right)\right\|^4 \\ &\quad + 4(1-\zeta_{\min}^2 + 2\alpha_I^2\zeta_M^2)\left\| \underset{\tau,\gamma}{E}\left(\tilde{W}_{I,k}\right)\right\|^2 \Delta\varepsilon_{IM}^2 + 4\omega\Delta\varepsilon_{IM}^4, \quad k = 0,1,\ldots,N\end{aligned} \tag{7B.6}$$

Additionally, according to Equation 7.33 and Theorem 7.2, the first difference of $L_{B,k}$ can be represented as

$$\begin{aligned}\Delta L_{B,k} &= \rho \left\| \underset{\tau,\gamma}{E}\left(\tilde{W}_{V,k+1}\right)\right\|^4 - \rho \left\| \underset{\tau,\gamma}{E}\left(\tilde{W}_{V,k}\right)\right\|^4 \\ &\le -2\rho\alpha_V(2-\chi-(\chi+5)\alpha_V)[2-2\alpha_V(2-\chi-(\chi+5)\alpha_V)]\left\| \underset{\tau,\gamma}{E}\left(\tilde{W}_{V,k}\right)\right\|^4 \\ &\quad + [1-2\alpha_V(2-\chi-(\chi+5)\alpha_V)]\rho\left\| \underset{\tau,\gamma}{E}\left(\tilde{W}_{V,k}\right)\right\|^2 \varepsilon_{TV}^2 + \rho\varepsilon_{TV}^4, \quad k = 0,1\ldots,N\end{aligned} \tag{7B.7}$$

with $0 < \chi = ((\varphi_{\min}^2 + \Delta\varphi_{\min}^2 + 2)/((\varphi_{\min}^2 + 1)(\Delta\varphi_{\min}^2 + 1))) < 2$ defined in Theorem 7.2.

Eventually, combining Equations 7A.2, 7B.3 through 7B.7, the first difference of $L_{CL,k}$ can be derived for $k = 0,1,\ldots,N$ as

$$\begin{aligned}\Delta L_{CL,k} &= \Delta L_{z,k} + \Delta L_{I,k} + \Delta L_{V,k} + \Delta L_{u,k} + \Delta L_{A,k} + \Delta L_{B,k} \\ &\le -(1-2l_o)\kappa\left\| \underset{\tau,\gamma}{E}(z_k)\right\|^2 - \frac{\alpha_u(1-3\alpha_u)\vartheta_{\min}^2}{\vartheta_M^2+1}\left\| \underset{\tau,\gamma}{E}\left(\tilde{W}_{uk}\right)\right\|^2 \\ &\quad - \frac{\alpha_d(1-3\alpha_d)\psi_{\min}^2}{\psi_M^2+1}\left\| \underset{\tau,\gamma}{E}\left(\tilde{W}_{dk}\right)\right\|^2 - (1-\alpha_I^2)\|\Lambda\|\left\| \underset{\tau,\gamma}{E}(e_{I,k})\right\|^2 \\ &\quad - \frac{W_{VM}^2(\varphi_{VM}')^2\zeta_M^2[\lambda_{\max}^2(R_z^{-1})(1+2\alpha_u^2)+\lambda_{\max}^2(S_z^{-1})(1+2\alpha_d^2)]}{2}\left\| \underset{\tau,\gamma}{E}\left(\tilde{W}_{I,k}\right)\right\|^2 \\ &\quad - \frac{(\varphi_{VM}')^2[(1+4\alpha_u^2)\lambda_{\max}^2(R_z^{-1})G_M^2 + (1+4\alpha_d^2)\lambda_{\max}^2(S_z^{-1})H_M^2]}{2}\left\| \underset{\tau,\gamma}{E}\left(\tilde{W}_{Vk}\right)\right\|^2 \\ &\quad - \frac{(\varphi_{VM}')^2[\lambda_{\max}^2(R_z^{-1})(1+4\alpha_u^2)+\lambda_{\max}^2(S_z^{-1})(1+4\alpha_d^2)]}{4}\left\| \underset{\tau,\gamma}{E}\left(\tilde{W}_{Vk}\right)\right\|^4 \\ &\quad - \frac{(\varphi_{VM}')^2\zeta_M^4[\lambda_{\max}^2(R_z^{-1})(1+4\alpha_u^2)+\lambda_{\max}^2(S_z^{-1})(1+4\alpha_d^2)]}{4}\left\| \underset{\tau,\gamma}{E}\left(\tilde{W}_{I,k}\right)\right\|^4 + \varepsilon_{CLM}\end{aligned} \tag{7B.8}$$

where $0<\varphi_{\min}<\|\varphi(z)\|$, $0<\Delta\varphi_{\min}<\|\Delta\varphi(z_k,N-k)\|$, and $0<\vartheta_{\min}<\|\vartheta(z)\|$, $0<\psi_{\min}<\|\psi(z)\|$ can be held by the PE condition given in Remark 7.2. In addition, ε_{CLM} is defined as

$$\varepsilon_{CLM}=8g_M^2\kappa\varepsilon_{uM}^2+8h_M^2\kappa\varepsilon_{dM}^2+2\|\Lambda\|\Delta\varepsilon_{I,k}^2+\|\Pi\|\varepsilon_{TV}^2+\varepsilon_{uM}+\varepsilon_{dM}$$

$$+4\omega\Delta\varepsilon_{IM}^4+\rho\varepsilon_{TV}^4 \quad \text{for} \quad k=0,1,\ldots,N$$

Recalling Theorems 7.1 and 7.2, while the identifier, critic, and two action NNs tuning parameters, α_I, α_V, α_u, are selected as $0<\alpha_I<\min\left\{(1/2\zeta_M),\left(\zeta_{\min}/\sqrt{2}\zeta_M\right)\right\}$, $0<\alpha_V<(2-\chi)/(\chi+5)$, and $0<\alpha_u<1$, $0<\alpha_d<1$, the terms $-(1-l_o)\kappa\left\|\underset{\tau,\gamma}{E}(z_k)\right\|^2$,

$$-(1-\alpha_I^2)\|\Lambda\|\left\|\underset{\tau,\gamma}{E}(e_{I,k})\right\|^2, -\frac{\alpha_d(1-3\alpha_d)\psi_{\min}^2}{\psi_M^2+1}\left\|\underset{\tau,\gamma}{E}\left(\tilde{W}_{dk}\right)\right\|^2,$$

$$-\frac{W_{VM}^2(\varphi'_{VM})^2\zeta_M^2[\lambda_{\max}^2(R_z^{-1})(1+2\alpha_u^2)+\lambda_{\max}^2(S_z^{-1})(1+2\alpha_d^2)]}{2}\left\|\underset{\tau,\gamma}{E}\left(\tilde{W}_{I,k}\right)\right\|^2,$$

$$-\frac{(\varphi'_{VM})^2[(1+4\alpha_u^2)\lambda_{\max}^2(R_z^{-1})G_M^2+(1+4\alpha_d^2)\lambda_{\max}^2(S_z^{-1})H_M^2]}{2}\left\|\underset{\tau,\gamma}{E}\left(\tilde{W}_{Vk}\right)\right\|^2,$$

$$-\frac{(\varphi'_{VM})^2[\lambda_{\max}^2(R_z^{-1})(1+4\alpha_u^2)+\lambda_{\max}^2(S_z^{-1})(1+4\alpha_d^2)]}{4}\left\|\underset{\tau,\gamma}{E}\left(\tilde{W}_{Vk}\right)\right\|^4 \quad \text{and}$$

$$-\frac{(\varphi'_{VM})^2\zeta_M^4[\lambda_{\max}^2(R_z^{-1})(1+4\alpha_u^2)+\lambda_{\max}^2(S_z^{-1})(1+4\alpha_d^2)]}{4}\left\|\underset{\tau,\gamma}{E}\left(\tilde{W}_{I,k}\right)\right\|^4$$

are all less than zeros.

Assume the initial system state is bounded as $\left\|\underset{\tau,\gamma}{E}(z_0)\right\|^2\le B_{z,0}$ with $B_{z,0}$ being a positive constant (Xu and Jagannathan 2013), the initial identification error is bounded as $\left\|\underset{\tau,\gamma}{E}(e_{I,0})\right\|^2\le B_{eI,0}$ with $B_{eI,0}$ being a positive constant (Xu and Jagannathan 2013), and the initial NN weight estimation errors of identifier, critic, and actions NNs are bounded as $\left\|\underset{\tau,\gamma}{E}\left(\tilde{W}_{I,0}\right)\right\|^2\le B_{WI,0}$, $\left\|\underset{\tau,\gamma}{E}\left(\tilde{W}_{V,0}\right)\right\|^2\le B_{WV,0}$, $\left\|\underset{\tau,\gamma}{E}\left(\tilde{W}_{u,0}\right)\right\|^2\le B_{Wu,0}$, and $\left\|\underset{\tau,\gamma}{E}\left(\tilde{W}_{d,0}\right)\right\|^2\le B_{Wd,0}$ with $B_{WI,0}$, $B_{WV,0}$, $B_{Wu,0}$, and $B_{Wd,0}$, being positive constants (Xu and Jagannathan 2013).

Further, according to standard Lyapunov theory (Sarangapani 2006) and geometric sequence theory (Brockett et al. 1983; Dankert et al. 2006), the Lyapunov function candidate $L_{CL,k}$, $k=0,1,\ldots,N$ in Equation 7B.1 can be derived as

$$L_{CL,k} = \Delta L_{CL,k-1} + \Delta L_{CL,k-2} + \cdots + \Delta L_{CL,0} + L_{CL,0} = \sum_{j=0}^{k-1} (\Delta L_{CL,j}) + L_{CL,0}$$
$$= \sum_{j=0}^{k-1} (\Delta L_{z,j} + \Delta L_{I,j} + \Delta L_{V,j} + \Delta L_{u,j} + \Delta L_{A,j} + \Delta L_{B,j}) + L_{CL,0} \tag{7B.9}$$

Using Equation 7A.9, Equation 7A.10 can be expressed as

$$L_{CL,k} = \kappa \left\| \underset{\tau,\gamma}{E}(z_k) \right\| + \|\Lambda\| \left\| \underset{\tau,\gamma}{E}(e_{I,k}) \right\|^2 + \|\Lambda\| \left\| \underset{\tau,\gamma}{E}\left(\tilde{W}_{I,k}\right) \right\|^2 + \|\Pi\| \left\| \underset{\tau,\gamma}{E}\left(\tilde{W}_{V,k}\right) \right\|^2 + \left\| \underset{\tau,\gamma}{E}\left(\tilde{W}_{u,k}\right) \right\|^2$$
$$+ \left\| \underset{\tau,\gamma}{E}\left(\tilde{W}_{d,k}\right) \right\|^2 + \omega \left\| \underset{\tau,\gamma}{E}\left(\tilde{W}_{I,k}\right) \right\|^4 + \rho \left\| \underset{\tau,\gamma}{E}\left(\tilde{W}_{V,k}\right) \right\|^4 \tag{7B.10}$$

$$\le l_o^k \kappa B_{x,0} + \left[1 - \frac{\alpha_u(1-3\alpha_u)\vartheta_{\min}^2}{\vartheta_M^2 + 1}\right]^k B_{u,0} + \left[1 - \frac{\alpha_d(1-3\alpha_d)\psi_{\min}^2}{\psi_M^2 + 1}\right]^k B_{d,0}$$
$$+ \left(1 - \frac{W_{VM}^2(\varphi'_{VM})^2\zeta_M^2[\lambda_{\max}^2(R_z^{-1})(1+2\alpha_u^2) + \lambda_{\max}^2(S_z^{-1})(1+2\alpha_d^2)]}{2}\right)^k B_{WI,0}$$
$$+ \left(1 - \frac{(\varphi'_{VM})^2\zeta_M^4[\lambda_{\max}^2(R_z^{-1})(1+4\alpha_u^2) + \lambda_{\max}^2(S_z^{-1})(1+4\alpha_d^2)]}{4}\right)^k B_{WI,0}^2$$
$$+ \left(1 - \frac{(\varphi'_{VM})^2[(1+4\alpha_u^2)\lambda_{\max}^2(R_z^{-1})g_M^2 + (1+4\alpha_d^2)\lambda_{\max}^2(S_z^{-1})h_M^2]}{2}\right)^k B_{WV,0} + \alpha_I^{2k}\|\Lambda\| B_{eI,0}$$
$$+ \left(1 - \frac{(\varphi'_{VM})^2[\lambda_{\max}^2(R_z^{-1})(1+4\alpha_u^2) + \lambda_{\max}^2(S_z^{-1})(1+4\alpha_d^2)]}{4}\right)^k B_{WV,0}^2 + \frac{1-\varpi^k}{1-\varpi}\varepsilon_{CLM}$$
$$\le \varpi^k B_{CL,0} + \frac{1-\varpi^k}{1-\varpi}\varepsilon_{CLM}, \quad k = 0,1,\ldots,N$$

with closed-loop NNCS two-player zero-sum game initial condition

$$B_{CL,0} = \kappa B_{z,0} + B_{u,0} + B_{d,0} + \alpha_I^2 B_{eI,0} + B_{WI,0} + B_{WI,0}^2 + B_{WV,0} + B_{WV,0}^2, \text{ and}$$

$$\varpi = \max\left\{ l_0, \left[1 - \frac{\alpha_u(1-3\alpha_u)\vartheta_{\min}^2}{\vartheta_M^2 + 1}\right], \alpha_I^2, \left[1 - \frac{\alpha_d(1-3\alpha_d)\psi_{\min}^2}{\psi_M^2 + 1}\right], \right.$$

$$\left(1-\frac{(\varphi'_{VM})^2\zeta_M^4[\lambda_{\max}^2(R^{-1})(1+4\alpha_u^2)+\lambda_{\max}^2(S^{-1})(1+4\alpha_d^2)]}{4}\right),$$

$$\left(1-\frac{[\lambda_{\max}^2(R_z^{-1})(1+2\alpha_u^2)+\lambda_{\max}^2(S_z^{-1})(1+2\alpha_d^2)]}{2}W_{VM}^2(\varphi'_{VM})^2\zeta_M^2\right),$$

$$\left(1-\frac{(\varphi'_{VM})^2[(1+4\alpha_u^2)\lambda_{\max}^2(R^{-1})g_M^2+(1+4\alpha_d^2)\lambda_{\max}^2(S^{-1})h_M^2]}{2}\right),$$

$$\left.\left(1-\frac{(\varphi'_{VM})^2[(1+4\alpha_u^2)\lambda_{\max}^2(R^{-1})g_M^2+(1+4\alpha_d^2)\lambda_{\max}^2(S^{-1})h_M^2]}{2}\right)\right\}$$

Therefore, the bounds for NNCS two-player zero-sum game in terms of the state, identification error, NN identifier weight estimation error, and critic, two action NNs weight estimation errors can be represented as

$$\left\|\underset{\tau,\gamma}{E}(z_k)\right\| \le \sqrt{\frac{\varpi^k}{\kappa}B_{CL,0}+\frac{1-\varpi^k}{\kappa(1-\varpi)}\varepsilon_{CLM}} \equiv B_{z,k}$$

or

$$\left\|\underset{\tau,\gamma}{E}(e_k)\right\| \le \sqrt{\frac{\varpi^k}{\alpha_I}B_{CL,0}+\frac{1-\varpi^k}{\alpha_I(1-\varpi)}\varepsilon_{CLM}} \equiv B_{eI,k} \tag{7B.11}$$

or

$$\left\|\underset{\tau,\gamma}{E}\left(\tilde{W}_{I,k}\right)\right\| \le \max\left\{\sqrt{\varpi^k B_{CL,0}+\frac{1-\varpi^k}{(1-\varpi)}\varepsilon_{CLM}},\ \sqrt[4]{\varpi^k B_{CL,0}+\frac{1-\varpi^k}{(1-\varpi)}\varepsilon_{CLM}}\right\} \equiv B_{WI,0}$$

or

$$\left\|\underset{\tau,\gamma}{E}\left(\tilde{W}_{V,k}\right)\right\| \le \max\left\{\sqrt{\varpi^k B_{CL,0}+\frac{1-\varpi^k}{(1-\varpi)}\varepsilon_{CLM}},\ \sqrt[4]{\varpi^k B_{CL,0}+\frac{1-\varpi^k}{(1-\varpi)}\varepsilon_{CLM}}\right\} \equiv B_{WV,0}$$

or

$$\left\|\underset{\tau,\gamma}{E}\left(\tilde{W}_{u,k}\right)\right\| \le \sqrt{\varpi^k B_{CL,0}+\frac{1-\varpi^k}{(1-\varpi)}\varepsilon_{CLM}} \equiv B_{u,k}$$

or

$$\left\| E\left(\tilde{W}_{d,k}\right) \right\|_{\tau,\gamma} \le \sqrt{\varpi^k B_{CL,0} + \frac{1-\varpi^k}{(1-\varpi)} \varepsilon_{CLM}} \equiv B_{d,k}, \quad k = 0,1,\ldots,N$$

Since $0 < l_o < 1$, $0 < \alpha_I < \min\left\{(1/2\zeta_M),\left(\zeta_{\min}/\sqrt{2}\zeta_M\right)\right\}$, $0 < \alpha_V < (2-\chi)/(\chi+5)$, and $0 < \alpha_u < 1/3$, $0 < \alpha_d < 1/3$, we have $0 < [1-((\alpha_u(1-3\alpha_u)\vartheta_{\min}^2)/(\vartheta_M^2+1))] < 1$ $0 < [1-((\alpha_d(1-3\alpha_d)\psi_{\min}^2)/(\psi_M^2+1))] < 1$, $0 < [1-2\alpha_V(2-\chi-(\chi+5))] < 1$, and $0 < [1-\alpha_V(2-\chi)] < 1$. Moreover, ϖ satisfies $0 < \varpi < 1$. Therefore, the term ϖ^k will decrease when kT_s increases. Additionally, since all provided initial bounds $B_{z,0}$, $B_{eI,0}$, $B_{WI,0}$, $B_{WV,0}$, and $B_{Wu,0}$, $B_{Wd,0}$ are positive constants, the closed-loop NNCS two-player zero-sum game initial condition $B_{CL,0}$ will also be a positive constant. Therefore, the bounds $B_{z,k}$, $B_{eI,k}$, $B_{WI,k}$, $B_{WV,k}$, and $B_{Wu,k}$, $B_{Wd,k}$ will decrease when the time instant kT_s increases.

Moreover, when the final time instant NT_s increases, all the signals will not only be *bounded* in the mean, but all bounds will also decrease with time. Recalling Theorems 7.1 and 7.2, when the time goes to infinity, that is, $k \to \infty$, the NNCS two-player zero-sum game system state, identification error, and weights estimation errors of three NNs will decrease close to the minimal bounds and the proposed finite-horizon stochastic optimal design for control and disturbance will approach infinite-horizon stochastic optimal control and disturbance policies closely.

REFERENCES

Al-Tamimi, A., Abu-Khalaf, M., and Lewis, F.L. 2007. Adaptive critic designs for discrete-time zero-sum games with application to H-infinite control. *IEEE Transactions on Systems, Man, and Cybernetics, Part B: Cybernetics*, 37, 240–247.

Basar, T. and Olsder, G.J. 1995. *Dynamic Non-Cooperative Game Theory*. 2nd edition, Academic, New York.

Bertsekas, D.P. and Tsitsiklis, J. 1996. *Neuro-Dynamics Programming*. Athena Scientific, Boston, MA, USA, 15–23.

Brockett, R.W., Millman, R.S., and Sussmann, H.J. (eds.) 1983. *Differential Geometric Control Theory*. Proceedings of the conference held at Michigan Technological University, Houghton, MI, June 28–July 1, Vol. 27. Birkhauser, USA.

Burgin, G.H. 1969. On playing two-person zero-sum games against nonminimax players. *Transactions on Systems Science and Cybernetics*, 5(4), 369–370.

Dankert, J., Yang, L., and Si, J. 2006. A performance gradient perspective on approximate dynamic programming and its application to partially observable Markov decision process. *Proceedings of the International Symposium on Intelligent Control*, Munich, Germany, pp. 458–463.

Dierks, T. and Jagannathan, S. 2012. Online optimal control of affine nonlinear discrete-time systems with unknown internal dynamics by using time-based policy update. *IEEE Transactions on Neural Networks and Learning Systems*, 23, 1118–1129.

Ge, S.S., Hong, F., and Lee, T.H. 2003. Adaptive neural network control of nonlinear system with unknown time delays. *IEEE Transactions on Automatic Control*, 48(11), 2004–2010.

Hespanha, J., Naghshtabrizi, P., and Xu, Y. 2007. A survey of recent results in networked control systems. *Proceedings of the IEEE*, 95, 138–162.

Heydari, A. and Balakrishnan, S.N. 2011. Finite-horizon input-constrained nonlinear optimal control using single network adaptive critics. *Proceedings of the American Control Conference*, San Francisco, USA, pp. 3047–3052.

Hong, F., Ge, S.S., and Lee, T.H. 2005. Practical adaptive neural control of nonlinear systems with unknown time delays. *IEEE Transactions on System, Man, and Cybernetics B, Cybernetics*, 35, 849–854.

Hu, S.S. and Zhu, Q.X. 2003. Stochastic optimal control and analysis of stability of networked control systems with long delay. *Automatica*, 39, 1877–1884.

Jia, X., Zhang, D., and Zheng, N. 2009. Fuzzy H-infinite control for nonlinear networked control systems in T-S fuzzy model. *IEEE Transactions on System, Man, and Cybernetics B, Cybernetics*, 39, 1073–1079.

Lewis, F.L. and Syrmos, V.L. 1995. *Optimal Control*. 2nd edition, Wiley, New York.

Liu, G., Xia, Y., Rees, D., and Hu, W. 2007. Design and stability criteria of networked predictive control systems with random network delay in the feedback channel. *IEEE Transactions on System, Man, and Cybernetics C. Applications and Reviews*, 37, 173–184.

Sarangapani, J. 2006. *Neural Network Control of Nonlinear Discrete-Time Systems*. CRC Press, Boca Raton, Florida.

Seiffertt, J., Snayal, S., and Wunsch, D.C. 2008. Hamilton–Jacobi–Bellman equations and approximate dynamic programming. *IEEE Transactions on System, Man, and Cybernetics B, Cybernetics*, 38, 918–923.

Stallings, W. 2002. *Wireless Communications and Networks*. 1st edition, Prentice-Hall, Upper Saddle River, NJ.

Stengel, R.F. 1986. *Stochastic Optimal Control: Theory and Application*. Wiley-Interscience, New York.

Vamvoudakis, K. and Lewis, F.L. 2010. Online solution of nonlinear two-player zero-sum games using synchronous policy iteration. *Proceedings of the Control Decision Conference*, pp. 3040–3047.

Walsh, G.C., Beldiman, O., and Bushnell, L.G. 2001. Asymptotic behavior of nonlinear networked control systems. *IEEE Transactions on Automatic Control*, 46, 1093–1097.

Wang, F.Y., Jin, N., Liu, D., and Wei, Q.L. 2011. Adaptive dynamic programing for finite horizon optimal control of discrete-time nonlinear system of ε-error bound. *IEEE Transactions on Neural Networks and Learning Systems*, 22, 24–36.

Wang, Z., Yang, F., Ho, D., and Liu, X. 2007. Robust H-infinite control for networked systems with random packet losses. *IEEE Transactions on Systems, Man, and Cybernetics B, Cybernetics*, pp. 916–924.

Werbos, P.J. 1983. A menu of designs for reinforcement learning over time. *Journal of Neural Networks Control*, 3, 835–846.

Wouw, N.V.D., Nesic, D., and Heemels, W.P.H. 2010. Stability analysis for nonlinear networked control systems: A discrete-time approach. *Proceedings of the IEEE Control Decision Conference*, Atlanta, GA, USA, pp. 7557–7563.

Xu, H. and Jagannathan, S. 2013. Stochastic optimal controller design for uncertain nonlinear networked control system via neuro dynamic programming. *IEEE Transactions on Neural Networks and Learning Systems*, 24, 471–484.

Xu, H., Jagannathan, S., and Lewis, F.L. 2012. Stochastic optimal control of unknown linear networked control system in the presence of random delays and packet losses. *Automatica*, 48, 1017–1030.

8 Distributed Joint Optimal Network Scheduling and Controller Design for Wireless Networked Control Systems

Wireless network is now being utilized in control systems, making the systems distributed in contrast to traditional dedicated control systems. These novel distributed systems are known as wireless networked control systems (WNCS) (Nilsson et al. 1998; Branicky et al. 2000; Baillieul and Antsaklis 2007; Schenato et al. 2007). Practical examples of such systems include smart power grid, network-enabled manufacturing, water distribution, traffic, and so on. In WNCS, wireless communication packets carry sensed data and control commands from different physical systems (or plants) and remote controllers. Although WNCS can reduce system wiring, ease of system diagnosis and maintenance, and increase agility, the uncertainty caused by shared wireless network and its protocols bring many challenging issues for both wireless network protocol and controller designs. As observed in the previous chapters network imperfections can transform a linear time-invariant system to an uncertain and stochastic system, which can become unstable unless a proper controller is designed. The wireless network is more unreliable due to channel issues such as path losses and fading when compared to a wired network as in controller area network (CAN) in automotive applications.

The issues for control design (Nilsson et al. 1998; Lian et al. 2001, 2003) include wireless network latency and packet loss (Lian et al. 2001, 2003), which are dependent on wireless communication channel quality and network protocol design, respectively. In general, wireless network latency and packet losses can destabilize the real-time control system and in many cases can result in safety concerns. However, the recent literature (Nilsson et al. 1998; Branicky et al. 2000; Baillieul and Antsaklis 2007; Schenato et al. 2007) only focused on stability or optimal controller design by assuming wireless network latency and packet losses to be a constant or random ignoring the real behavior of the wireless network component.

Similarly, the current wireless network protocol designs (Dai and Prabhakar 2000; IEEE Standard 802.11a 2003) ignore the effects of the real-time nature of the control system or the application making them unsuitable for real-time control applications. Thus, a truly cross-layer WNCS codesign that optimizes not only the performance of the wireless network but also the controlled system is necessary.

Toward this end, the distributed scheduling scheme is critical in wireless network protocol design (IEEE Standard 802.11a 2003) if a communication network is shared by many dynamic systems. Compared with traditional centralized scheduling (Dai and Prabhakar 2000), the main advantage with distributed scheduling is that it does not need a central processor to deliver the schedules after collecting information from all the systems.

In IEEE 802.11 standard (2003), carrier sense multiple access (CSMA) protocol is introduced to schedule wireless users in a distributed manner where a wireless node wishing to transmit does so only if it does not hear an ongoing transmission. Meanwhile, fairness is a nonnegligible factor in distributed scheduling design. Vaidya et al. (2005) proposed distributed fairness scheduling in packet switched network and wireless LAN network. Different users wishing to share a wireless channel can be allocated bandwidth in proportion to their weights, which ensured fairness among different users (Vaidya et al. 2005). However, since the random access scheme is used in most CSMA-based distributed scheduling (Vaidya et al. 2005; Li and Negi 2010) and these schemes (Bennett and Zhang 1996; Vaidya et al. 2005; Zheng et al. 2009) focus on improving the performance of the link layer alone, which in turn increases wireless network latency, these protocols are unsuitable for WNCS since they can cause degradation in the performance of WNCS.

The article by Xu and Jagannathan (2013) presents a cross-layer approach that considers controller design from the application layer and wireless network performance from the MAC layer to derive a stochastic optimal controller and distributed scheduling schemes for WNCS. The proposed stochastic control design can generate optimal control policies through the certainty-equivalent stochastic value function estimator by relaxing system dynamics, wireless network latency, and packet losses. In addition, the cross-layer distributed scheduling protocol optimizes not only the wireless network performance but also the system (or plant) performance by maximizing the utility function generated from both the wireless network and the plant.

First, Section 8.1 presents the background of WNCS. Subsequently, Section 8.2 develops WNCS codesign. An example is covered in this section. Conclusions are given in Section 8.3.

8.1 BACKGROUND OF WIRELESS NETWORKED CONTROL SYSTEMS

The basic structure of WNCS with multiple users is shown in Figure 8.1 where users include controller–plant pair (or system) sharing a common wireless network. Each WNCS pair includes five main components: (1) real-time physical system or plant to be controlled; (2) a sensor that measures system outputs from the plant; (3) a controller that generates commands in order to maintain a desired plant performance; (4) wireless network to facilitate communication among plants with their controllers; and (5) actuators that change plant states based on commands received from the controller. Figure 8.2 illustrates the structure of a WNCS pair. It is important to note that $\tau_{sc}(t)$ represents wireless network latency between the sensor and the controller, $\tau_{ca}(t)$ is wireless latency between the controller and actuator, and $\gamma(t)$ is the packet loss indicator.

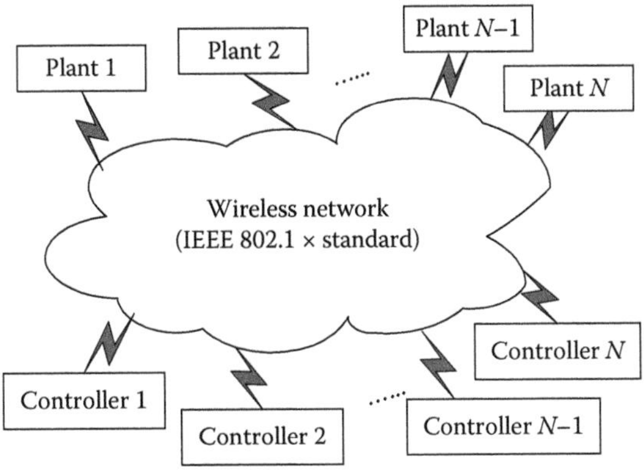

FIGURE 8.1 Multiple WNCS pairs.

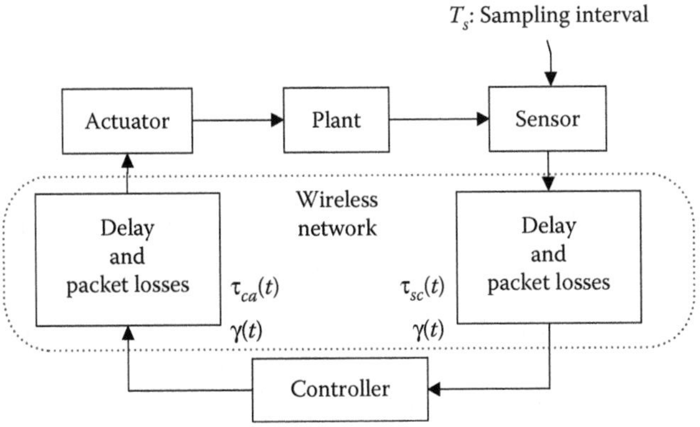

FIGURE 8.2 Structure of one WNCS pair.

8.2 WIRELESS NETWORKED CONTROL SYSTEMS CODESIGN

8.2.1 Overview

In our algorithm, the control design and scheduling protocols are implemented into all WNCS pairs that share the wireless network. Each WNCS pair tunes its stochastic optimal controller under wireless imperfections (e.g., wireless network latency and packet losses) caused by current distributed scheduling design, estimates its stochastic value function (Stengel 1986) based on tuned control design, and transmits that information (i.e., value function) to the link layer. The link layer tunes the distributed scheduling scheme based on throughput from the link layer and the stochastic value function value received from the application layer (i.e., controlled plant). The cross-layer WNCS codesign framework is shown in Figure 8.3.

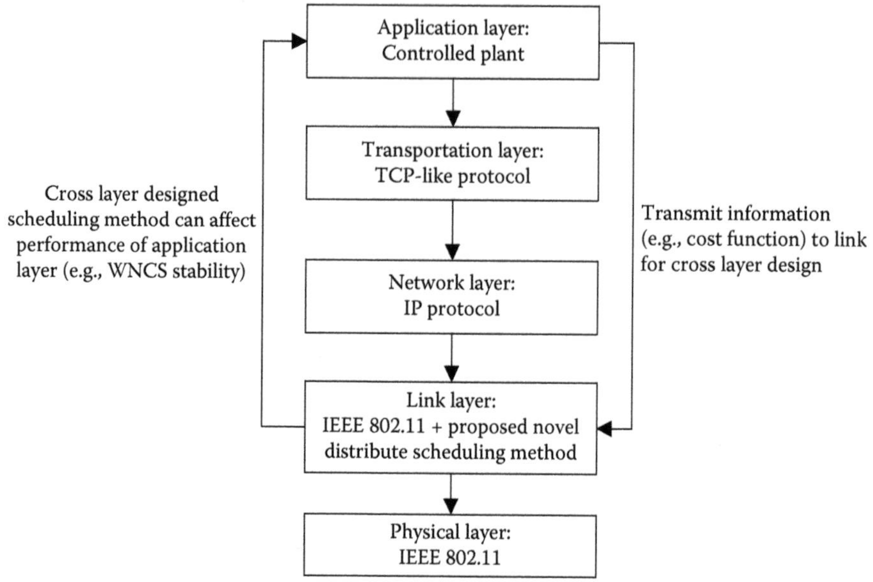

FIGURE 8.3 Network structure for the cross-layer design.

8.2.2 Plant Model

Suppose each WNCS pair is described as a linear time-invariant continuous-time system $\dot{x}^l(t) = A^l x(t) + B^l u^l(t), \forall l = 1,2,\ldots,N$ with system dynamics A^l, B^l for the lth WNCS pair, and sampling interval T_s. For stochastic optimal control design, the wireless network latency for every WNCS pair has to be bounded by $\tau^l \le \bar{d}T_s, \forall l = 1,\ldots,N$, which needs to be ensured by the cross-layer distributed scheduling protocol. Considering wireless network latency and packet losses, the lth WNCS pair dynamics can be represented as (Xu and Jagannathan 2011)

$$x_{k+1}^l = A_s^l x_k^l + B_k^{l1} u_{k-1}^{la} + B_k^{l2} u_{k-2}^{la} + \cdots + B_k^{l\bar{d}} u_{k-\bar{d}}^{la} + B_k^{i0} u_k^{la}$$

$$u_{k-i}^{la} = \gamma_{k-i}^l u_{k-i}^l \quad \forall i = 0,1,2,\ldots,\bar{d}, \quad \forall k = 0,1,2,\ldots \tag{8.1}$$

where u_k^{la} is the actual control input received by the lth WNCS actuator at time kT_s, u_k^l is the control input computed by the lth WNCS controller at time kT_s, and stochastic variables γ_k^l model the packet losses for the lth WNCS at time kT_s, which follows the Bernoulli distribution with $P(\gamma_k^l = 1) = \bar{\gamma}^l$. $A_s^l, B_k^{l1},\ldots,B_k^{l\bar{d}}$ represent the augment system model dynamics resulting from wireless network latency and packet losses for the lth WNCS at time kT_s (*Note*: The definition of these dynamics is given in the previous chapters and from Xu and Jagannathan (2011), and are omitted here). By defining the augment state $z_k^l = [(x_k^l)^T \; (u_{k-1}^l)^T \; (u_{k-2}^l)^T \; \ldots \; (u_{k-\bar{d}}^l)^T]^T$ of the lth WNCS pair at time kT_s, the plant dynamics (8.1) can be rewritten as

$$z_{k+1}^l = A_{zk}^l z_k^l + B_{zk}^l u_k^l \tag{8.2}$$

where the time-varying augmented system matrices are given

$$A_{zk}^{l} = \begin{bmatrix} A_s^l & \gamma_{k-1}^l B_k^{l1} & \cdots & \gamma_{k-i}^l B_k^{lk} & \cdots & \gamma_{k-\bar{d}}^l B_k^{l\bar{d}} \\ 0 & 0 & \cdots & \cdots & \cdots & 0 \\ 0 & I_m & \cdots & \cdots & 0 & 0 \\ \vdots & 0 & I_m & \cdots & \cdots & 0 \\ \vdots & \vdots & & \ddots & & \vdots \\ 0 & 0 & \cdots & \cdots & I_m & 0 \end{bmatrix}$$

$$B_{zk}^{l} = \begin{bmatrix} \gamma_k^l B_k^{l0} \\ I_m \\ 0 \\ 0 \\ \vdots \\ 0 \end{bmatrix}$$

It is important to note that there are two main challenges in Equation 8.2 for the codesign. In practical WNCS, since wireless imperfections are not known before hand, system representation (8.2) is uncertain, and stochastic optimal control of WNCS has to be designed without knowing the system dynamics, which is the first challenge for the control part of the codesign. Second, stochastic optimal control is designed based on the constraints of wireless imperfections. However, these wireless imperfections depend upon the network scheduling scheme. Therefore, designing an optimal distributed scheduling protocol that not only optimizes wireless network performance but also satisfies its network imperfection constraints from different WNCS pairs is another challenge for the codesign. Based on these challenges, the stochastic optimal controller and cross-layer distributed scheduling schemes are proposed.

8.2.3 Stochastic Optimal Control Design

In this part, a novel stochastic optimal control scheme is proposed for uncertain plant dynamics due to wireless imperfections. Without loss of generality, the lth WNCS pair is chosen for convenience for the optimal control development. Based on optimal control theory (Stengel 1986), the lth WNCS stochastic value function can be defined as

$$V_k^l = \underset{\tau,\gamma}{E}\{(z_k^l)^T P_k^l z_k^l\} \tag{8.3}$$

where $P_k^l \geq 0$ is the solution of the SRE for the lth WNCS pair and $\underset{\tau,\gamma}{E}\{\cdot\}$ is the expect operator of $\{(z_k^l)^T P_k^l z_k^l\}$. Stochastic optimal control for the lth WNCS pair at time kT_s

can be solved by minimizing the value function, that is, $(u_k^l)^* = \arg\min V_k^l, \forall k = 1,2,\ldots$. Similar to Xu and Jagannathan (2011), the value function (8.3) can be expressed as

$$V_k^l = \underset{\tau,\gamma}{E}\{[(z_k^l)^T \; u_k^{lT}] H_k^l [(z_k^l)^T \; u_k^{lT}]^T\} = (\bar{h}_k^l)^T \bar{\chi}_k^l \tag{8.4}$$

where

$$\bar{H}_k^l = \underset{\tau,\gamma}{E}(H_k^l) = \begin{bmatrix} \bar{H}_k^{lzz} & \bar{H}_k^{lzu} \\ \bar{H}_k^{luz} & \bar{H}_k^{luu} \end{bmatrix}$$

$$= \begin{bmatrix} S_z^l + \underset{\tau,\gamma}{E}[(A_{zk}^l)^T P_{k+1}^l A_{zk}^l] & \underset{\tau,\gamma}{E}[(A_{zk}^l)^T P_{k+1}^l B_{zk}^l] \\ \underset{\tau,\gamma}{E}[(B_{zk}^l)^T P_{k+1}^l A_{zk}^l] & R_z^l + \underset{\tau,\gamma}{E}[(B_{zk}^l)^T P_{k+1}^l B_{zk}^l] \end{bmatrix}$$

$\bar{h}_k^l = vec(\bar{H}_k^l)$, $\chi_k^l = [(z_k^l)^T \; u^T(z_k^l)]^T$ and $\bar{\chi}_k^l$ is the Kronecker product quadratic polynomial basis vector of the lth WNCS pair and $\bar{h}_k^l = vec(\bar{H}_k^l)$ with the vector function acting on square matrices, thus yielding a column vector (*Note*: The vec(·) function is constructed by stacking the columns of the matrix into one column vector with the off-diagonal elements that can be combined as $H_{mn}^l + H_{nm}^l$ (Xu and Jagannathan 2011). According to Stengel (1986), the optimal control of the lth WNCS pair can be expressed by using the H^l matrix (8.4) as

$$\begin{aligned} \underset{\tau,\gamma}{E}[(u_k^l)^*] &= -[R_z^l + \underset{\tau,\gamma}{E}(B_{zk}^{lT} P_{k+1}^l B_{zk}^l)]^{-1} \underset{\tau,\gamma}{E}(B_{zk}^{lT} P_{k+1}^l A_{zk}^l) z_k^l \\ &= -\underset{\tau,\gamma}{E}\left[(\bar{H}_k^{luu})^{-1} \bar{H}_k^{luz} z_k^l\right] \end{aligned} \tag{8.5}$$

Therefore, if the H^l matrix is obtained for the lth WNCS pair, then the stochastic optimal control is solved. However, since the system dynamics is unknown, the H^l matrix cannot be solved directly. Similar to Xu and Jagannathan (2011), we adaptively estimate the stochastic value function H^l matrix and obtain the certainty-equivalent optimal control.

The value function estimation error e_{hk}^l of the lth WNCS pair at time kT_s can be defined and expressed as $\hat{V}_k^l - \hat{V}_{k-1}^l + z_k^{lT} S_z^l z_k^l + u_k^{lT} R_z^l u_k^l = e_{hk}^l$, where $\hat{V}_l$ is the estimated stochastic value function of the lth WNCS pair at time kT_s, and S_z^l, R_z^l are positive definite matrix and positive semidefinite matrix of the lth WNCS pair, respectively. Then, the update law for the parameter vector for the value function can be given as

$$\underset{\tau,\gamma}{E}\left(\hat{\bar{h}}_{k+1}^l\right) = \underset{\tau,\gamma}{E}\left(\hat{\bar{h}}_k^l + \alpha_h^l \frac{\Delta\bar{\chi}_k^l (e_{hk}^l - z_k^{lT} S_z^l z_k^l - u_k^{lT} R_z^l u_k^l)^T}{\Delta\bar{\chi}_k^{lT} \Delta\bar{\chi}_k^l + 1}\right) \tag{8.6}$$

where $\bar{\chi}_k^l$ is the regression function of the lth WNCS pair, $\Delta\bar{\chi}_k^l$ is defined as $\Delta\bar{\chi}_k^l = \bar{\chi}_k^l - \bar{\chi}_{k-1}^l$, and α_h^l is the learning rate of the value function estimator for the lth WNCS pair.

Based on the estimated H^l matrix, the stochastic optimal control policy for the lth WNCS pair can be expressed as

$$\underset{\tau,\gamma}{E}\left(\hat{u}_k^l\right) = -\underset{\tau,\gamma}{E}\left[\left(\hat{\bar{H}}_k^{luu}\right)^{-1}\hat{\bar{H}}_k^{luz} z_k^l\right] \tag{8.7}$$

Algorithms 8.1 and 8.2 (Xu and Jagannathan 2013) represent the proposed stochastic optimal control design, while Theorem 8.1 shows that the value function estimation errors are asymptotically stable in the mean. Further, the estimated control inputs will also converge to the optimal control signal asymptotically.

Algorithm 8.1

Stochastic optimal control for the lth WNCS pair

1. **Initialize:** $\hat{\bar{h}}_0^l = \mathbf{0}$ and implementing admissible control u_0^l
2. **while** $\{kT_s \le t < (k+1)T_s\}$ **do**
3. **Calculate** the value function estimation error e_{hk}^l
4. **Update** the parameters of the value function estimator
5. $\underset{\tau,\gamma}{E}\left(\hat{\bar{h}}_{k+1}^l\right) = \underset{\tau,\gamma}{E}\left(\hat{\bar{h}}_k^l + \alpha_h^l \dfrac{\Delta\bar{\chi}_k^l (e_{hk}^l - z_k^{lT} S_z^l z_k^l - u_k^{lT} R_z^l u_k^l)^T}{\Delta\bar{\chi}_k^{lT}\Delta\bar{\chi}_k^l + 1}\right)$
6. **Update** control input based on the estimated H^l matrix
7. $\underset{\tau,\gamma}{E}\left(\hat{u}_k^l\right) = -\underset{\tau,\gamma}{E}\left[\left(\hat{\bar{H}}_k^{luu}\right)^{-1}\hat{\bar{H}}_k^{luz} z_k^l\right]$
8. **end while**
9. **Go to** the next time interval $[(k+1)T_s, (k+2)T_s)$ (i.e., $k = k+1$), and go back to line 2.

Algorithm 8.2

Plant of the lth WNCS pair

1. **Initialize:** lth WNCS pair states z_k^l
2. **while** $\{kT_s \le t < (k+1)T_s\}$ **do**
3. **Receive and implement** control inputs from controller
4. **if** multiple control inputs have been received at the plant on the same time **then**
5. **Apply** the recent control input to the plant and discard other control inputs [7]
6. **else if** old control input arrived after new control input being received at the plant **then**
7. **Discard** old control inputs and apply newer control inputs to the plant
8. **else** the control input is received by plant at different time and keep in order **then**

9. **Apply** the control inputs $\{u_{k-1}^l, u_{k-2}^l, \ldots, u_{k-\bar{d}}^l\}$ received during this time interval to the plant sequentially
10. **end if**
11. **end if**
12. **end while**
13. **Go to** next time period $[(k+1)T_s, (k+2)T_s]$ (i.e., $k = k+1$), and go back to while loop (line 2).

Theorem 8.1

Given the initial state z_0^l for the lth WNCS pair, estimated value function, and value function vector $\hat{\bar{h}}_k^l$ of the lth WNCS pair be bounded in the set **S**, let u_{0k}^l be any initial admissible control policy for the lth WNCS pair at the time kT_s (8.1) with wireless imperfections satisfying latency constraints (i.e., $\tau < \bar{d}T_s$) caused by the distributed scheduling protocol. Let the value function parameters be tuned and the estimated control policy be provided by Equations 8.6 and 8.7, respectively. Then, there exists positive constant α_h^l such that the system state z_k^l and stochastic value function parameter estimation errors $\tilde{\bar{h}}_k^l$ are all asymptotically stable in the mean. In other words, as $k \to \infty$, $z_k^l \to 0$, $\tilde{\bar{h}}_k^l \to 0$, $\hat{V}_k^l \to V_k^l$, and $\underset{\tau,\gamma}{E}\left(\hat{u}_k^l\right) \to \underset{\tau,\gamma}{E}\left[(u_k^l)^*\right] \forall l$.

Proof: Follow steps similar to Xu et al. (2012).

8.2.4 Optimal Cross-Layer Distributed Scheduling Scheme

In this section, we focus on a novel utility-optimal distributed scheduling design, which is mainly at the link layer. Therefore, without loss of generality, traditional wireless ad hoc network protocol (Jagannathan 2007) is applied to the other layers. For optimizing the performance of WNCS and satisfying the constraints from the proposed stochastic optimal control design in the application layer, a novel optimal cross-layer distributed scheduling algorithm is proposed here by using controlled plane information from the application layer.

First, the utility function for the lth WNCS pair is defined as

$$Utility_k^l = 2^{(R_l + \Delta V_k^l)} \tag{8.8}$$

Since the performance of the controlled plant and wireless network are considered in wireless network protocol design, the utility function includes two parts: (1) a value function from the lth WNCS pair's controlled plant at time kT_s is $\Delta V_k^l = (z_k^{lT} S_z^l z_k^l + u_k^{lT} R_z^l u_k^l) - (z_k^{lT} S_z^l z_k^l + u_k^{lT} R_z^l u_k^l)$; and (2) the throughput of the lth WNCS pair, R_l, which can be represented as Equation 8.9 by using Shannon theory as

$$R_l = B_{WNCS} \log_2\left(1 + \frac{P_l d_l^{-2}}{n_0^l B_{WNCS}}\right) \tag{8.9}$$

where B_{wncs} is the bandwidth of the entire wireless network, P_l is the transmitting power of the lth WNCS pair, d_l is the distance between the plant and the controller of the lth WNCS pair, and n_0^l is the constant noise density of lth WNCS pair.

Next, the optimal distributed scheduling problem can be formulated as maximizing the following utility function:

$$\text{maximize} \sum_{l=1}^{N} Utility_k^l = \text{maximize} \sum_{l=1}^{N} 2^{(R_l + \Delta V_k^l)}$$
$$\text{subject to}: \tau_k^l \le \bar{d}T_s \quad \forall l = 1,2,\ldots,N; \quad \forall k = 0,1,2,\ldots \tag{8.10}$$

where τ_k^l is the wireless network latency of the lth WNCS pair at kT_s.

It is important to note that the wireless network latency constraints in Equation 8.10 represent the proposed optimal control design constraints. Cross-layer distributed scheduling not only maximizes the sum of all WNCS pairs but also satisfies the wireless network latency requirement for every plant–controller pair, which in turn ensures that the proposed control design can optimize the controlled plant properly.

The main idea of the proposed distributed scheduling scheme is to separate the transmission time of different WNCS pairs by using a back-off interval (BI) (Jagannathan 2007) based on the utility function in a distributed manner. The proposed distributed scheduling framework is shown in Figure 8.4. For solving optimal scheduling problem (8.10) by different WNCS pairs, the BI is designed as

$$BI_k^l = \xi * \frac{\sum_{j=1}^{N} 2^{(R_j + \Delta V_k^j)}}{\beta_k^l \sum_{j=1}^{N} 2^{(R_j + \Delta V_k^j)} + 2^{(R_l + \Delta V_k^l)}}, \quad \forall l = 1,2,\ldots,N \tag{8.11}$$

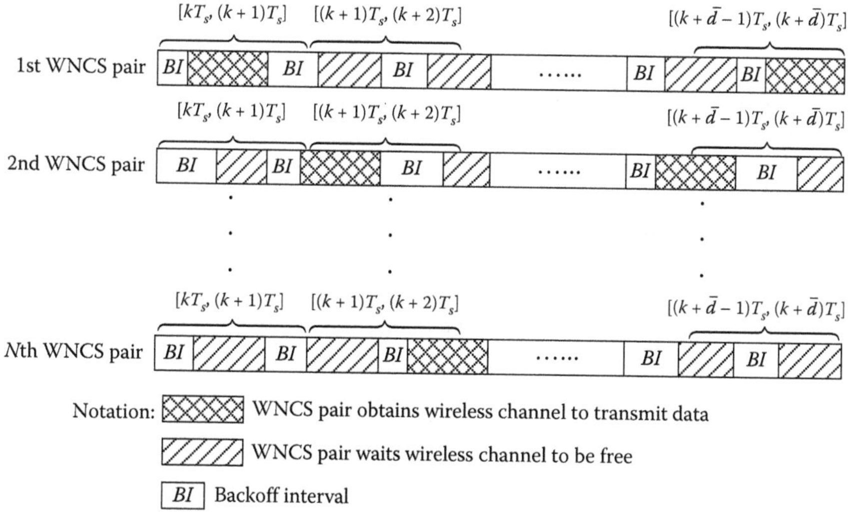

FIGURE 8.4 The proposed cross-layer distributed scheduling framework.

where ξ is the scaling factor and β_k^l is the balancing parameter of the lth WNCS pair at time kT_s, which is equal to the index of the first unsent packet stored in the transmission buffer of the lth WNCS pair. It is important to note that the balancing parameter is used to satisfy the latency constraints in Equation 8.10, which is illustrated in Theorem 8.2.

Algorithm 8.3

Novel utility-optimal cross-layer distributed scheduling scheme

1. **Initialize:** The balancing parameters are initialized as $\beta_0^l = 0, \forall l = 1,2,\ldots,N$, and each WNCS pair broadcasts its utility function value $2^{(R_l+\Delta V_0^l)}$ and receives utility function values of other pairs to calculate the network utility function value (i.e., $\sum_{l=1}^{N} 2^{(R_l+\Delta V_0^l)}$)
2. **While** $\{kT_s \le t \le (k+1)T_s\}$ **do**
3. **Calculate** back-off interval (BI) by different WNCS pair

$$BI_k^l = \xi * \frac{\sum_{j=1}^{N} 2^{(R_j+\Delta V_k^j)}}{\beta_k^l \sum_{j=1}^{N} 2^{(R_j+\Delta V_k^j)} + 2^{(R_l+\Delta V_k^l)}}, \quad \forall l = 1,2,\ldots,N$$

4. **Contend** wireless resource
5. **If** lth WNCS pair has the smallest BI **then**
6. **Schedule** lth WNCS pair and transmit lth WNCS pair's data through wireless network
7. **If** transmission is over, **then**
8. **Update** the scheduled WNCS pair's balancing parameter β_k^l
9. **Broadcast** the message to notify all the users that wireless channel is free
10. **end if**
11. **else**
12. **Update** entire wireless network utility $\sum_{l=1}^{N} 2^{(R_l+\Delta V_k^l)}$ and WNCS pairs' balancing parameters $\beta_k^i, \forall i,k$
13. **Wait** for wireless channel to be free
14. **end if**
15. **Update** time stamp: $t = t + BI_k^l + T_k^l$ (BI_k^l is the back-off interval of the scheduled WNCS pair. T_k^l is the transmission time of the scheduled WNCS pair)
16. **end while**
17. **Update and broadcast** utility function $2^{(R_l+\Delta V_k^l)}$ from all WNCS pairs
18. **Go to** next time period $[(k+1)T_s, (k+2)T_s)$ (i.e., $k = k+1$), and go back to line 2.

Remark 8.1

Since every WNCS pair decides its schedule by using local information, the proposed novel optimal cross-layer scheduling scheme is distributed. In this chapter, we assume that every WNCS pair broadcasts its utility function periodically in order to calculate the entire wireless network utility.

Remark 8.2

Compared with other distributed scheduling schemes (Vaidya et al. 2005; Zheng et al. 2009; Li and Negi 2010), the proposed algorithm designs the BI intelligently by optimizing the utility function instead of selecting it randomly (Vaidya et al. 2005; Zheng et al. 2009; Li and Negi 2010).

Theorem 8.2

The proposed distributed scheduling protocol based on cross-layer design delivers the desired performance in terms of satisfying the delay constraints $\tau_k^l \leq \bar{d}T_s$, $\forall l = 1,2,\ldots,N$; $\forall k = 0,1,2,\ldots$ (i.e., during $[kT_s,(k+\bar{d})T_s)$, every WNCS pair should be scheduled at least once).

Proof: If there exists an WNCS pair i that is not scheduled and another WNCS pair n has been scheduled a second time, then $\exists\, i \in [1,N]$, $\exists\, n \in [1,N]$, and $n \neq i$ such that $\beta_{k+s}^i - \beta_{k+s}^n > 1$,

$$BI_{k+s}^i = \frac{\sum_{j=1}^N 2^{(R_j+\Delta V_{k+s}^j)}}{\beta_{k+s}^i \sum_{j=1}^N 2^{(R_j+\Delta V_{k+s}^j)} + 2^{(R_i+\Delta V_{k+s}^i)}}$$

and $BI_{k+s}^n < BI_{k+s}^i$, where $s \in (0,\bar{d})$. Therefore,

$$BI_{k+s}^i = \frac{\sum_{j=1}^N 2^{(R_j+\Delta V_{k+s}^j)}}{\beta_{k+s}^i \sum_{j=1}^N 2^{(R_j+\Delta V_{k+s}^j)} + 2^{(R_i+\Delta V_{k+s}^i)}}$$

$$> \frac{\sum_{j=1}^N 2^{(R_j+\Delta V_{k+s}^j)}}{\beta_{k+s}^n \sum_{j=1}^N 2^{(R_j+\Delta V_{k+s}^j)} + 2^{(R_n+\Delta V_{k+s}^n)}} = BI_{k+s}^n \tag{8.12}$$

In other words,

$$-2^{(R_i+\Delta V_{k+s}^i)} > \sum_{j=1,j\neq n}^{N} 2^{(R_j+\Delta V_{k+s}^j)} \tag{8.13}$$

Since according to the definition $2^{(R_j+\Delta V_{k+s}^j)} > 0, \forall j = 0,1,\ldots,N$, $\forall k \sum_{j=1,j\neq n}^N 2^{(R_j+\Delta V_{k+s}^j)} > 0 > -2^{(R_i+\Delta V_{k+s}^i)}$, which is contrary to Equation 8.12. Therefore, if WNCS pair i has not been scheduled once, then WNCS pair n cannot be scheduled twice. Namely, there is no WNCS pair that can be scheduled twice before every WNCS pair has been scheduled once.

Theorem 8.3

When priorities of different WNCS pairs are equal, the proposed scheduling protocol can render best performance schedules for each WNCS pair.

Proof: Based on the definition of BI design (8.11), the priority term is $\beta_k^i \sum_{j=1}^N 2^{(R_j+\Delta V_k^j)}$. If it is same for any WNCS pair, it indicates

$$\beta_k^i \sum_{j=1}^{N} 2^{(R_j+\Delta V_k^j)} = \beta_k^l \sum_{j=1}^{N} 2^{(R_j+\Delta V_k^j)}, \quad \forall i, l \in [1, N] \text{ and } i \neq l \tag{8.14}$$

Therefore, for $\forall i, l \in [1, N]$ and $i \neq l$, the BI of different WNCS pair should satisfy

$$\frac{BI_k^i}{BI_k^l} = \frac{\left(\sum_{j=1}^N 2^{(R_j+\Delta V_k^j)}\right) \Big/ \left(\beta_k^i \sum_{j=1}^N 2^{(R_j+\Delta V_k^j)} + 2^{(R_i+\Delta V_k^i)}\right)}{\left(\sum_{j=1}^N 2^{(R_j+\Delta V_k^j)}\right) \Big/ \left(\beta_k^l \sum_{j=1}^N 2^{(R_j+\Delta V_k^j)} + 2^{(R_l+\Delta V_k^l)}\right)} = \frac{\beta_k^l \sum_{j=1}^N 2^{(R_j+\Delta V_k^j)} + 2^{(R_l+\Delta V_k^l)}}{\beta_k^i \sum_{j=1}^N 2^{(R_j+\Delta V_k^j)} + 2^{(R_i+\Delta V_k^i)}} > 1 \tag{8.15}$$

If and only if

$$2^{(R_l+\Delta V_k^l)} > 2^{(R_i+\Delta V_k^i)} \tag{8.16}$$

Next, proof for the proposed cross-layer distributed scheduling algorithm is given. First of all, the utility function of the whole WNCS is defined as

$$Utility^{tot} = \sum_{j=1}^{N} 2^{(R_j+\Delta V_k^j)} \tag{8.17}$$

Assume that set $\mathbf{S}_1$ is the unscheduled WNCS pairs set and $\mathbf{S}_2$ is the scheduled WNCS pairs set. $\forall l \in \mathbf{S}_1$, $\forall n \in \mathbf{S}_2$ such that $BI_k^l > BI_k^n$ and $2^{(R_l+\Delta V_k^l)} < 2^{(R_n+\Delta V_k^n)}$.

Therefore, $U_{S_2} = \sum_{j=\mathbf{S}_2} 2^{(R_j+\Delta V_k^j)}$ and $U_{S_2}^1$ can be derived as

$$\begin{aligned} U_{S_2}^1 &= \sum_{j=\mathbf{S}_2, j\neq n} 2^{(R_j+\Delta V_k^j)} + 2^{(R_l+\Delta V_k^l)} \\ &= \sum_{j=\mathbf{S}_2} 2^{(R_j+\Delta V_k^j)} + \left[2^{(R_l+\Delta V_k^l)} - \left(2^{(R_n+\Delta V_k^n)}\right)\right] \\ &< \sum_{j=\mathbf{S}_2} 2^{(R_j+\Delta V_k^j)} = U_{S_2} \end{aligned} \tag{8.18}$$

Thus, when all WNCS pair priorities are the same, the utility function of WNCS pairs based on the proposed scheduling algorithm achieves the maximum, which illustrates the optimality.

Remark 8.3

Fairness is an important factor to evaluate the performance of the scheduling algorithm. For the proposed cross-layer distributed scheduling, a fairness index (FI) (Bennett and Zhang 1996) is defined as

$$FI = \frac{\left(\sum_{i=1}^{N} R_i \Big/ \left(2^{(R_i+\Delta V^l)}\right)\right)^2}{N^* \sum_{i=1}^{N} \left(R_i \Big/ \left(2^{(R_i+\Delta V^l)}\right)\right)^2}$$

to measure the fairness among different WNCS pairs.

8.2.5 Numerical Simulations

To evaluate the cross-layer codesign, the wireless network includes 10 pairs of physical plant and remote controllers, which are located within a 150 m*150 m square area randomly. Since the batch reactor is considered as a benchmark example for WNCS (Xu and Jagannathan 2011), all 10 pairs use it.

EXAMPLE 8.1

The continuous-time model is

$$\dot{x} = \begin{bmatrix} 1.38 & -0.2077 & 6.715 & -5.676 \\ -0.5814 & -4.29 & 0 & 0.675 \\ 1.067 & 4.273 & -6.654 & 5.893 \\ 0.048 & 4.273 & 1.343 & -2.104 \end{bmatrix} x + \begin{bmatrix} 0 & 0 \\ 5.679 & 0 \\ 1.136 & -3.146 \\ 1.136 & 0 \end{bmatrix} u \tag{8.19}$$

First, the performance of the proposed stochastic optimal control algorithm is shown in Figures 8.5 and 8.6. Owing to page limitation, and without loss of generality, an average value of 10 different state regulation errors is shown in Figure 8.5. The results indicate that the stochastic optimal control under wireless network latency with unknown dynamics can make the state regulation errors converge to zero quickly while ensuring that all WNCS are stable. Note that there are some overshoots observed at the beginning because the optimal control tuning needs time.

Second, the performance of the proposed cross-layer distributed scheduling is evaluated. For comparison embedded round robin (ERR) (Jagannathan 2007) and Greedyscheduling (Dai and Prabhakar 2000) have been added. In Figure 8.7, the wireless network latency of each WNCS pair is shown. At the beginning, based on the different value of the utility function defined in Equation (8.8), one WNCS pair contends the wireless resource to communicate. Meantime, since the unscheduled WNCS pairs have to wait, the wireless network latencies of the other WNCS pairs

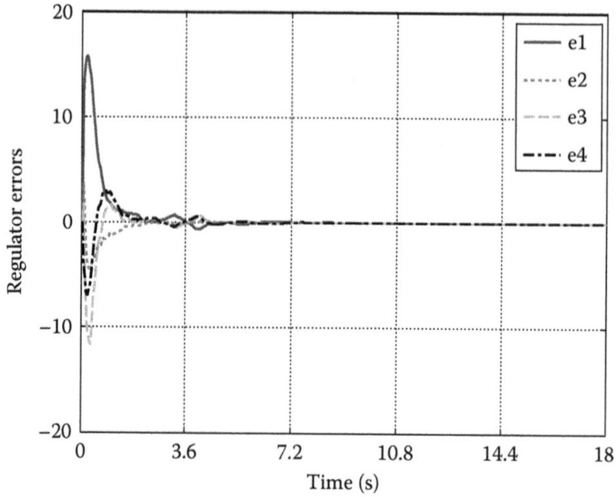

FIGURE 8.5 State regulation errors with stochastic optimal control.

have been increased. However, with wireless network latencies increasing, their BI values have to be decreased and the scheduled WNCS pair's BI need to be increased based on the proposed cross-layer distributed algorithm (8.11). Therefore, when the BI of the unscheduled WNCS pair is smaller than the scheduled BI, it can access the wireless resource to transmit. It is important to note that wireless network latency of all 10 WNCS pairs have never been increased beyond $\bar{d}T_s$ (*Note*: $\bar{d} = 2, T_s = 0.034\,\text{s}$), which indicates that the wireless network latency constraints of all 10 WNCS pairs have been satisfied.

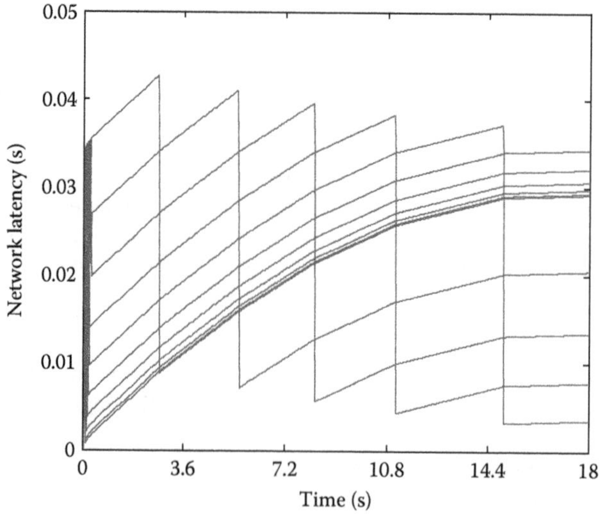

FIGURE 8.6 Fairness comparison.

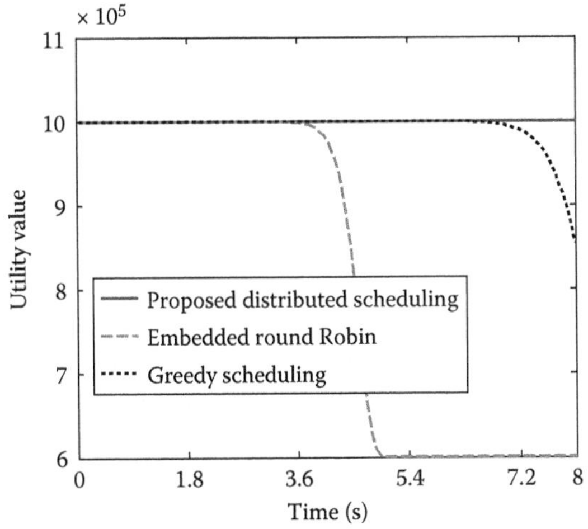

FIGURE 8.7 Performance of novel optimal cross-layer distributed scheduling algorithm: wireless network latency.

Third, the utility function of WNCS with three different scheduling schemes is compared. As shown in Figure 8.8, the proposed cross-layer scheduling maintains a high value while the utilities of WNCS with ERR and Greedy scheduling are much less than the proposed scheduling. It indicates that the proposed scheduling scheme can improve the performance of WNCS better than ERR and Greedy scheduling. It is important to note that since Greedy scheduling only optimizes the link layer performance and the utility function of WNCS is defined from both the link layer and the application layer, it cannot optimize the WNCS performance.

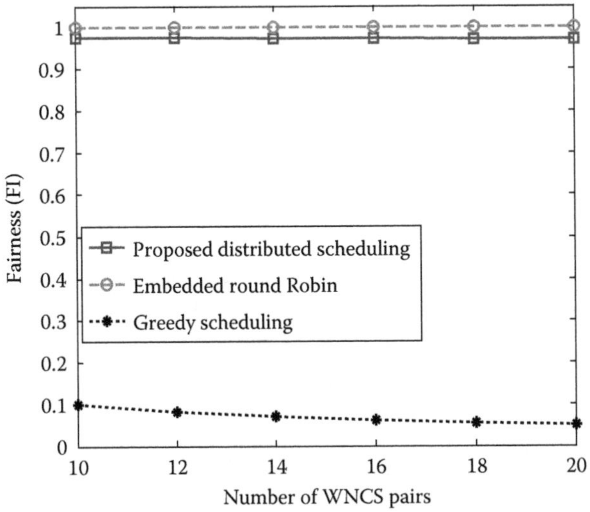

FIGURE 8.8 Utility comparison.

Eventually, the fairness of different scheduling algorithms with different number of WNCS pairs has been compared. As shown in Figure 8.8, fairness indices of the proposed cross-layer distributed scheduling and ERR scheduling schemes are close and equal to one, whereas that of Greedy scheduling is much less than one, thus indicating a fair allocation of wireless resource for the proposed one. According to the above results, the proposed cross-layer WNCS codesign optimizes the performance of both wireless network and plant.

8.3 CONCLUSIONS

In this chapter, through a novel utility-optimal cross-layer codesign, it is demonstrated that the proposed algorithm can optimize not only the performance of the controller but also the wireless network. The stochastic optimal control scheme does not require system dynamics and wireless network latency and packet losses, which are quite useful for hardware implementation, and the scheduling algorithm is utility-optimal and distributed, which is simpler and requires less computation when compared to centralized scheduling algorithms.

PROBLEMS

Section 8.4

Problem 8.4.1: Repeat Example 8.1 with 15 user pairs.

Problem 8.4.2: Repeat Example 8.1 with 10 user pairs and different initial conditions.

REFERENCES

Baillieul, J. and Antsaklis, P.J. 2007. Control and communication challenges in networked real-time systems. *Proceedings of the IEEE*, 95, 9–28.

Bennett, J.C.R. and Zhang, H. 1996. WF2Q: Worst-case fair weighted fair queuing. *Proceedings of the 15th IEEE International Joint Conference of Computer Societies and Networking of the Next Generation*, IEEE Press, San Francisco, USA, pp. 120–128.

Branicky, M.S., Phillips, S.M., and Zhang, W. 2000. Stability of networked control systems: Explicit analysis of delay. *Proceedings of the 2000 American Control Conference*, IEEE Press, Chicago, USA, pp. 2352–2357.

Dai, J.G. and Prabhakar, B. 2000. The throughput of data switch with and without speedup. *Proceedings of the 31th IEEE International Conference on Computer Communications*, IEEE Press, Tel-Aviv, Israel, pp. 556–564.

IEEE Standard 802.11a. 1999. R2003. Wireless LAN Medium Access Control (MAC) and Physical Layer (PHY) Specifications, High-Speed Physical Layer in the 5 GHz Band.

Jagannathan, S. 2007. *Wireless Ad-Hoc and Sensor Networks: Protocols, Performance, and Control*. CRC Press, Boca Raton, FL.

Li, Q. and Negi, R. 2010. Greedy maximal scheduling in wireless networks. *Proceedings of 2010 IEEE Global Communication Conference*, IEEE Press, Miami, USA, pp. 1–5.

Lian, F., Moyne, J., and Tilbury, D. 2001. Analysis and modeling of networked control systems: MIMO case with multiple time delays. *Proceedings of the American Control Conference*, Arlington, VA, USA, pp. 4306–4312.

Lian, F., Moyne, J., and Tilbury, D. 2003. Modeling and optimal controller design of networked control systems with multiple delays. *International Journal of Control*, 76, 591–606.

Nilsson, J., Bernhardsson, B., and Wittenmark, B. 1998. Stochastic analysis and control of real-time systems with random delays. *Automatica*, 34, 57–64.

Schenato, L., Sinopoli, B., Franceschetti, M., Polla, K., and Sastry, S. 2007. Foundations of control and estimation over lossy networks. *Proceedings of the IEEE*, 95, 163–187.

Stengel, R.F. 1986. *Stochastic Optimal Control: Theory and Application*. Wiley-Interscience, New York.

Vaidya, N., Dugar, A., Gupta, S., and Bahl, P. 2005. Distributed fair scheduling in wireless LAN. *IEEE Transactions on Mobile Computing*, 4, 616–629.

Xu, H. and Jagannathan, S. 2011. Stochastic optimal control of unknown linear networked control system using Q-learning methodology. *Proceedings of the 2011 American Control Conference*, IEEE Press, San Francisco, USA, pp. 2819–2824.

Xu, H. and Jagannathan, S. 2013. Optimal adaptive distributed power allocation for enhanced cognitive radio network in the presence of channel uncertainties. *International Journal of Computer Networks and Communications*, 5(1), 1–20.

Xu, H., Jagannathan, S., and Lewis, F.L. 2012. Stochastic optimal control of unknown networked control systems in the presence of random delays and packet losses. *Automatica*, 48, 1017–1030.

Zheng, D., Ge, W., and Zhang, J. 2009. Distributed opportunistic scheduling for ad hoc networks with random access: An optimal stopping approach. *IEEE Transactions on Information Theory*, 55, 205–222.

9 Event-Sampled Distributed Networked Control Systems

Embedded intelligent control (Tian et al. 2010) and cross-layer network protocol designs (Xia et al. 2012) have been considered as the two of the fastest-growing research areas in recent years. As mentioned by Xia et al. (2011), combining these two areas can introduce significant advantages for both modern control and networking communities. This novel class of systems is referred to as NCS (Day and Zimmermann 1983; Srivastava and Motani 2005; Nie and Comaniciu 2006; Guan et al. 2011).

Since the control and communication aspects are tightly coupled in NCS (Liu et al. 2005), one needs to consider these aspects together for efficient NCS design. Therefore, incorporating fixed communication network effects, Day and Zimmermann (1983) proposed a control scheme to maintain the stability of the control system part of the NCS in the presence of deterministic packet dropout rate. Srivastava and Motani (2005), from the communication network protocol side, implemented a stable control scheme and evaluated the performance of the widely used IEEE 802.11 protocol (e.g., carrier sense multiple access CSMA/CS) for DNCS.

However, research efforts by Day and Zimmermann (1983) and Srivastava and Motani (2005) have not considered the real-time interaction between the control and communication subsystems. A revolutionary scheme for DNCS should utilize the real-time interaction between the physical and cyber layers to optimize the performance of the NCS. From the well-known open system interconnection (OSI) architecture perspective (Liu et al. 2005), the physical systems of an NCS are connected through the communication network protocol that resides in the network layers. Therefore, cross-layer designs (Dai and Prabhakar 2000; IEEE Standard 2003), considering the interaction properly among different layers is necessary.

In IEEE Standard (2003), the authors have shown that cross-layer design can attain performance gains by exploiting the dependence among protocol layers when compared to the traditional individual layered protocol designs. However, most cross-layer designs are implemented for network layer and cyber layer (Dai and Prabhakar 2000) wherein the cyber layer is neglected. For DNCS, the controller and the IEEE 802.11 network protocol design for communication have to be considered jointly.

Among the different protocols utilized for communication networks, the distributed scheduling (Dai and Prabhakar 2000) scheme is critical for the communication protocol design when there are multiple systems sharing a common network. Compared to a traditional centralized scheduling (Vaidya et al. 2005), the main advantage of distributed scheduling is that it does not require a central processor to deliver the schedules after collecting information from all the communication links in the network. According to the IEEE 802.11 standard (Jiang and Walrand 2010), CSMA protocol is introduced to schedule communication links in a distributed manner where a communication link by requesting the network resources transmits the message only if it does not hear an ongoing transmission from the shared network.

Further, Jiang and Walrand (2010) derived a throughput-optimal distributed scheduling algorithm and proved that even a distributed scheduling scheme can still maximize the throughput. Moreover, a random access scheme is widely incorporated in most CSMA-based distributed scheduling (Dai and Prabhakar 2000; Zheng et al. 2009; Jiang and Walrand 2010) schemes. However, these protocols (Dai and Prabhakar 2000; Zheng et al. 2009; Jiang and Walrand 2010) are neither optimal nor suitable for NCS since they focus on improving network layer performance alone, which in turn neglects the effects on the control system in the cyber layer. In particular, the feedback signals and the control inputs are packetized and transmitted to the controller and actuator, respectively, via the communication network. Neglecting the effects of the network layer on the control scheme can result in significant deterioration in performance.

At the cyber layer, the control of such NCS has to be carefully dealt with since existing periodically sampled schemes (Jagannathan 2007) are unsuitable to transmit the feedback signals through over the communication network. Periodic sampling of feedback signals and transmission increases network resources causing potential congestion. Such NCS requires an event-sampled control system design framework wherein the state vector is transmitted to the controller in an aperiodic manner in order to save communication network resources unlike the stochastic optimal control design using periodic sampling as covered in the previous chapters. Specifically, the optimal event-triggered control design is necessary for NCS in order to minimize network resources.

It is important to note that the traditional optimal control schemes require periodically sampled state and control vector requiring significant network resources such as bandwidth for communication between the sensor and the controller, which could potentially cause congestion. In contrast, in the event-triggered design, the control input updates are nonperiodic whereas this creates challenges in terms of stability and convergence. Controlling an LNCS in an optimal way is quite challenging, especially in the presence of uncertain system dynamics (Guinaldo et al. 2012).

The article by Xu et al. (2014) proposes a novel cross-layer scheme for each system in a DNCS, which includes an optimal event-triggered controller design and a distributed scheduling scheme for the communication network in the network layers. The cross-layer design in the event-sampled domain brings several challenges in terms of selecting a utility function for both the layers, design of the optimal policy and network protocol, overall stability, and so on. Owing to the cross-layer design, significant benefits are expected.

Therefore, the article by Xu et al. (2014) will cover (1) an adaptive optimal event-triggered control scheme, which is designed in a forward-in-time manner without using the knowledge of linear dynamics of each system whereas most existing designs are either backward-in-time or require system dynamics, and (2) the distributed scheduling via cross-layer approach, which improves the performance of the

NCS by minimizing the cost function from both the network layer and physical systems. Compared with time-based periodic sampling, the proposed event-triggered scheme designed via the cross-layer approach utilizing event-based sampling saves significant computational and network resources (Sahoo et al. 2014).

First, Section 9.1 presents the background of DNCS. Subsequently, Section 9.2 develops optimal adaptive event-triggered control. Eventually, Section 9.3 provides a novel cross-layer distributed scheduling scheme for NCS.

9.1 DISTRIBUTED NETWORKED CONTROL SYSTEMS

During the past several decades, great strides have been made such that novel communication network and control designs provide benefits such as robustness, optimality, and flexibility. Modern control theory has been developing quickly during the past few decades to address challenges in such systems with constrained resources. Of all the methods, optimal control and event-triggered control are two important and relevant ones for discussion.

Owing to an explosive growth in these areas, an interesting idea is to combine the network protocol design and control theory together to harvest the benefits. This novel class of NCS is a first step to attain the cyber physical system (Xia et al. 2011, 2012). The basic structure of a DNCS is described in Figure 9.1, where numerous systems communicate to their corresponding controllers through a common communication network such as an IEEE 802.1x. It is considered as a DNCS since multiple subsystems are distributed without any interconnection among them in the physical

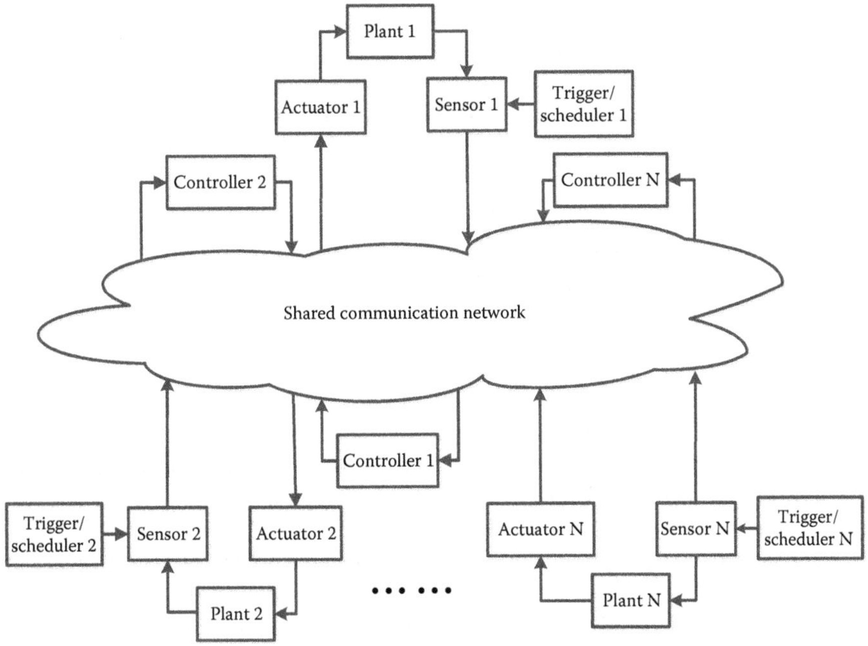

FIGURE 9.1 Distributed NCS.

layer. It is clear that the shared communication network will affect the performance of the control system. For example, when the shared communication network gets congested due to incorrect scheduling, the NCS subsystem under consideration cannot maintain stability due to lack of transmission of information from the physical system to its controller. Therefore, a novel cross-layer design is necessary for this DNCS where the design incorporates the control and distributed scheduling simultaneously. Without loss of generality, numerous coupled homogeneous systems are considered as a part of the DNCS in this chapter.

Without loss of generality, consider each system in the overall NCS described as a LTI continuous-time system described by $\dot{x}_l(t) = A_l x(t) + B_l u_l(t), \forall l = 1,2,\ldots,N$, where A_l, B_l represents the dynamics of the lth system and T_s is the sampling interval. For the optimal control design, the communication network delay for each system is assumed to be unknown but bounded as $\tau_l \le \bar{d}T_s, \forall l = 1,\ldots,N$ and this needs to be ensured by selecting the proper cross-layer distributed network scheduling protocol. Then, incorporating the network delay and packet losses, the dynamics of the lth system can be represented (Wei et al. 2002) as a stochastic time-varying discrete-time system described by

$$
\begin{aligned}
x_{l,k+1} &= A_{l,s}x_{l,k} + B_{l1,k}u_{la,k-1} + B_{l2,k}u_{la,k-2} + \cdots + B_{l\bar{d},k}u_{la,k-\bar{d}} + B_{l0,k}u_{la} \\
u_{la,k-i} &= \gamma_{l,k-i}u_{l,k-i} \quad \forall l = 0,1,2,\ldots,\bar{d}, \forall k = 0,1,2,\ldots
\end{aligned}
\tag{9.1}
$$

where $u_{la,k}$ is the actual control policy received by the actuator of the lth system at the time instant kT_s, $u_{l,k}$ is the control input computed by the lth controller at the time instant kT_s, $\gamma_{l,k}$ models the stochastic variable representing the packet losses for the lth system at the time instant kT_s, which follows the Bernoulli distribution, with $P(\gamma_{l,k} = 1) = \bar{\gamma}_l$, $A_{l,s}, B_{l1,k},\ldots,B_{l\bar{d},k}$ denoting the augment system dynamics resulting by incorporating the communication network delay and packet losses for the lth system at the time instant kT_s.

Defining the augment state vector to be $z_{l,k} = [(x_{l,k})^T \; (u_{l,k-1})^T \; (u_{l,k-2})^T \ldots (u_{l,k-\bar{d}})^T]^T$ for the lth system at the time instant kT_s, the system dynamics (9.1) can be rewritten as a linear stochastic time-varying system described by

$$
z_{l,k+1} = A_{l,zk}z_{l,k} + B_{l,zk}u_{l,k}, \quad \forall l = 0,1,2,\ldots,\bar{d}, \quad \forall k = 0,1,2,\ldots \tag{9.2}
$$

Note that the detailed time-varying augmented system matrices are given in Guinaldo et al. (2012). After incorporating the random delays and packet losses from the network, the continuous-time linear system becomes an uncertain stochastic linear discrete-time system as in Equation 9.2.

It is well known that event-sampled or event-triggered control (Dimarogonas and Johanson 2009; Eqtami et al. 2010; Mazo Jr. and Tabuada 2011) can bring in significant advantages for DNCS, in terms of saving computational cost and reducing network traffic enormously. Further, the event-triggered control framework is incorporated into the DNCS design. However, compared to stabilizing event-triggered control, optimality is more preferred (Xu et al. 2014). Therefore, a novel optimal adaptive event-triggered control is developed in the next section.

9.2 OPTIMAL ADAPTIVE EVENT-SAMPLED CONTROL

In this section, a novel optimal adaptive zero-order-hold (ZOH)-based event-triggered control of DNCS is developed. First, the general ZOH event-triggered control system is introduced. Subsequently, optimal adaptive ZOH event-triggered control scheme is derived for DNCS, which can maintain stability even when the system dynamics are unknown and in the presence of event-based sampling of system state and control input vector. Without loss of generality, the lth system is selected to describe the optimal adaptive event-triggered control as follows.

9.2.1 ZOH-Based Event-Triggered Control System

Recently, ZOH-based event-triggered control of linear systems has been a topic of significant interest to the NCS community due to its network benefits (Wang and Lemmon 2011). In Figure 9.2, the structure of a ZOH-based event-triggered control system is shown. Compared to time-driven control systems, a trigger mechanism is included at the sensor node of each system to decide when to transmit the system state information z_k over the communication network to the controller.

According to Equations 9.1 and 9.2, the lth system is described as $z_{l,k+1} = A_{l,zk}z_{l,k} + B_{l,zk}u_{l,k} + w_{l,k}$, where $w_{l,k}$ is a random disturbance input with zero mean. Here, a ZOH event-triggered controller will hold the last received system state vector until a new state vector is received. In order to improve the performance of the ZOH event-triggered system, the optimal control theory (Stengel 1986) has to be incorporated into the event-triggered control design. However, since network imperfections can bring uncertainty into system dynamics (9.2), a novel ADP-based stochastic optimal event-triggered control (Wang and Lemmon 2011) scheme will be introduced as follows.

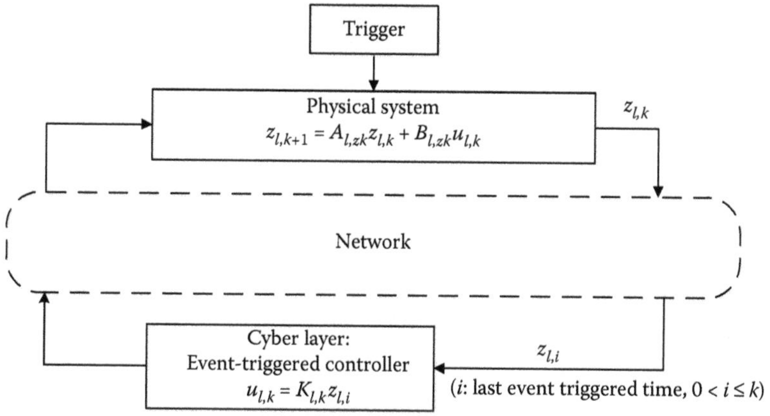

FIGURE 9.2 ZOH event-triggered system with controller implemented at the cyber layer.

9.2.2 Optimal Adaptive ZOH-Based Event-Triggered Control

In this section, optimal ZOH event-triggered control system design is given in a detailed manner. First, a certainty-equivalent value function is introduced and estimated adaptively. Then, the optimal ZOH event-triggered control scheme is developed by utilizing the estimated value function.

9.2.2.1 Value Function Setup

According to stochastic optimal control theory (Stengel 1986), the certainty-equivalent stochastic value function for the lth system can be represented as

$$V_{l,k} = \sum_{k=0}^{\infty} (z_{l,k}^T Q_{l,z} z_{l,k} + u_{l,k}^T S_{l,z} u_{l,k}) = \underset{\tau,\gamma}{E}\{z_{l,k}^T P_{l,k} z_{l,k}\} \tag{9.3}$$

where $Q_{l,z}$, $S_{l,z}$ are positive semidefinite and positive definite matrices, respectively, $P_{l,k} \geq 0$ is the solution to the certainty-equivalent SRE for the lth system with $\underset{\tau,\gamma}{E}\{\cdot\}$ the expectation operator of $\{(z_{l,k})^T P_{l,k} z_{l,k}\}$. Then, the Hamiltonian for the lth system can be represented as

$$H(z_{l,k}, u_{l,k}) = r(z_{l,k}, u_{l,k}) + V(z_{l,k+1}, u_{l,k+1}) - V(z_{l,k}, u_{l,k}) \tag{9.4}$$

where $r(z_{l,k}, u_{l,k}) = z_{l,k}^T Q_{l,z} z_{l,k} + u_{l,k}^T S_{l,z} u_{l,k}$ is a one-step cost-to-go function. Utilizing the standard optimal control theory (Lewis and Syrmos 1995), the stochastic optimal control signal of the lth system at the time instant kT_s can be solved by minimizing this stochastic value function (9.3) (i.e., $(u_{l,k})^* = \arg\min V_{l,k}$, $\forall l = 1,2,\ldots,N$, $\forall k = 1,2,\ldots$), which yields

$$\underset{\tau,\gamma}{E}(u_{l,k}^*) = \underset{\tau,\gamma}{E}(K_{l,k}^* z_{l,k}) = -\left[R_{l,z} + \underset{\tau,\gamma}{E}(B_{l,zk}^T P_{l,k+1} B_{l,zk})\right]^{-1} \underset{\tau,\gamma}{E}(B_{l,zk}^T P_{l,k+1} A_{l,zk}) z_{l,k} \tag{9.5}$$

Remark 9.1

It is important to note that computing the optimal control policy in Equation 9.5 requires the system dynamics that is not usually known for an NCS as mentioned in the previous chapters.

On the other hand, the stochastic optimal action-dependent value function with event-sampled state vector can be expressed as

$$V_{l,k} = \underset{\tau,\gamma}{E}\{[(z_{l,k})^T \; u_{l,k}^T] H_{l,k} [(z_{l,k})^T \; u_{l,k}^T]^T\} = \bar{h}_{l,k}^T \bar{\chi}_{l,k} \tag{9.6}$$

where

$$\bar{H}_{l,k}=\underset{\tau,\gamma}{E}(H_{l,k})=\begin{bmatrix}\bar{H}_{l,k}^{zz} & \bar{H}_{l,k}^{zu}\\ \bar{H}_{l,k}^{uz} & \bar{H}_{l,k}^{uu}\end{bmatrix}=\begin{bmatrix}Q_{l,z}+\underset{\tau,\gamma}{E}[(A_{l,zk})^T P_{l,k+1}A_{l,zk}] & \underset{\tau,\gamma}{E}[(A_{l,zk})^T P_{l,k+1}B_{l,zk}]\\ \underset{\tau,\gamma}{E}[(B_{l,zk})^T P_{l,k+1}A_{zk}^{l}] & S_{l,z}+\underset{\tau,\gamma}{E}[(B_{l,zk})^T P_{l,k+1}B_{l,zk}]\end{bmatrix}$$

$\chi_{l,k}=[(z_{l,k})^T\ u^T(z_{l,k})]^T$, and $\bar{\chi}_{l,k}$ is the Kronecker product quadratic polynomial basis vector of the lth system and $\bar{h}_{l,k}=\text{vec}(\bar{H}_{l,k})$ with the vector function acting on square matrices thus yielding a column vector. Note that the vec(·) function is constructed by stacking the columns of the matrix into one column vector with the off-diagonal elements that can be combined as $H_{l,mn}+H_{l,nm}$ (Xu et al. 2012).

Next, according to Xu et al. (2014), the optimal control gain of the lth system can be represented by using the H_l matrix as

$$\underset{\tau,\gamma}{E}(K_{l,k}^{*})=-[R_{l,z}+\underset{\tau,\gamma}{E}(B_{l,zk}^{T}P_{l,k+1}B_{l,zk})]^{-1}\underset{\tau,\gamma}{E}(B_{l,zk}^{T}P_{l,k+1}A_{l,zk})=-(\bar{H}_{l,k}^{uu})^{-1}\bar{H}_{l,k}^{uz} \quad (9.7)$$

Therefore, if H_l matrix is solved for the lth system, the stochastic optimal control gain can be obtained by utilizing Equation 9.5. However, since the system dynamics are unknown, H_l cannot be solved directly. Similar to Wang and Lemmon (2011), a novel adaptive estimator will be proposed to estimate the value function, H matrix (i.e., H_l), and the stochastic optimal control gain.

9.2.2.2 Model-Free Online Tuning of Value Function

Recalling the ZOH event-triggered control scheme in Section 9.2.1 and the novel ADP technique in Donkers and Heemels (2012), the value function can be estimated when an event is triggered as

$$\hat{V}_{l,i}=\underset{\tau,\gamma}{E}\left\{[(z_{l,i})^T\ u_{l,i}^{T}]\hat{H}_{l,i}[(z_{l,i})^T\ u_{l,i}^{T}]^T\right\}=\hat{\bar{h}}_{l,i}^{T}\bar{\chi}_{l,i} \quad (9.8)$$

where iT_s is the triggering time instant with T_s denoting the sampling time. According to traditional optimal control theory (Lewis and Syrmos 1995), the Bellman equation is represented in terms of the value function as

$$V_{l,i+1}-V_{l,i}+\underset{\tau,\gamma}{E}[r(z_{l,i},u_{l,i})]=0 \quad (9.9)$$

However, Equation 9.9 cannot be guaranteed with the estimated $\hat{H}_{l,i}$. After incorporating the estimated $\hat{H}_{l,i}$ into the Bellman equation, Equation 9.9 can be presented as

$$\hat{V}_{l,i+1}-\hat{V}_{l,i}+\underset{\tau,\gamma}{E}[r(z_{l,i},u_{l,i})]=\underset{\tau,\gamma}{E}(e_{l,i+1}) \quad (9.10)$$

where $e_{l,i+1}$ denotes the temporary difference (TD) error (Xu et al. 2012) in the Bellman equation. Moreover, substituting Equation 9.8 into Equation 9.10, Equation 9.10 can be represented as

$$\begin{aligned} \underset{\tau,\gamma}{E}(e_{l,i+1}) &= \underset{\tau,\gamma}{E}[r(z_{l,i},u_{l,i})] + \hat{V}_{l,i+1} - \hat{V}_{l,i} = \underset{\tau,\gamma}{E}[r(z_{l,i},u_{l,i})] + \hat{\bar{h}}_{l,i+1}^{T}\bar{\chi}_{l,i+1} - \hat{\bar{h}}_{l,i+1}^{T}\bar{\chi}_{l,i} \\ &= \underset{\tau,\gamma}{E}[r(z_{l,i},u_{l,i})] + \hat{\bar{h}}_{l,i+1}^{T}(\bar{\chi}_{l,i+1} - \bar{\chi}_{l,i}) = \underset{\tau,\gamma}{E}[r(z_{l,i},u_{l,i})] + \hat{\bar{h}}_{l,i+1}^{T}\Delta\bar{\chi}_{l,i} \end{aligned} \tag{9.11}$$

with $\Delta\bar{\chi}_{l,i} = \bar{\chi}_{l,i+1} - \bar{\chi}_{l,i}$. Now, select the parameter tuning law for $\hat{H}_{l,i}$ as

$$\underset{\tau,\gamma}{E}(\hat{\bar{h}}_{l,i+1}) = \underset{\tau,\gamma}{E}\left(\hat{\bar{h}}_{l,i} + \alpha_l \frac{\Delta\bar{\chi}_{l,i} e_{l,i+1}^{T}}{\Delta\bar{\chi}_{l,i}^{T}\Delta\bar{\chi}_{l,i} + 1}\right) \tag{9.12}$$

where the TD error with event-sampled inputs drives the parameter update. Define the parameter estimation error as $\tilde{\bar{h}}_{l,i} = \bar{h}_l - \hat{\bar{h}}_{l,i}$, and the parameter estimation error dynamics can be expressed as

$$\underset{\tau,\gamma}{E}(\tilde{\bar{h}}_{l,i+1}) = \underset{\tau,\gamma}{E}\left(\tilde{\bar{h}}_{l,i} - \alpha_l \frac{\Delta\bar{\chi}_{l,i} e_{l,i+1}^{T}}{\Delta\bar{\chi}_{l,i}^{T}\Delta\bar{\chi}_{l,i} + 1}\right) \tag{9.13}$$

Remark 9.2

It is observed that the estimated stochastic value function will no longer be tuned when the system state vector converges to zero. It can be seen as a PE (Green and Moore 1989) requirement for the input to the value function estimator whereas the optimal adaptive ZOH event-triggered control system state vector must be persistently exciting long enough for the estimator to learn the value function. The PE condition is well known in adaptive control theory and can be ensured by adding exploration noise (Xu and Jagannathan 2012). This PE condition is satisfied only when there is a sufficient number of triggered events such that the adaptive estimator generates the H matrix.

Next, the estimation of the optimal adaptive ZOH event-triggered control is derived. According to Donkers and Heemels (2012), the optimal ZOH event-triggered control can be attained by minimizing the value function. Recalling Equation 9.7, the optimal control gain can be designed by using the estimated parameter $\hat{H}_{l,i}$ as

$$\underset{\tau,\gamma}{E}(\hat{u}_{l,i}) = \underset{\tau,\gamma}{E}(\hat{K}_{l,i}z_{l,i}) = -(\hat{H}_{l,i}^{uu})^{-1}\hat{H}_{l,i}^{ux}\underset{\tau,\gamma}{E}(z_{l,i}) \tag{9.14}$$

Obviously, the optimal adaptive ZOH event-triggered control gain can be obtained in terms of the $\hat{H}_{l,i}$ matrix, which is solved by estimating the value function with an event-sampled state vector. It is important to note that the proposed design (9.12) and (9.14) not only relaxes the requirement of the system dynamics but also eliminates the value and policy iterations.

Next, the event-triggering condition that facilitates the events is derived. Since the controller will hold the latest received system state (i.e., x_i, $0 < i \le k$) at the time instant kT_s, the measurement error e_k^{ZOH} or also referred to as event-trigger error can be represented in terms of the state vector as

$$\underset{\tau,\gamma}{E}(e_k^{ZOH}) = \underset{\tau,\gamma}{E}(x_k - x_i), \quad 0 < i \le k \tag{9.15}$$

When the measurement error e_k^{ZOH} exceeds the threshold defined as the event-trigger condition, the sensor will transmit a new state vector to the controller. Upon receiving the state vector, the controller generates the control input and resets the event-trigger error e_k^{ZOH} to zero. Moreover, the ZOH-based event-triggered closed-loop control system dynamics can be represented as

$$x_{k+1} = (A_{zk} + B_{zk}K_k)x_k - B_{zk}K_k e_k^{ZOH} \tag{9.16}$$

with

$$e_k^{ZOH} = \begin{cases} 0 & \text{event is initiated} \\ x_k - x_i & \text{event is not initiated} \end{cases}$$

Next, to maintain the stability of the ZOH event-triggered control system, a novel event-trigger condition can be represented as

$$\underset{\tau,\gamma}{E}\left\| e_k^{ZOH} \right\| \le \sigma(A_{zk}, B_{zk}, K_k)\underset{\tau,\gamma}{E}\left\| x_k \right\| \tag{9.17}$$

where $\sigma(A_{zk}, B_{zk}, K_k)$ is the state-dependent threshold of the event-trigger condition to be designed. The event will be triggered and the system state vector will be transmitted to the controller when the event-trigger error, e_k^{ZOH}, fails to meet the event-triggering condition given by Equation 9.17. On the other hand, an event will not be triggered and the controller will hold the previous received system state vector when the event-triggering condition (9.17) is satisfied. The design of the event-trigger condition is very challenging since the system matrices, A_{zk}, B_{zk} are unknown. Next, the stability of the event-triggered control scheme is introduced.

Theorem 9.1: ZOH-Based Optimal Event-Trigger Control Scheme

Consider the linear discrete-time systems in a DNCS (9.2) along with the adaptive estimator (9.12) and adaptive estimator-based optimal control (9.14). Let $u_0(x_{l,k})$ be an initial stabilizing control policy for the system (9.2). Let the adaptive estimator parameter $\hat{\bar{h}}_{l,0}$ be initialized in $\mathcal{D} \subset \Re^n$. Further, there exists a positive constant α_l such that the closed-loop event-triggered augmented system state vector, $z_{l,k}$, and adaptive estimator parameter estimation errors $\tilde{\bar{h}}_{l,k}$, are asymptotically stable in the mean, provided the system states are transmitted to the controller and the adaptive estimator parameters are updated using Equation 9.12 through the violation of an event-trigger condition defined by

$$D(\underset{\tau,\gamma}{E}\left\|e_{l,k}^{ZOH}\right\|) \le \sigma_{l,k} \underset{\tau,\gamma}{E}\left\|z_{l,k}\right\|$$

where $\sigma_{l,k} = \sqrt{(1-2l_o^2)\Gamma_l / 4B_{l,M}^2 \left\|\hat{K}_{l,k}\right\|^2}$ with $0 < l_o < 1/\sqrt{2}$, $0 < \Gamma_l < l$, $B_{l,M}$ is the upper bound on the control coefficient matrix $B_{l,zk}$, and $\hat{K}_{l,k}$ is the estimated control gain defined in Equation 9.14.

9.2.3 Cross-Layer Distributed Scheduling Design

In this section, a novel cross-layer design (Shakkottai et al. 2003; Reena and Jacob 2007) is proposed for DNCS. First, the main idea of a cross-layer design is provided. Subsequently, a novel distributed scheduling (Zheng et al. 2009) is proposed to optimize the DNCS by minimizing the cost function, which includes information from both the cyber and network layers. To the best knowledge of the authors, it is the first time that both the ZOH-based optimal adaptive event-triggered control design in the application layer and the distributed scheduling scheme design in the network layer are designed for the DNCS via a cross-layer approach.

9.2.3.1 Cross-Layer Design

In the proposed scheme, the novel event-triggered control design and distributed scheduling protocol are implemented into the cyber layer and network layer, respectively, for an NCS with a shared communication network. Since the NCS system dynamics and the scheduling scheme—as part of the communication network protocol—can affect each other, a cross-layer approach will be introduced to jointly optimize the NCS and communication network based on the information from both the cyber and network layers.

Each NCS will tune its stochastic optimal controller in the presence of network imperfections such as network-induced delay and packet losses resulting from the current distributed scheduling design, estimate its value function (Xu et al. 2012) based on tuned control design, and transmit that information (i.e., value function) to the network layer. The network layer tunes the distributed scheduling scheme based on throughput from the network layer and the value function information received from the physical systems. The cross-layer NCS design framework is shown in Figure 9.3.

9.2.3.2 Distributed Scheduling

In this section, optimal distributed scheduling is developed at the network layer. Without loss of generality, traditional wireless ad hoc network protocol

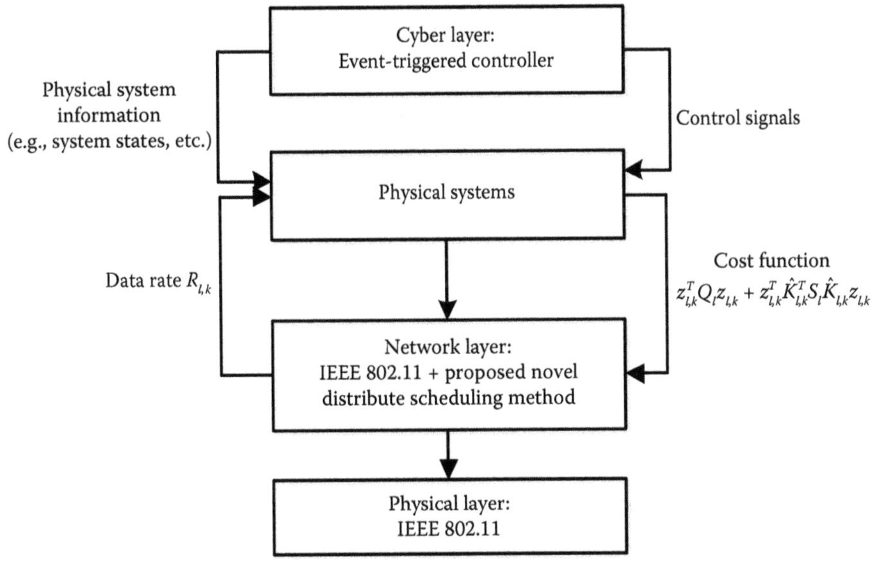

FIGURE 9.3 Framework for the cross-layer design.

(Jagannathan 2007) is implemented at the other protocol layers. In order to optimize the performance of the overall NCS, which includes the performance of both physical systems and network layers, a novel optimal cross-layer distributed scheduling algorithm is derived by incorporating control system information from the physical systems. Similar to Section 9.2.1, the NCS lth system is considered here.

First, to design the distributed scheduling and optimal control jointly, the cost function for the lth system can be represented as

$$J_{l,k} = \underset{\tau,\gamma}{E}(z_{l,k}^T Q_{l,z} z_{l,k} + u_{l,k}^T S_{l,z} u_{l,k} + \beta_l R_{l,k}) \tag{9.18}$$

where $R_{l,k}$ denotes the lth system average traffic payload during $[kT_s,(k+1)T_s]$ and β_l is the weight parameter of the average traffic payload. When β_l becomes large, it indicates that the average traffic payload will affect the total cost. Subsequently, the entire cost function for the DNCS can be expressed as

$$J_k = \sum_{l=1}^{M} J_{l,k} = \sum_{l=1}^{M} \underset{\tau,\gamma}{E}(z_{l,k}^T Q_{l,z} z_{l,k} + u_{l,k}^T S_{l,z} u_{l,k} + \beta_l R_{l,k}) \tag{9.19}$$

with M denoting the number of NCS systems. Then, the optimal design of the DNCS should minimize the cost function (9.19) to obtain

$$J_k^* = \min_{u,\pi} \sum_{l=1}^{M} \underset{\tau,\gamma}{E}(z_{l,k}^T Q_{l,z} z_{l,k} + u_{l,k}^T S_{l,z} u_{l,k} + \beta_l R_{l,k}) \tag{9.20}$$

where u is the control design and π is the scheduling policy.

Minimizing the cost function (9.20) requires the design of (1) a novel distributed scheduling, and an (2) optimal control scheme. For the control input design, the optimal adaptive ZOH event-triggered control scheme has been derived in Section 9.2. For the distributed scheduling part, a novel scheme will be developed in this section. Moreover, each NCS system has two options for scheduling its transmission of event-sampled state vector either to schedule or not to schedule. It is important to note that whether or not each NCS system is scheduled depends on which option will reduce the cost function. For example,

Case 1: lth system is scheduled

$$\begin{aligned} J_{l,k}^{S,1} &= \underset{\tau,\gamma}{E}(z_{l,k}^T Q_{l,z} z_{l,k} + z_{l,k}^T \hat{K}_{l,k}^T S_{l,z} \hat{K}_{l,k} z_{l,k} + \beta_l R_{l,k}^Y) \\ &= \underset{\tau,\gamma}{E}(z_{l,k}^T Q_{l,z} z_{l,k} + z_{l,k}^T \Lambda_{l,z} z_{l,k} + \beta_l R_{l,k}^Y) \end{aligned} \tag{9.21}$$

where $\Lambda_{l,z} = \hat{K}_{l,k}^T S_{l,z} \hat{K}_{l,k}$ and $R_{l,k}^Y$ is the average traffic payload when the NCS lth system has been scheduled.

Case 2: lth system is not scheduled

$$\begin{aligned} J_{l,k}^{S,2} &= \underset{\tau,\gamma}{E}(z_{l,k}^T Q_{l,z} z_{l,k} + z_{l,i}^T \hat{K}_{l,i}^T S_{l,z} \hat{K}_{l,i} z_{l,i} + \beta_l R_{l,k}^N) \\ &= \underset{\tau,\gamma}{E}(z_{l,k}^T Q_{l,z} z_{l,k} + z_{l,i}^T \Lambda_{l,i} z_{l,i} + \beta_l R_{l,k}^N) \end{aligned} \tag{9.22}$$

where $\Lambda_{l,i} = \hat{K}_{l,i}^T S_{l,z} \hat{K}_{l,i}$ with $\hat{K}_{l,i}$ and $z_{l,i}$ being held control gain and system state vector, which are defined in Section 9.2, and $R_{l,k}^N$ is the average traffic payload when the lth NCS system has not been scheduled.

Next, the difference between these two cases can be considered in the utility function and represented as

$$\begin{aligned} \Delta J_{l,k}^S &= \pi_{l,k}(J_{l,k}^{S,2} - J_{l,k}^{S,1}) \\ &= \pi_{l,k} \underset{\tau,\gamma}{E}(z_{l,k}^T Q_l x_{l,k} + z_{l,i}^T \Lambda_{l,i} z_{l,i} + \beta_l R_{l,k}^N) \\ &\quad -\pi_{l,k} \underset{\tau,\gamma}{E}(z_{l,k}^T Q_l z_{l,k} + z_k^T \Lambda_{l,k} z_{l,k} + \beta_l R_{l,k}^Y) \\ &= \underset{\tau,\gamma}{E}(\pi_{l,k} \phi(e_{l,k}^{ZOH}) - \pi_{l,k} \beta_l D_{l,k}) \end{aligned} \tag{9.23}$$

where $D_{l,k} = R_{l,k}^Y - R_{l,k}^N$ is the difference of average traffic payload for the lth NCS system between the two scenarios, which is expressed as

$$D_{l,k} = R_{l,k}^Y - R_{l,k}^N = \left(\frac{N_k + 1}{kT_s} - \frac{N_k}{kT_s} \right) N_{l,bit} = \frac{1}{kT_s} N_{l,bit} \tag{9.24}$$

where $N_{l,bit}$ is the number of bits for packetizing the sensed event of the lth NCS system, $\pi_{l,k}$ is the schedule indicator of the lth NCS system (i.e.,

$$\pi_{l,k}=\begin{cases}1, & \text{Scheduled}\\ 0, & \text{Not scheduled}\end{cases} \quad e_{l,k}^{ZOH}=z_{l,k}-z_{l,i} \quad \text{and} \quad \phi(e_{l,k}^{ZOH})=(z_{l,i}^{T}\Lambda_{l,i}z_{l,i}-z_{l,k}^{T}\Lambda_{l,k}z_{l,k})$$

Obviously, when $\Delta J_{l,k}^{S}>0$, it indicates that scheduling the NCS lth system to transmit over the network yields additional advantage. Otherwise, scheduling the NCS lth system will degrade the performance. Therefore, when $\Delta J_{l,k}^{S}>0$, this NCS system can be considered as scheduled. It is important to note that there are multiple NCS systems in a DNCS (i.e., M coupled NCS systems) eligible to transmit, and probably the utility functions of several subsystems are higher than zero, which indicates that all of these NCS systems have to be scheduled. However, according to the literature on networking (Dai and Prabhakar 2000), only one NCS system can access the network.

In order to optimize the performance of the network, the optimal scheduling policy should maximize the total utility function given by

$$(\Delta J_{k}^{S})^{*}=\max_{\pi}\sum_{l\in\mathbf{G}_{k}}\Delta J_{l,k}^{S} \tag{9.25}$$

where $\mathbf{G}_k$ is the subsystem set with positive value of the utility function at time instant kT_s, that is, $\Delta J_{l,k}^{S}>0$ for $l\in\mathbf{G}_k$. Obviously, for a centralized scheduler design, the optimal schedule has to select a subsystem that has the maximum value of $\Delta J_{l,k}^{S}$. However, in the centralized scheduling scheme, finding the maximum value $\Delta J_{l,k}^{S}$ requires significant information from every system, which might overload the network. Therefore, the novel distributed scheduling scheme is needed to solve this drawback.

In this chapter, the main idea of the scheduling algorithm as shown in Figure 9.4 is to separate the transmission time of different systems by using the BI (Jagannathan 2007), which is designed based on a related utility function in the distributed manner. To solve the optimal scheduling problem (9.25) for the DNCS, the BI can be designed as

$$BI_{l,k}=\varsigma\times(e^{-\Delta J_{l,k}}+n_{l,k}) \quad \text{for} \quad l\in\mathbf{G}_{k} \tag{9.26}$$

where ς is the scaling factor and $n_{l,k}$ is a random variable satisfying the Gaussian distribution (i.e., $n_{l,k}\sim L^{*}N(0,\sigma^{2})$), and $L=\min_{l,j\in G_k}(e^{-\Delta J_{l,k}}-e^{-\Delta J_{j,k}})$ is the range of the random value $n_{l,k}$. Next, the proposed novel distributed scheduling algorithm is introduced.

Algorithm

Novel optimal distributed scheduling scheme

1. **Initialize:** The utility functions are initialized as $\Delta J_{l,0} = 0$, $\forall l = 1,2,\ldots,M$
2. **While** $\{kT_s \le t < (k+1)T_s\}$ **do**
3. **Calculate** the back-off interval (BI) for different subsystems of the overall NCS adaptive
   ZOH event-triggered control system $BI_{l,k} = \varsigma \times (e^{-\Delta J_{l,k}} + n_{l,k})$ for $l \in \mathbf{G}_k$.
4. **Contend** the shared communication network resource.
5. **If** the lth system of the DNCS has the smallest BI **then**
6. **Schedule** event-triggered pair l and transmit the lth system's data through the shared communication network.
7. **If** transmission is over, **then**
8. **Update** the scheduled system's utility function $\Delta J_{l,k}$.
9. **end if**
10. else
11. **Wait** for shared communication network channel to be free.
12. end if
13. **Update** time stamp: $t = t + BI_{l,k} + T_{l,k}$ ($BI_{l,k}$ is the back-off interval of the scheduled DNCS. $T_{l,k}$ is the transmission time of scheduled system.)
14. **end while**
15. **Update and broadcast** utility function $\Delta J_{l,k}$ from each system of the overall NCS.
16. **Go to** next time period $[(k+1)T_s, (k+2)T_s)$ (i.e., $k = k+1$), and go back to line 2.

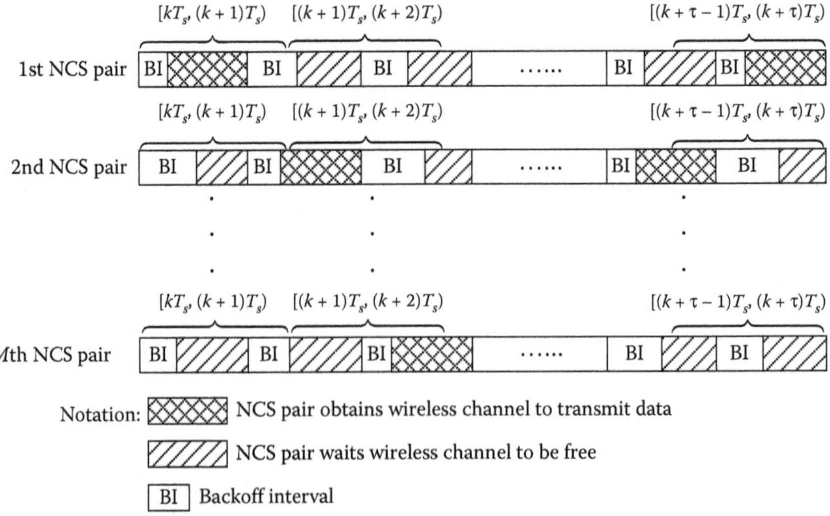

FIGURE 9.4 Framework of the proposed cross-layer distributed scheduling scheme.

Remark 9.3

Since each NCS system decides its scheduling policy by using local information from the physical and network layers, the proposed novel cross-layer scheduling scheme is distributed.

Remark 9.4

Compared with other distributed scheduling schemes (Vaidya et al. 2005; Jiang and Walrand 2010), the proposed algorithm generates the BI intelligently by optimizing a utility function instead of selecting it at random as in Vaidya et al. (2005) and Jiang and Walrand (2010), which can be considered as the main contribution of this distributed scheduling algorithm. Next, the optimality of the proposed novel distributed scheduling is shown in Theorem 9.2.

Theorem 9.2: Optimal Distributed Scheduler Performance

Given the multiple systems in an NCS and the adaptive optimal event-triggered control scheme given by Equation 9.14, the proposed distributed scheduling scheme selects the optimal adaptive ZOH-based event-triggered system of the NCS with the highest utility function value since it has the shortest BI and the highest priority to access the shared communication network. In addition, the proposed algorithm can render best performance schedules for each system of the overall NCS.

Proof: Assume the NCS lth system has the highest utility function value (i.e., $\Delta J_{l,k} = \max_{i \in \mathbf{G}_k} \Delta J_{i,k}$). Then we have $\Delta J_{l,k} > \Delta J_{i,k}$ for any $i = \mathbf{G}_k$, $i \neq l$. Therefore,

$$e^{-\Delta J_{l,k}} < e^{-\Delta J_{i,k}} \quad i = \mathbf{G}_k,\ i \neq l \tag{9.27}$$

Next, for any $i = \mathbf{G}_k$, $i \neq l$, the BI can be expressed as

$$\begin{aligned} BI_{l,k} &= \varsigma \times (e^{-\Delta J_{l,k}} + n_{l,k}) \\ &< \varsigma \times [e^{-\Delta J_{l,k}} + \min_{l,j \in [1,M]} (e^{-\Delta J_{l,k}} - e^{-\Delta J_{j,k}})] \\ &< \varsigma \times (e^{-\Delta J_{l,k}} + e^{-\Delta J_{i,k}} - e^{-\Delta J_{l,k}}) \\ &< \varsigma \times e^{-\Delta J_{i,k}} < \varsigma \times (e^{-\Delta J_{i,k}} + n_{i,k}) < BI_{i,k} \end{aligned} \tag{9.28}$$

Hence, $BI_{l,k} < BI_{i,k}$ for any $i = \mathbf{G}_k$, $i \neq l$. Based on the proposed distributed scheduling algorithm, the NCS lth system can be scheduled to use shared communication network due to its shortest BI. Next, the cost-optimality of the proposed scheme will be proven by using the contradiction method.

Assume that there exists another system j, and scheduling it can render better performance than scheduling the lth system even if the lth system has the shortest BI (i.e., cost function value $J_k^{sj} < J_k^{sl}$, but $BI_{l,k} < BI_{j,k}$). According to the definition of cost function (9.19), J_k^{sj} can be defined as

$$J_k^{sj} = \left(\sum_{i=1}^{M} J_{i,k} \right) - \Delta J_{j,k} \quad \text{for} \quad j = 1, 2, \ldots, M \tag{9.29}$$

Since $BI_{l,k} < BI_{j,k}$ is given in the assumption above, we have $\Delta J_{j,k} < \Delta J_{l,k}$ by using Equations 9.26 and 9.28. Meanwhile, the cost function of the jth NCS can be derived as

$$J_k^{sj} = \left(\sum_{i=1}^{M} J_{i,k} \right) - \Delta J_{j,k} > \left(\sum_{i=1}^{M} J_{i,k} \right) - \Delta J_{l,k} = J_k^{sl} \tag{9.30}$$

It is important to note that $J_k^{sj} > J_k^{sl}$ in Equation 9.30 contradicts the assumption $J_k^{sj} > J_k^{sl}$. By contradiction, there is no other system that can obtain a better performance than scheduling the lth system, which has the shortest BI. In other words, the proposed distributed scheduling scheme can render the best performance by scheduling the lth system with the shortest BI.

On the other hand, any system with negative utility function value, $\Delta J_{i,k} < 0$, should not contend the shared communication network resource since it will degrade the performance. Next, the proof is given in detail. Assume the pth system with a negative utility function, $\Delta J_{p,k} < 0$ is scheduled. Then the cost function with this decision can be expressed as

$$J_k^{sp} = \sum_{i=1}^{M} J_{i,k} - \Delta J_{p,k} > \sum_{i=1}^{M} J_{i,k} = J_k (\text{no pair is scheduled}) \tag{9.31}$$

Therefore, scheduling an NCS system with negative utility function will degrade the performance.

Remark 9.5

Fairness is an important factor to evaluate the performance of scheduling schemes. For the proposed distributed scheduling algorithm, an index given by

$$FI = \frac{\left(\sum_{i=1}^{M} \frac{R_i}{\sum_{j=0}^{\infty} (x_{i,j}^T Q_i x_{i,j} + u_{i,j}^T R_i u_{i,j}) + \beta_i R_i} \right)^2}{M * \sum_{i=1}^{M} \left(\frac{R_i}{\sum_{j=0}^{\infty} (x_{i,j}^T Q_i x_{i,j} + u_{i,j}^T R_i u_{i,j}) + \beta_i R_i} \right)^2}$$

is defined to measure the fairness among different systems.

Until now, the novel distributed scheduling is proposed to improve the performance of both the physical systems and network layers since the utility function included information from both layers. Combining with the proposed optimal adaptive ZOH event-triggered control scheme in Section 9.2, the proposed cross-layer approach is completed.

9.3 SIMULATION

Consider an NCS with six systems that are located within a 300 m*300 m square area at random. For maintaining the homogeneous property, all six systems are using similar control systems as Garcia and Antsaklis (2013).

EXAMPLE 9.1

The discrete-time representation is given as

$$x_{k+1} = \begin{bmatrix} 1.1138 & -0.0790 \\ 0.0592 & 0.8671 \end{bmatrix} x(t) + \begin{bmatrix} 0.2033 \\ 0.1924 \end{bmatrix} u(t) \tag{9.32}$$

Moreover, the sampling interval is selected as $T_s = 0.15$ s, the number of bits for the six quantized sensed data for the systems are defined as $N_{bit} = [10\ 8\ 6\ 7\ 8\ 4]$. Further, the cost function weight matrices are selected as $Q_l = \begin{bmatrix} 1 & 0 \\ 0 & 1 \end{bmatrix}$, $R_l = 0.1$. The initial system state vectors are given by $x_{1,0} = [20 \quad 7]^T$, $x_{2,0} = [12 \quad 5]^T$, $x_{3,0} = [10 \quad 3]^T$, $x_{4,0} = [8 \quad 4]^T$, $x_{5,0} = [10 \quad 5]^T$, and $x_{6,0} = [0.1 \quad 6]^T$.

a. Design the event-triggering condition and estimated optimal control.
b. Design the distributed scheduling scheme.
c. Plot the DNCS system state response.
d. Compared with ERR and Greedy scheduling schemes, plot the fairness of the optimal distributed scheduling and DNCS cost function performance.

The augment state is generated as $z_{l,k} = [x_{l,k} \quad u_{l,k-1} \quad u_{l,k-2}]^T \in \Re^4$ or $\chi_{l,k} = [(z_{l,k})^T\ u^T\ (z_{l,k})]^T \in \Re^6$. Moreover, the regression function for the adaptive estimator is $\{\chi_{l,1}^2, \chi_{l,1}\chi_{l,2}, \chi_{l,1}\chi_{l,3}, \ldots, \chi_{l,2}^2, \ldots, \chi_{l,5}^2, \ldots, \chi_{l,6}^2\}$ as per Equation 9.6. Further, the tuning parameter for the adaptive estimator is selected as $\alpha_{l,h} = 10^{-6}$ while the initial parameters for the adaptive estimator are set to zero at the beginning of the simulation. The parameter for the ZOH event-trigger condition is chosen as $\Gamma_l = 0.9$.

First, the performance of the proposed optimal adaptive ZOH event-triggered control is shown. Without loss of generality, an average value of state regulation errors for the six subsystems is shown in Figure 9.5. The results indicate that the proposed optimal adaptive ZOH event-triggered control design can force the regulation errors to converge to zero asymptotically while ensuring that all systems are stable.

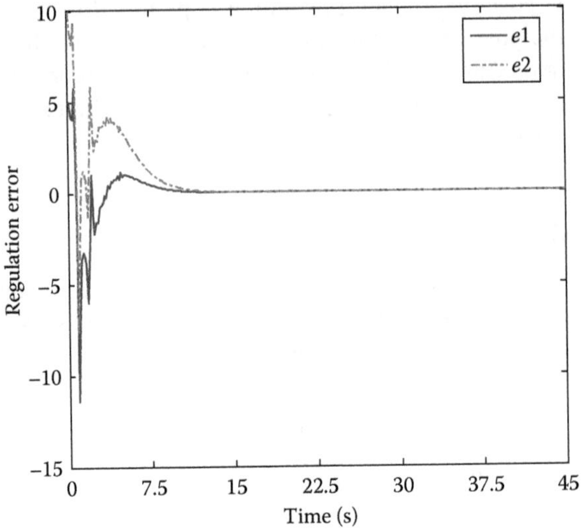

FIGURE 9.5 State regulation errors.

It is important to note the overshoots observed at the beginning take place because the value function estimator tuning needs a brief adaptation phase.

Then, the performance of the optimal adaptive ZOH event-triggered control system is demonstrated. Without loss of generality, we pick the 5th optimal adaptive ZOH event-triggered subsystem to evaluate the performance. As shown in Figure 9.6,

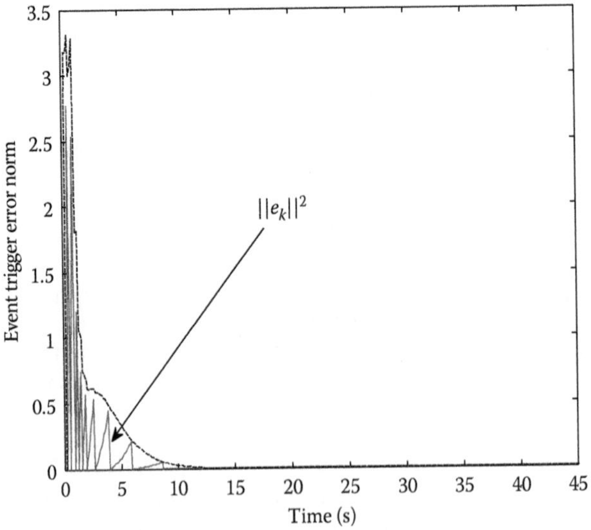

FIGURE 9.6 Event-trigger error norm $\|e_k\|^2$.

when an error event is triggered and scheduled, the estimation error will be reset to zero since the actual subsystem state will be received at the controller. On the other hand, if an error event is not triggered and scheduled, the estimation error increases due to the held system state in the ZOH event-triggered control system. Once the parameters of the value function estimator are assessed accurately, the estimation errors converge to zero. It is important to note that error events are triggered and scheduled more frequently at the beginning since the value function estimator needs to be tuned often. After a short adaptation phase, the error events become less frequent and reduce network traffic as the number of transmissions of state and control input vector reduces significantly.

Next, the performance of the proposed cross-layer distributed scheduling has been evaluated. For comparison, classical ERR (Jagannathan 2007) and Greedy scheduling (Jagannathan 2007) schemes are considered. In Figure 9.7, the cost function of multiple subsystems of the DNCS with three different scheduling schemes is compared. The proposed novel cross-layer distributed scheduling scheme maintains the lowest value while the cost function of the DNCS with ERR and Greedy scheduling is much more than the proposed scheduling scheme.

This indicates that the proposed distributed scheduling scheme can improve the performance of the DNCS much better than the widely used ERR and Greedy scheduling. It is important to note that: (1) Since ERR only guarantees that each NCS system can have the same probability to access the shared communication network resource, it could not assign communication network resources based on practical usage of each system of the DNCS, which might waste precious network resources. Therefore, ERR cannot optimize the multiple subsystems of DNCS performance. (2) Since Greedy scheduling only assigns network resources to the system with the

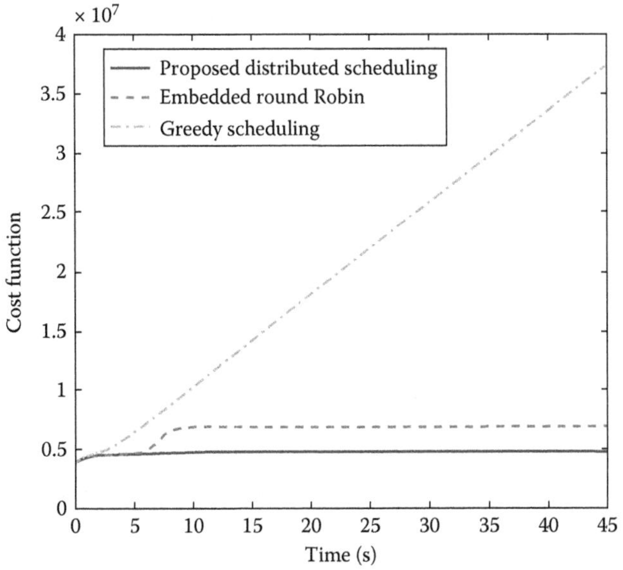

FIGURE 9.7 Cost function comparison.

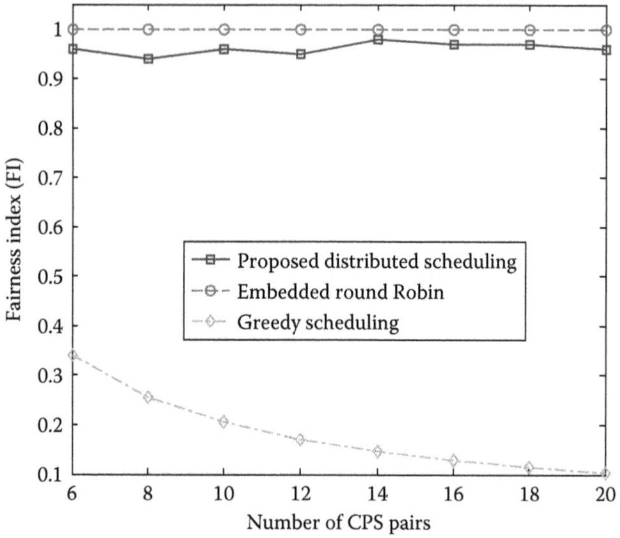

FIGURE 9.8 Fairness comparison.

best network layer performance whereas the physical system performance of each system in the DNCS is overlooked, it also cannot optimize the performance of the DNCS.

Eventually, the fairness of different scheduling schemes has been evaluated. As shown in Figure 9.8, fairness indices of the proposed cross-layer distributed scheduling and widely used ERR schemes are close and equal to one, whereas that of Greedy scheduling is much less than one, thus indicating a fair allocation of shared communication network resource for the proposed one while meeting the overall performance. The ERR method is fair but has a higher cost than the proposed one. According to the results illustrated in Figures 9.5 through 9.8, the proposed cross-layer codesign included an optimal adaptive ZOH event-triggered control and a cross-layer distributed scheduling scheme to optimize the performance of both the communication network and the subsystems.

9.4 CONCLUSIONS

In this study through a novel cross-layer codesign for DNCS, it is demonstrated that the proposed scheme can optimize not only the performance of the control system but also the shared communication network. The novel optimal adaptive ZOH-based event-triggered control scheme does not require the system dynamics of each system and uses the event-based sampling of state and control policy instead of an inefficient periodic time-driven sampling, reducing transmission cost. The parameter tuning for the event-triggered control is aperiodic. The novel scheduling algorithm is distributed and simple, and requires less computation than centralized scheduling algorithms.

PROBLEMS

Section 9.5

Problem 9.5.1: Repeat Example 9.1 and show the reduction in the number of transmissions and plot with periodic sampling.

Problem 9.5.2: Repeat Example 9.1 with different initial conditions.

REFERENCES

Dai, J.G. and Prabhakar, B. 2000. The throughput of data switches with and without speedup. *Proceedings of the 31st IEEE International Conference on Computer Communications*, Tel Aviv, Israel, pp. 556–564.

Day, J.D. and Zimmermann, H. 1983. The OSI reference model. *Proceedings of the IEEE*, 71, 1334–1340.

Dimarogonas, D.V. and Johanson, K.H. 2009. Event-triggered control for multi-agent systems. *Proceedings of the 48th IEEE Conference on Decision and Control*, Shanghai, China, pp. 7131–7136.

Donkers, M. and Heemels, W. 2012. Output-based event-triggered control with guaranteed $\mathcal{L}_\infty$-gain and improved and decentralised event-triggering. *IEEE Transactions on Automatic Control*, 57, 1362–1376.

Eqtami, A., Dimarogonas, D.V., and Kyriakopoulos, K.J. 2010. Event-triggered control for discrete-time systems. *Proceedings of the American Control Conference*, Baltimore, MD, pp. 4719–4724.

Garcia, E. and Antsaklis, P.J. 2011. Model-based event-triggered control with time-varying network delays. *Proceedings of the 50th IEEE Conference on Decision and Control*, Orlando, FL, USA, pp. 1650–1655.

Garcia, E. and Antsaklis, P.J. 2013. Model-based event-triggered control for systems with quantization and time-varying network delays. *IEEE Transactions on Automatic Control*, 58, 422–434.

Green, M. and Moore, J.B. 1989. Persistency of excitation in linear systems. *System & Control Letters*, 7, 351–360.

Guan, X., Wu, H., and Bi, S. 2011. A game theory based obstacle avoidance routing protocol for wireless sensor networks. *Sensor*, 11, 9327–9343.

Guinaldo, M., Lehmann, D., Sanchez, J., Dormido, S., and Johansson, K.H. 2012. Distributed event-triggered control with network delays and packet losses. *Proceedings of the Control and Decision Conference*, Maui, HI, USA, pp. 1–6.

IEEE Standard 802.11a. 1999. Wireless LAN medium access control (MAC) and physical layer (PHY) specification. High-speed physical layer in 5 GHz band.

Jagannathan, S. 2007. *Wireless Ad-Hoc and Sensor Network: Protocols, Performance, and Control*. CRC Press, Boca Raton, FL.

Jiang, L. and Walrand, J. 2010. A distributed CSMA algorithm for throughput and utility maximization in wireless networks. *IEEE/ACM Transactions on Networking*, 18, 960–972.

Lewis, F.L. and Syrmos, V.L. 1995. *Optimal Control*. 2nd edition. Wiley, New York.

Liu, Q., Zhou, S., and Giannakis, G.B. 2005. Queuing with adaptive modulation and coding over wireless links: Cross-layer analysis and design. *IEEE Transactions on Wireless Communication*, 4, 1142–1153.

Mazo Jr., M. and Tabuada, P. 2011. Decentralized event-triggered control over wireless sensor/actuator networks. *IEEE Transactions on Automatic Control*, 56, 2456–2461.

Nie, N. and Comaniciu, C. 2006. Adaptive channel allocation spectrum etiquette for cognitive radio networks. *Journal of Mobile Networks and Applications*, 11, 779–797.

Reena, P. and Jacob, L. 2007. Joint congestion and power control in UWB based wireless sensor network. *Proceedings of the 32nd IEEE Conference on Local Computer Network (LCN)*, pp. 911–918.

Sahoo, A., Xu, H., and Jagannathan, S. 2014. Event-based optimal regulator design for nonlinear networked control systems. *Proceedings of the Adaptive Dynamic Programming and Reinforcement Learning*, Orlando, FL, USA, 295–300.

Shakkottai, S., Rappaport, T.S., and Karlsson, P.C. 2003. Cross-layer design for wireless networks. *IEEE Communication Magazine*, 41, 74–80.

Srivastava, V. and Motani, M. 2005. Cross-layer design: A survey and the road ahead. *IEEE Communication Magazine*, 43, 112–119.

Stengel, R.F. 1986. *Stochastic Optimal Control: Theory and Application*. Wiley-Interscience, New York.

Tian, G., Tian, Y.C., and Fidge, C. 2010. Performance analysis of IEEE 802.11 DCF based WNCS networks. *Proceedings of Local Computer Networks*, Denver, CO, USA, pp. 512–519.

Vaidya, N., Dugar, A., Gupta, S., and Bahl, P. 2005. Distributed fair scheduling in wireless LAN. *IEEE Transactions on Mobile Computing*, 4, 616–629.

Wang, X. and Lemmon, M. 2011. Event-triggered in distributed networked control systems. *IEEE Transactions on Automatic Control*, 56, 586–601.

Wei, Y., Heidemann, J., and Estrin, D. 2002. An energy-efficient MAC protocol for wireless sensor network. *Proceedings of INFOCOM 2002*, pp. 1567–1576.

Xia, F., Kong, X., and Xu, Z. 2012. Cyber-physical control over wireless sensor and actuator networks with packet loss. *Wireless Networking Based Control*. Springer, New York, USA.

Xia, F., Vinel, A., Gao, R., Wang, L., and Qiu, T. 2011. Evaluating IEEE 802.15.4 for cyber-physical systems. *EURASIP Journal of Wireless Communication and Networking*, 2, 1–14.

Xu, H., Jagannathan, S., and Lewis, F.L. 2012. Stochastic optimal control of unknown networked control systems in the presence of random delays and packet losses. *Automatica*, 48, 1017–1030.

Xu, H., Sahoo, A., and Jagannathan, S. 2014. Stochastic adaptive event-triggered control and network scheduling protocol co-design for distributed networked systems. *IET Control Theory and Applications*, 8(18), 2253–2265.

Zheng, D., Ge, W., and Zhang, J. 2009. Distributed opportunistic scheduling for ad hoc networks with random access: an optimal stopping approach. *IEEE Transactions on Information Theory*, 55, 205–222.

10 Optimal Control of Uncertain Linear Control Systems under a Unified Communication Protocol

In the previous chapters, the optimal control of NCS in the presence of network imperfections is covered when disturbances are included provided the state vector of NCS is perfectly measured. While Halevi and Ray (1988), Zhang et al. (2001), Nilsson et al. (1998), and Lian et al. (2002) did not include the behavior of the communication protocol that causes these network imperfections, Schenato et al. (2007) studied the effect of communication protocols such as TCP and UDP by using LNCS with known dynamics and network imperfections and suggested observer-based optimal controller via the Riccati equation.

However, it is demonstrated in Chapter 3 (Xu et al. 2012) that incorporating network imperfections, which are not known in advance, into a known time-invariant linear system, will result in a stochastic LNCS with uncertain system dynamics. Consequently, Riccati equation-based optimal control schemes cannot be utilized for such systems since they require system matrices.

As presented in Chapter 5, ADP techniques, on the contrary, proposed by Werbos (1983), intend to solve optimal controller design for unknown linear and nonlinear system in a forward-in-time manner instead of traditional optimal control schemes that generate backward-in-time control inputs. A suite of ADP schemes have been covered in the previous chapters. However, the behavior of a communication protocol (e.g., TCP, UDP) is not studied (Xu et al. 2012).

The addition of a communication network protocol will necessitate an output feedback (Schenato et al. 2007) scheme that is more involved than the state feedback-based ADP schemes. An observer can complicate the optimal controller design (Xu et al. 2012) and stability analysis especially when the system dynamics are uncertain. Meanwhile, a traditional full acknowledgment-based communication protocol such as TCP can provide a more reliable feedback loop while it requires more bandwidth due to its acknowledgment feature but suffers from retransmissions, which is unsuitable for real-time control. In some cases, full acknowledgment may not be possible due to limited network bandwidth and in the worst case, no acknowledgments are possible as in UDP.

Hence, the effect of a communication protocol with full and intermittent acknowledgments with and without retransmissions is studied under a new unified framework from Xu et al. (2012) in this chapter. It is important to note that the proposed unified framework includes three scenarios (i.e., Case 1: TCP with full acknowledgments,

Case 2: TCP with intermittent acknowledgments, and Case 3: UDP with no acknowledgments). Traditional TCP and UDP (Stevens 1995) therefore can be considered as two cases of the controller design under the influence of the unified communication protocol framework for an uncertain LNCS.

In this chapter, after first incorporating the network imperfections into linear time-invariant systems, the resulting stochastic system is expressed in an input–output form (equations 1.9 and 1.10 in Chapter 1). Subsequently, to implement the output feedback under the unified communication protocol, a novel observer is introduced to estimate the system states when the dynamics are unknown. Next, by using the observed system states and an initial stabilizing control, the stochastic value function is estimated (Dierks and Jagannathan 2012) and its parameter vector is tuned online and forward-in-time by using the Bellman equation (Stengel 1986). Eventually, certainty-equivalent stochastic control inputs that optimize the value function can be calculated based on parameters provided by the value function estimator.

Compared with traditional optimal control theory that requires the knowledge of system dynamics to solve the SRE, the proposed novel observer and value function estimator relax the need for system state vector and dynamics, and information on network-induced delays and packet losses, respectively, for NCS under the unified communication protocol, and yield optimal control without using value or policy iterations. In the case of no acknowledgment, the protocol becomes UDP and the performance of the optimal adaptive controller is evaluated. TCP with full acknowledgment is considered as another special case of the unified communication protocol. The net result is the stochastic optimal adaptive output feedback control technique for LNCS in discrete-time under the influence of the unified communication protocol, unknown system dynamics, and network imperfections. Here, the outputs have to be measured perfectly. Lyapunov stability is introduced.

The background of NCS is given in Chapter 1. Next, Section 10.1 presents the optimal adaptive control for LNCS under the unified communication protocol. Subsequently, the closed-loop stability is given in Section 10.2. Eventually, Section 10.3 presents the simulation results to demonstrate the effectiveness of the developed method. The conclusions are given in Section 10.4.

10.1 OPTIMAL CONTROL DESIGN UNDER UNIFIED COMMUNICATION PROTOCOL FRAMEWORK

In this section, observers (Schenato et al. 2007) and ADP (Werbos 1992) are used to derive stochastic optimal adaptive output-feedback control of an NCS under the unified communication protocol, unknown system dynamics due to network imperfections. First, a novel observer is designed online to estimate the augment system state vector at the controller by assuming that the output vector can be measured perfectly. Second, we estimate the unknown value function for the NCS with network imperfections under the unified communication protocol. Third, a model-free online tuning of the parameters of observer and value function estimator by using the ADP method incorporating the observed augment system states is proposed. Eventually, the convergence proof is given. The block diagram representation of the NCS with protocol effects is illustrated in Figure 10.1.

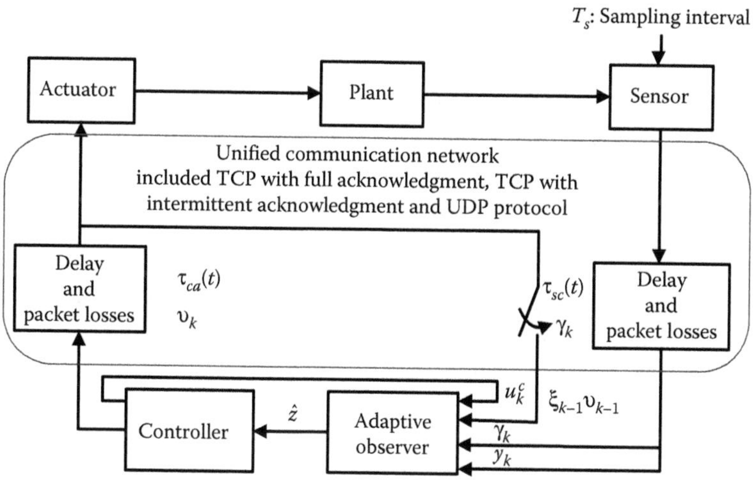

FIGURE 10.1 Networked control system under the communication protocol.

10.1.1 Observer Design

An observer or estimator is normally utilized when certain states are unavailable for measurement. The observer design for an NCS requires the knowledge of system dynamics (Schenato et al. 2007) and therefore traditional observers cannot be employed for NCS. Therefore, in this section, a novel observer is proposed to estimate the system states for the NCS under the unified communication protocol by relaxing the need for system dynamics due to network imperfections. Next, the details of the novel observer design are given.

The system state vector, z_k, can be estimated at time instant, kT_s, as

$$\hat{z}_{k|k} = E\left[\hat{\vartheta}_k^T M_{\xi,k}\hat{s}_{k-1} \,\middle|\, n_{k-1}\right] + \kappa_o E[\tilde{y}_{k-2}^o \,|\, \phi_{k-2}] \tag{10.1}$$

where

$$\bar{M}_{\xi,k} = \begin{bmatrix} I^n/2 & 0 & \cdots & 0 & 0 \\ 0 & (1-\xi_{k-1})\bar{\upsilon}+\xi_{k-1}\upsilon_{k-1} & \cdots & 0 & 0 \\ \vdots & \vdots & \ddots & \vdots & \vdots \\ 0 & 0 & \cdots & (1-\xi_{k-b})\bar{\upsilon}+\xi_{k-b}\upsilon_{k-b} & 0 \\ 0 & 0 & \cdots & 0 & (1-\xi_k)\bar{\upsilon}+\xi_k\upsilon_k \end{bmatrix}$$

$$\tilde{y}_k^o = \begin{bmatrix} \tilde{y}_k \\ \tilde{y}_{k-1} \\ \vdots \\ \tilde{y}_{k-N+1} \end{bmatrix}$$

and I^n is an $n \times n$ identity matrix, the output estimation error being $\tilde{y}_k = y_k - \hat{y}_k$, the state estimation error at $(k-1)T_s$ is denoted as $\tilde{z}_{k-1|k-1} = z_{k-1} - \hat{z}_{k-1|k-1}$, s_k is given by $\left[z_k^T, u_k^T\right]^T$, and $\hat{s}_{k-1} = E\left[\left[\hat{z}_{k-1|k-1}^T\ u_{k-1}^T\right]^T \middle| \phi_{k-1}\right] + n_{k-1}$ with $\|\hat{s}_{k-1}\| < \Psi_M$ since $E\left[\hat{z}_{k-1|k-1}^T \middle| \phi_{k-1}\right]$ and $M_{\xi,k}$ are known at time kT_s, $\hat{\vartheta}_k$ is the estimated parameters for the observer and n_k is independent and identically distributed white Gaussian noise (i.e., $n_k \sim N(0, \sigma_0^2)$ where σ_0 is the variance of white Gaussian noise with $\sigma_0 \neq 0$), and κ_o is a design constant matrix.

It is important to note that if acknowledgment ξ_k is received, then υ_k is known and used in the observer. Otherwise, when ξ_k, υ_k are unknown as in the case of UDP, the mean value (i.e., $\bar{\upsilon}$) can be utilized in the novel observer. For example, one element $M_{\xi,k}^{bb}$ of matrix $M_{\xi,k}$ can be expressed as

$$
\begin{aligned}
M_{\xi,k}^{bb} &= \xi_k \upsilon_k + (1 - \xi_k)\bar{\upsilon} \\
&= \begin{cases} \upsilon_k & \text{if acknowledgment is received (i.e., } \xi_k = 1) \\ \bar{\upsilon} & \text{if acknowledgment is not received (i.e., } \xi_k = 0) \end{cases}
\end{aligned}
$$

Then, the observer design is detailed as follows. First, the system state vector in the next time step $\hat{z}_{k+1|k}$ can be predicted as

$$
E\left[\hat{z}_{k+1|k} \middle| \phi_k\right] = E\left[\hat{\vartheta}_k^T M_{\xi,k} \hat{s}_k \middle| \phi_k\right] + \kappa_o E\left[\tilde{y}_{k-1}^o \middle| \phi_{k-1}\right] \tag{10.2}
$$

Moreover, the prediction error $E\left[\tilde{z}_{k+1|k} \middle| \phi_k\right]$ can be derived as

$$
\begin{aligned}
E\left[\tilde{z}_{k+1|k} \middle| \phi_k\right] &= E\left[z_{k+1} - \hat{z}_{k+1|k} \middle| \phi_k\right] \\
&= [U_z - \kappa_o (D_y^o)^+] E\left[\tilde{z}_{k|k} \middle| \phi_k\right] + [1 - \kappa_o (D_y^o)^+] E\left[\tilde{\vartheta}_k^T M_{\xi,k} \hat{s}_k \middle| \phi_k\right] \\
&\quad + [1 - \kappa_o (D_y^o)^+] E\left[\vartheta^T (M_k - \bar{M}_{\xi,k}) \hat{s}_k \middle| \phi_k\right]
\end{aligned} \tag{10.3}
$$

where

$$
U_z = diag\{A_z, I_m, \ldots, I_m\} \in \mathbb{R}^{(n+(\bar{d}-1)m)\times(n+(\bar{d}-1)m)},
$$

and

$$
M_k = \begin{bmatrix} I^n/2 & 0 & \cdots & 0 & 0 \\ 0 & \upsilon_{k-1} & \cdots & 0 & 0 \\ \vdots & \vdots & \ddots & \vdots & \vdots \\ 0 & 0 & \cdots & \upsilon_{k-b} & 0 \\ 0 & 0 & \cdots & 0 & \upsilon_k \end{bmatrix}
$$

Note that: (1) when acknowledgment packets are transmitted perfectly, $\xi_k = 1$ for every $k \in \mathbb{N}$ and Equation 10.3 is an NCS under the TCP with full acknowledgment; (2) when acknowledgment packets are transmitted intermittently, then Equation 10.3 is an NCS under TCP with intermittent acknowledgment; and (3) while no acknowledgment can be transmitted correctly, $\xi_k = 0$ for every $k \in \mathbb{N}$ and Equation 10.3 is an NCS under UDP. Meanwhile, since the unified communication protocol includes all these three cases, the general parameter update law needs to be determined for the NCS under the unified communication framework.

Define the update law for the parameter vector $E\left[\hat{\vartheta}_k \middle| \phi_k\right]$ of the observer as

$$E\left[\hat{\vartheta}_{k+1} \middle| \phi_{k+1}\right] = E\left[\hat{\vartheta}_k \middle| \phi_k\right] + \alpha_o E\left[\frac{M_{\xi,k}\hat{s}_k\left(y_{k+1} - \Gamma_{k+1}\hat{z}_{k+1|k}\right)^T}{(M_{\xi,k}\hat{s}_k)^T M_{\xi,k}\hat{s}_k + 1} \middle| \phi_k\right] \tag{10.4}$$

Next, substituting the NCS dynamics given in Chapter 1 in Equation 10.4, $\hat{\vartheta}_{k+1}$ can be expressed as

$$E\left[\hat{\vartheta}_{k+1} \middle| \psi_k\right] = E\left[\hat{\vartheta}_k \middle| \phi_k\right] + \alpha_o \gamma_{k+1} E\left[\frac{M_{\xi,k}\hat{s}_k \Gamma_{k+1}\tilde{z}_{k+1|k}^T}{(M_{\xi,k}\hat{s}_k)^T M_{\xi,k}\hat{s}_k + 1} \middle| \phi_k\right] \tag{10.5}$$

where α_o is the tuning parameter satisfying $0 < \alpha_o < 1$. Meanwhile, observer parameter estimation error dynamics $E\left[\tilde{\vartheta}_k \middle| \phi_k\right]$ can be represented as

$$E\left[\hat{\vartheta}_{k+1} \middle| \psi_k\right] = E\left[\hat{\vartheta}_k \middle| \phi_k\right] - \alpha_o \gamma_{k+1} E\left[\frac{M_{\xi,k}\hat{s}_k \Gamma_{k+1}\tilde{z}_{k+1|k}^T}{(M_{\xi,k}\hat{s}_k)^T M_{\xi,k}\hat{s}_k + 1} \middle| \phi_k\right] \tag{10.6}$$

Eventually, at time $(k + 1)T_s$, the observed state $E\left[\hat{z}_{k+1|k+1} \middle| \phi_{k+1}\right]$ and the estimation error dynamics $E\left[\tilde{z}_{k+1|k+1} \middle| \phi_{k+1}\right]$ can be expressed as

$$\begin{aligned}
E\left[\hat{z}_{k+1|k+1} \middle| \phi_{k+1}\right] &= E\left[\hat{\vartheta}_{k+1}^T M_{\xi,k}\hat{s}_k \middle| n_k\right] + \kappa_o E\left[\tilde{y}_{k-1}^o \middle| \phi_{k-1}\right] \\
E\left[\tilde{z}_{k+1|k+1} \middle| \phi_{k+1}\right] &= E\left[z_{k+1} - \hat{z}_{k+1|k+1} \middle| \phi_{k+1}\right] \\
&= [U_z - \kappa_o (D_y^o)^+] E\left[\tilde{z}_{k|k} \middle| \phi_k\right] + [1 - \kappa_o (D_y^o)^+] E\left[\tilde{\vartheta}_{k+1}^T M_{\xi,k}\hat{s}_k \middle| \phi_k\right] \\
&\quad + [1 - \kappa_o (D_y^o)^+] E\left[\vartheta^T (M_k - \bar{M}_{\xi,k})\hat{s}_k \middle| \phi_k\right]
\end{aligned} \tag{10.7}$$

Next the following assumption and definitions are needed.

Assumption 10.1: Observability

In order to meet the observability criterion, critical arrival probability of packets carrying the output feedback between the sensor and the controller need to be in the region (Schenato et al. 2007) defined by $P(\gamma_k = 1) > (1/N_0)$, where $P(\gamma_k = 1)$ is the arrival probability at the controller and N_0 is a finite positive constant.

Remark 10.1

This assumption implies that a broken communication link is not present between the controller and the system, which in turn ensures that there exists at least one packet containing the output vector that traverses through the network so as to observe the system states.

Theorem 10.1

Let the proposed observer state, estimation errors, and parameter vector update be defined by Equations 10.1, 10.3, and 10.4, respectively. Under Assumption 10.1 and the unified communication protocol, there exists positive constants η and α_o satisfying

$$0 < \eta < \sqrt{\frac{1}{3((1-\frac{\chi_{\min}^2}{\chi_{\min}^2+1}\alpha_o)^2 - \alpha_o(1+\alpha_o))}}, \quad \frac{1}{\chi_{\min}^2+1} < \alpha_o < 1,$$

such that the estimation errors $E\left[\tilde{z}_{k|k}\,|\phi_k\right]$ and parameter estimation errors $E\left[\tilde{\vartheta}_k\,|\phi_k\right]$ (10.6) are *bounded* in the mean with bounds given by $\left\|E\left[\tilde{z}_{k|k}\,|\phi_k\right]\right\| \le B_{eo}$ and $\left\|E\left[\tilde{\vartheta}_k\,|\phi_k\right]\right\| \le B_{\vartheta}$.

Proof: Refer to Appendix 10A.

Remark 10.2

It is important to note that the communication acknowledgment indicator ξ_k, its mean $\bar{\xi}$, and variance σ_ξ^2 will affect the value of the bound ε_M^o and bounds B_{eo}, B_ϑ. For an NCS under TCP with full acknowledgment, bounds will tend to zero or be asymptotically stable in the mean. For an NCS under TCP with intermittent acknowledgment, the bounds will increase when mean $\bar{\xi}$ and variance σ_ξ^2 decrease. Eventually, for NCS under UDP, bounds will be become maximum in this unified communication framework.

10.1.2 Stochastic Value Function

Consider the NCS under the unified communication protocol and network imperfections represented in Chapter 1. Given the NCS under the unified communication protocol with a unique equilibrium point, $z = 0$, on a set Ω, minimizing the stochastic value function $V_k(z)$ renders the stochastic optimal control input as $u_k^{c*} = -K_k z_k$ where K_k is the optimal gain. According to Stengel (1986), the certainty-equivalent stochastic value function can be rewritten as

$$V^*(z_k) = E\left[z_k^T P_k z_k \middle| \phi_k\right] \tag{10.8}$$

where $P_k \geq 0$ is the solution of the SRE (Åstrom 1970; Stengel 1986). Then we can define the optimal action-dependent value function in terms of the expected value as

$$\begin{aligned} V^*(z_k) &= E\left\{[r(z_k,u_k)+V^*(z_{k+1})]\middle|\phi_k\right\} \\ &= E\left\{[z_k^T \; u_k^T]\Theta_k[z_k^T \; u_k^T]^T \middle|\phi_k\right\} \end{aligned} \tag{10.9}$$

with cost-to-go defined as $r(z_k,u_k) = z_k^T O z_k + u_k^T R u_k$. Then, similar to Xu et al. (2012), using the Bellman equation and stochastic value function, substituting the value function into the Bellman equation results in

$$\begin{aligned} \begin{bmatrix} z_k \\ u_k \end{bmatrix}^T E\left(\Theta_k \middle| \psi_k\right)\begin{bmatrix} z_k \\ u_k \end{bmatrix} &= E\left\{[r(z_k,u_k)+V^*(z_{k+1})]\middle|\phi_k\right\} \\ &= \begin{bmatrix} z_k \\ u_k \end{bmatrix}^T \begin{bmatrix} O+E\left(A_{zk}^T P_{k+1} A_{zk}\middle|\phi_k\right) & E\left(A_{zk}^T P_{k+1} B_{zk}\middle|\phi_k\right) \\ E\left(B_{zk}^T P_{k+1} A_{zk}\middle|\phi_k\right) & R+E\left(B_{zk}^T P_{k+1} B_{zk}\middle|\phi_k\right) \end{bmatrix}\begin{bmatrix} z_k \\ u_k \end{bmatrix} \end{aligned} \tag{10.10}$$

Therefore, $E\left(\Theta_k \middle| \psi_k\right)$ can be expressed as

$$\bar{\Theta}_k = \begin{bmatrix} \bar{\Theta}_k^{zz} & \bar{\Theta}_k^{zu} \\ \bar{\Theta}_k^{uz} & \bar{\Theta}_k^{uu} \end{bmatrix} = \begin{bmatrix} O+E(A_{zk}^T P_{k+1} A_{zk}|\phi_k) & E(A_{zk}^T P_{k+1} B_{zk}|\phi_k) \\ E(B_{zk}^T P_{k+1} A_{zk}|\phi_k) & R+E(B_{zk}^T P_{k+1} B_{zk}|\phi_k) \end{bmatrix} \tag{10.11}$$

Then, according to Åstrom (1970) and Equation 10.11, the optimal control gain for an NCS under the unified communication protocol can be represented in terms of value function parameters, $\bar{\Theta}_k$, as

$$K_k = \left[R+E\left(B_{zk}^T P_{k+1} B_{zk}\middle|\phi_k\right)\right]^{-1} E\left(B_{zk}^T P_{k+1} A_{zk}\middle|\phi_k\right) = \left(\bar{\Theta}_k^{uu}\right)^{-1}\bar{\Theta}_k^{uz} \tag{10.12}$$

It is important to note that even if the kernel matrix P_k is known, solving time-varying optimal control gain still requires slowly time-varying system matrices. However, if the parameter vector $\bar{\Theta}_k$ that is slowly varying can be estimated online, then system dynamics are not needed to calculate the optimal control gain.

10.1.3 Model-Free Online Tuning of Adaptive Estimator

In this section, when the value function (10.8) is estimated online, matrix $\bar{\Theta}_k$ is obtained, which in turn is used to derive stochastic optimal control inputs via Equation 10.12 without the knowledge of system matrices. By assuming that the value function, $V^*(z_k)$, can be represented as the LIP and according to Åstrom (1970) and Equation 10.8, the value function is given as

$$V^*(z_k) = \varphi_k^T \bar{\Theta}_k \varphi_k = \bar{\theta}_k^T \bar{\varphi}_k \tag{10.13}$$

where $\bar{\theta}_k = \text{vec}\left(\bar{\Theta}_k\right)$, $\varphi_k = E\left[[z_k^T \; u(z_k^T)]^T \mid \phi_k\right] \in \mathbb{R}^{n+bm=l}$, and $\bar{\varphi}_k = (\varphi_{k1}^2, \ldots, \varphi_{k1}\varphi_{kl}, \varphi_{k2}^2, \ldots, \varphi_{kl-1}\varphi_{kl}, \varphi_{kl}^2)$ is the Kronecker product quadratic polynomial basis vector (Werbos 1992) consisting of current state and past control inputs, vec(·) function is constructed by stacking the columns of matrix into one column vector with off-diagonal elements (Werbos 1992). Since matrix $\bar{\Theta}_k$ can be considered as slowly time varying, the value function can be represented as a function of the target unknown parameter vector and regression function $\bar{\phi}_k$.

Meanwhile, the value function can also be represented in terms of $\bar{\Theta}_k$ as

$$V^*(z_k) = \varphi_k^T \bar{\Theta}_k \varphi_k = \bar{\theta}_k^T \bar{\varphi}_k \tag{10.14}$$

Since the observed system states $\hat{z}_{k|k}$ are only available at the controller, the value function with observed system states can be expressed as

$$V_k^*(\hat{z}) = \varphi_k^{eT} \bar{\Theta}_k \varphi_k^e = \bar{\theta}_k^T \bar{\varphi}_k^e \tag{10.15}$$

where $\varphi_k^e = E\left[\left[\hat{z}_{k|k}^T \; u_k^T\right]^T \middle| \phi_k\right]$ and $\bar{\phi}_k^e$ is the Kronecker product quadratic polynomial basis vector of φ_k^e.

Next, we will derive the residual errors by using the Bellman equation. Normally, the Bellman equation can be rewritten as $V_{k+1}^*(z) - V_k^*(z) + r(z_k, u_k) = 0$. However, this relationship does not hold when we apply the observed states $\hat{z}_{k|k}$. Hence, substituting observed system states into the Bellman equation, the residual errors $V^*\left(\hat{z}_{k|k}\right) - V^*\left(\hat{z}_{k-1|k-1}\right) + r\left(\hat{z}_{k-1|k-1}, u_{k-1}\right) = E\left[e_k^o \middle| \phi_k\right]$. It is important to note that e_k^o is caused by observer error dynamics $E\left[\tilde{z}_{k|k} \middle| \phi_k\right]$ and can be derived as

$$\begin{aligned}
E\left[e_k^o \middle| \psi_k\right] &= E\Big[-\tilde{z}_{k-1|k-1}^T P_{k-1} \tilde{z}_{k-1|k-1} - 2\tilde{z}_{k-1|k-1}^T P_{k-1} z_{k-1} + 2\tilde{z}_{k|k}^T P_k z_k \\
&\quad - \tilde{z}_{k|k}^T P_k \tilde{z}_{k|k} + 2\tilde{z}_{k-1|k-1}^T Q z_{k-1} - \tilde{z}_{k-1|k-1}^T Q \tilde{z}_{k-1|k-1} \\
&\quad + 2(A_{zk}\tilde{z}_{k-1|k-1} - \tilde{z}_{k|k})^T P_k (A_{zk} - B_{zk} K_k)\hat{z}_{k-1} \\
&\quad + (A_{zk}\tilde{z}_{k-1|k-1} - \tilde{z}_{k|k})^T P_k (A_{zk}\tilde{z}_{k-1|k-1} - \tilde{z}_{k|k}) \Big| \phi_k\Big] \\
&= E\left[\tilde{z}_{k-1|k-1}^T (Q - P_{k-1} + A_{zk}^T P_k A_{zk}) \tilde{z}_{k-1|k-1} \middle| \phi_k\right] \\
&= E\left[\tilde{z}_{k-1|k-1}^T O_e \tilde{z}_{k-1|k-1} \middle| \phi_k\right]
\end{aligned} \tag{10.16}$$

Next, the value function estimation with observed system states will be considered. First, the value function with observed system states $V_k^*(\hat{z})$ can be expressed in terms of estimated parameters $\hat{\bar{\theta}}_k$ as

$$\hat{V}(\hat{z}_{k|k}) = \varphi_k^{eT}\hat{\bar{\Theta}}_k\varphi_k^e = \hat{\bar{\theta}}_k^T\bar{\varphi}_k^e \tag{10.17}$$

where $\hat{\bar{\theta}}_k$ is the estimated value of the target parameter vector $\bar{\theta}_k$ defined before.

Substituting Equation 10.17 and estimated system states into the Bellman equation, Equation 10.9 is not guaranteed to hold. By using delayed values for convenience, the residual error associated with Equation 10.17 can be expressed as

$$\begin{aligned} E\left[e_k^a + e_k^o \,|\varphi_k\right] &= \hat{V}\left(\hat{z}_{k|k}\right) - \hat{V}\left(\hat{z}_{k-1|k-1}\right) + r(\hat{z}_{k-1|k-1}, u_{k-1}) \\ &= E\left[r\left(\hat{z}_{k-1|k-1}, u_{k-1}\right) + \hat{\bar{\theta}}_k^T\Delta\varphi_{k-1}^e \,|\phi_k\right] \end{aligned} \tag{10.18}$$

where $\Delta\bar{\phi}_k^e = \bar{\phi}_k^e - \bar{\phi}_{k-1}^e$ is the first difference by using the regression function and e_k^a represent estimation errors.

The dynamics of Equation 10.18 can be expressed as $E\left[e_{k+1}^a + e_{k+1}^o \,|\phi_k\right] = E\left[r\left(\hat{z}_{k|k}, u_k\right) + \hat{\bar{\theta}}_{k+1}^T\Delta\varphi_k^e \,|\phi_k\right]$. Next, define the update law for the parameter vector $E\left[\hat{\bar{\theta}}_k \,|\phi_k\right]$ as

$$E\left[\hat{\bar{\theta}}_{k+1} \,|\phi_{k+1}\right] = E\left[\hat{\bar{\theta}}_k + \alpha_h \frac{\Delta\bar{\varphi}_k^e(e_k^o + e_k^a)^T}{\Delta\bar{\phi}_k^{eT}\Delta\bar{\phi}_k^e + 1} \,|\phi_k\right] \tag{10.19}$$

where $0 < \alpha_h < 1$ is the tuning parameter for value function estimation.

Remark 10.3

It is observed that the value function $V^*(z_k)$ and its estimation $\hat{V}(\hat{z}_{k|k})$ (10.17) will become zero only when $E[z_k \,|\phi_k] = 0$ and $E[\hat{z}_{k|k} \,|\phi_k] = 0$. Hence, when system states have converged to zero, the estimated system states $E[\hat{z}_{k|k} \,|\phi_k]$ will also converge to zero according to Theorem 10.2 and the value function estimation is no longer updated. It can be seen as a PE requirement for the inputs to the value function estimator wherein the system states must be persistently exciting long enough for the estimator to learn the optimal value function. Therefore, exploration noise is added to NCS in order to satisfy the PE condition (Xu et al. 2012).

Next, the dynamics of parameter estimation errors of the value function can be expressed as

$$E\left[\tilde{\bar{\theta}}_{k+1} \,|\phi_{k+1}\right] = E\left[\tilde{\bar{\theta}}_k - \alpha_h \frac{\Delta\bar{\varphi}_k^e(e_k^o + e_k^a)^T}{\Delta\bar{\varphi}_k^{eT}\Delta\bar{\varphi}_k^e + 1} \,|\phi_k\right] \tag{10.20}$$

Then, the convergence of the value function estimation errors with parameter error dynamics $E[\tilde{\bar{\theta}}_k|\phi_k]$ given by Equation 10.20 is demonstrated for an initial admissible control (Sarangapani 2006) policy. It is important to note that slowly time-varying (Dacic and Nesic 2007) LNCS is asymptotically stable in the mean if an initial admissible control can be implemented provided system matrices are known. However, the proposed estimated value function with observed system states results in estimation errors for the value function V_k, whose stability needs to be studied. Therefore, Theorem 10.2 will prove that the value function estimation errors converge while the overall closed-loop system stability is shown in Theorem 10.3.

Theorem 10.2

Given the initial parameter vector $E[\hat{\bar{\theta}}_0|\phi_0]$ for the value function estimator to be bounded in the set Ω, let u_{0k} be an initial admissible control policy for the LNCS under the unified communication protocol. Let the observer parameter update law be given by Equation 10.19. Then there exists positive constants α_h, α_o and η satisfying $0 < \alpha_h < ((2\Delta\phi_{\min}^2)/(3(\Delta\phi_{\min}^2 + 1)))$ with $0 < \Delta\phi_{\min} < \left\|\Delta\bar{\varphi}_k^e\right\|$, and $\Delta\phi_{\min}$ is the lower bound of $\left\|\Delta\bar{\varphi}_k^e\right\|$,

$$0 < \eta < \sqrt[4]{\frac{3\alpha_o^3}{(1+6\alpha_o+9\alpha_o^2+\alpha_o^3)(U_M-1)^4}}, \quad 0 < \alpha_o < \frac{(\chi_{\min}^2+1)}{21(\chi_{\min}^2+1)^2+3}$$

and that the value function estimation errors for NCS under the unified communication protocol are *bounded* in the mean with the bound given by $\left\|E\left[\tilde{\bar{\theta}}_k|\phi_k\right]\right\| \le B_\theta$.

Proof: Refer to Appendix 10B.

Similar to Remark 10.2, the communication acknowledgment indicator ξ_k will also affect ε_M^{AE} and the bound B_θ. For an NCS under TCP with full acknowledgment, the bound will tend to be zero or asymptotically stable in the mean. For an NCS under TCP with intermittent acknowledgment, the bound will increase when mean $\bar{\xi}$ and variance σ_ξ^2 decrease. For an NCS under UDP, bounds will become maximum. Under the framework, the stochastic optimal control signal is estimated by using the estimated $\bar{\Theta}_k$ matrix as

$$\hat{u}_k = -E\left[\hat{K}_k\hat{z}_{k|k}|\phi_k\right] = -E\left[\left(\hat{\bar{\Theta}}_k^{uu}\right)^{-1}\hat{\bar{\Theta}}_k^{uz}\hat{z}_{k|k}|\phi_k\right] \tag{10.21}$$

Next, the stability of the observation error, value function, control estimation, and adaptive estimation error dynamics are considered.

10.2 CLOSED-LOOP SYSTEM STABILITY

In this section, we will show that observer errors, slowly time-varying parameter, and value function estimation error dynamics are *bounded* in the mean. Moreover, the observed system states and estimated control inputs for NCS under the unified communication protocol will converge close to actual system states and optimal control signals and stay within the bound. It is important to note these bounds are due to imperfect acknowledgments. Whenever the acknowledgments are all transmitted correctly and on time (i.e., Case 2 in the unified communication protocol), these bounds will become zero and all the signals are asymptotically stable in the mean, and the communication protocol becomes TCP with full acknowledgment. On the other hand, when the acknowledgments are disabled, these bounds will attain maximum and the communication protocol becomes UDP. The flowchart of the proposed stochastic optimal regulator of LNCS under the unified communication protocol is shown in Figure 10.2.

Here, the initial system states are assumed to reside in the set Ω stabilized by using the initial admissible control input u_0 and they are deterministic. Then,

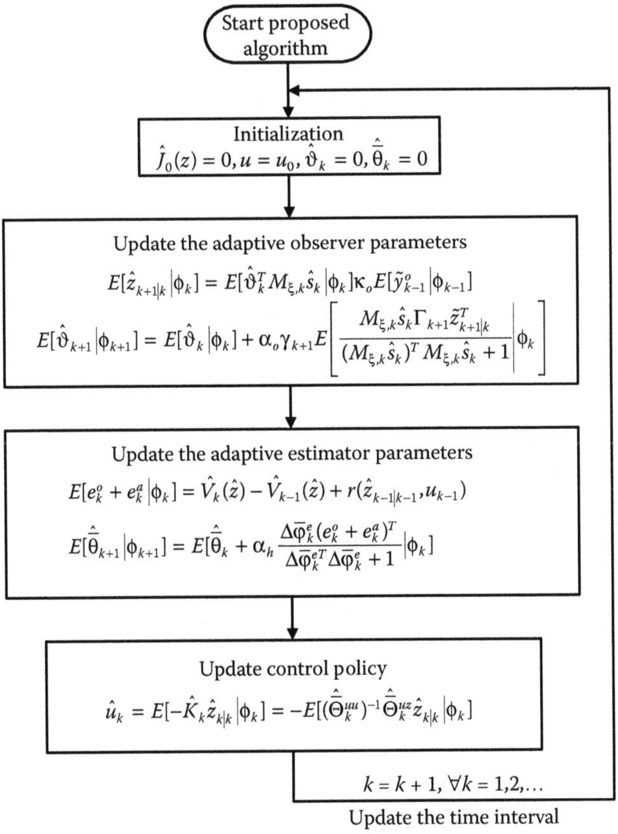

FIGURE 10.2 Stochastic optimal regulator for LNCS under the unified communication protocol.

sufficient conditions for the observer and value function estimator tuning gains α_o, α_h are derived to guarantee all the future states will remain *bounded* in the mean. Eventually, it can be shown that actual control inputs converge close to the optimal value. Before introducing the convergence proof, the following lemma is needed to establish the bounds in the mean on the optimal closed-loop dynamics while the optimal control is implemented on the NCS under the unified communication protocol with network and acknowledgment imperfections.

Lemma 10.1

Consider the NCS in the presence of the unified communication protocol, and then, there exists a set of optimal control policy such that the following inequality is satisfied

$$\left\| E\left[A_{zk} z_k + B_{zk} u_k^* \middle| \phi_k \right] \right\|^2 \le k_a \left\| E\left[z_k \middle| \phi_k \right] \right\|^2 \tag{10.22}$$

where $0 < k_a < 1/2$ is a constant.

Proof: Proof follows similar to Xu et al. (2012).

Next, the main result is stated.

Theorem 10.3

Given the initial conditions for the system state z_0, observer parameter estimation vector $E[\hat{\bar{\vartheta}}_0 | \phi_0]$, value function parameter vector $E[\hat{\bar{\theta}}_0 | \phi_0]$ are bounded in Ω, let u_0 be any initial admissible control policy for an NCS under the unified communication protocol in the presence of network and possible acknowledgment imperfections satisfying the bounds given by Equation 10.22. Let the observer, value function estimated parameters be tuned and the estimated control policy be provided by Equations 10.5, 10.19, and 10.21, respectively. Then, there exist positive constants α_o, η given by Theorems 10.1 and 10.2, and α_h given by Theorem 10.2 such that the system state vector z_k, observer parameter estimation error vector $E[\tilde{\bar{\vartheta}}_k | \psi_k]$, and value function parameter estimation error vector $E[\tilde{\bar{\theta}}_k | \psi_k]$ for an NCS under this unified communication protocol framework are all *bounded* in the mean with bounds given by $\left\| E\left[z_k \middle| \phi_k \right] \right\| \le b_z$, $\left\| E\left[\tilde{z}_{k|k} \middle| \phi_k \right] \right\| \le b_e$, $\left\| E\left[\tilde{\bar{\vartheta}}_k \middle| \phi_k \right] \right\| \le b_\vartheta$, and $\left\| E\left[\tilde{\bar{\theta}}_k \middle| \phi_k \right] \right\| \le b_\theta$.

Proof: Refer to Appendix 10C.

Remark 10.4

When an acknowledgment can be transmitted perfectly, the unified communication protocol becomes TCP with full acknowledgment, system states z_k, observer parameter estimation error vector $E[\tilde{\bar{\vartheta}}_k|\phi_k]$, and value function parameter estimation error vector $E[\tilde{\bar{\theta}}_k|\psi_k]$ are all asymptotically stable in the mean since $\bar{\xi}=1$, $\sigma_\xi^2=0$, and $\left(1-\bar{\xi}\right)^2+\sigma_\xi^2=0$, which makes $\varepsilon_{TM}=0$ and bounds b_z, b_e, b_θ, and b_ϑ are all zero. Meanwhile, when no acknowledgment can be transmitted correctly, the unified communication protocol becomes UDP. Then, system states z_k, observer parameter estimation error vector $E[\tilde{\bar{\vartheta}}_k|\phi_k]$, and value function parameter estimation error vector $E[\tilde{\bar{\theta}}_k|\phi_k]$ are all *bounded* in the mean with the bound reaching maximum since $\bar{\xi}=0$, $\sigma_\xi^2=0$, and $\left(1-\bar{\xi}\right)^2+\sigma_\xi^2=1$, and ε_{TM} along with the bounds b_z, b_e, b_θ, b_ϑ (10C.6) reach maximum values.

10.3 SIMULATION RESULTS

In this section, the performance of the proposed stochastic optimal controller for NCS under the unified communication protocol is evaluated with a single protocol at a time. Meanwhile, the standard optimal control of NCS under the unified communication protocol with forced assumption of known system dynamics, acknowledgment reception, and network imperfections are also simulated for comparison.

EXAMPLE 10.1

The continuous-time version of a batch reactor system dynamics are given as (Carnevale et al. 2007)

$$\dot{x}=\begin{bmatrix}1.38 & -0.2077 & 6.715 & -5.676\\ -0.5814 & -4.29 & 0 & 0.675\\ 1.067 & 4.273 & -6.654 & 5.893\\ 0.048 & 4.273 & 1.343 & -2.104\end{bmatrix}x+\begin{bmatrix}0 & 0\\ 5.679 & 0\\ 1.136 & -3.146\\ 1.136 & 0\end{bmatrix}u$$

$$y=\begin{bmatrix}1 & 0 & 1 & -1\\ 0 & 1 & 0 & 0\end{bmatrix}x \tag{10.23}$$

with $x\in\mathbb{R}^{4\times1}$ and $u\in\mathbb{R}^{2\times1}$. This batch reactor example has been developed over the years for NCS (Carnevale et al. 2007; Dacic and Nesic 2007).

The NCS parameters under the unified communication protocol are selected as (Xu et al. 2012)

1. The sampling time: $T_s=50$ ms.
2. The delay bound is selected as two, that is, $b=2$.
3. The random delays: $E(\tau_{sc})=35$ ms, $E(\tau)=75$ ms.
4. Packet losses follow the Bernoulli distribution with $\bar{\gamma}=0.3$ and $\bar{\upsilon}=0.2$.
5. The communication acknowledgment indicator follows the Bernoulli distribution with $\bar{\xi}=0.8$.
6. To obtain the ergodic performance, Monte Carlo simulation runs were executed with 1000 iterations.

10.3.1 Traditional Pole Placement Controller Performance with Network Imperfections

First, we consider the effect of network imperfections for an NCS under the unified communication protocol. In Figure 10.3, the standard control inputs

$$u_k = -\begin{bmatrix} 3.78 & 1.82 & 0.50 & 4.27 \\ -0.28 & 0.98 & -0.91 & 6.48 \end{bmatrix} x_k$$

designed by the pole placement method cannot maintain the system stability in the presence of network imperfection caused by the communication protocol. This controller renders acceptable performance with known network imperfections.

10.3.2 NCS under TCP with Intermittent Acknowledgment

Next, the proposed stochastic optimal controller and novel observer designs are applied on an NCS under Case 2 of the unified communication protocol (i.e., TCP with intermittent acknowledgment) with unknown system dynamics resulting from network imperfections and intermittent acknowledgment, respectively. The augment state z_k is derived as $z_k = [x_k \; u_{k-1} \; u_{k-2}]^T \in \mathbb{R}^{8\times 1}$ and $\varphi^e = [\hat{z} \; u]^T \in \mathbb{R}^{10\times 1}$. The initial admissible policy for proposed algorithm is selected as

$$u_0 = -\begin{bmatrix} 0.87 & 0.85 & -0.1 & 1.24 & 0.03 & 0 & 0.13 & 0.01 \\ -1.51 & 0.09 & -2.55 & 2.47 & 0 & 0.08 & -0.05 & 0.52 \end{bmatrix} E\left[\hat{z}_{k|k} \middle| \phi_k\right]$$

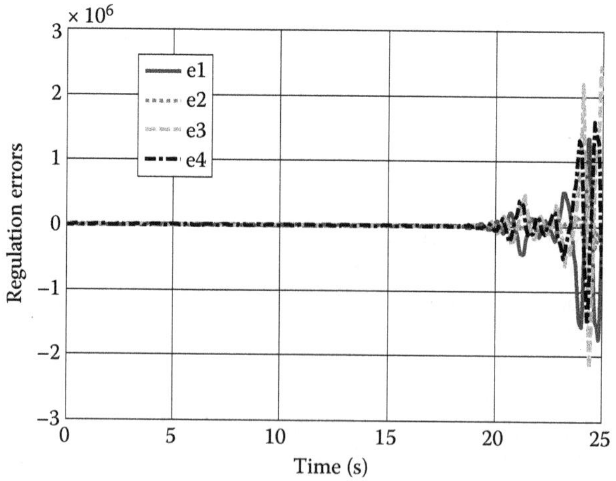

FIGURE 10.3 State regulation errors of standard control when network imperfections are present for NCS.

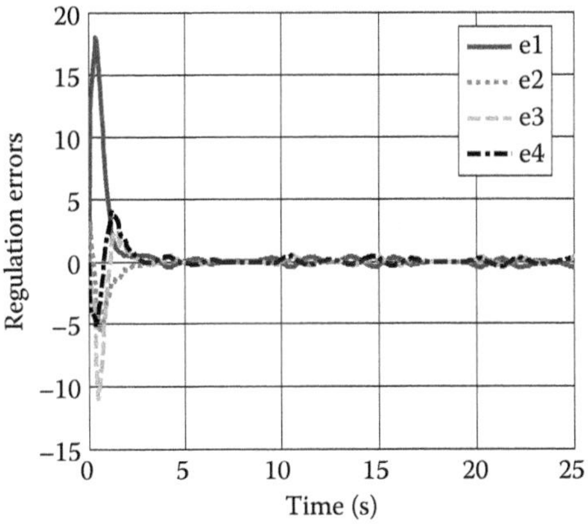

FIGURE 10.4 Performance of the proposed stochastic optimal control: State regulation errors of NCS under Case 2 of the unified communication protocol.

while regression functions for value function estimation is generated as $\{\varphi_1^{e2}, \varphi_1^e\varphi_2^e, \varphi_1^e\varphi_3^e, \ldots, \varphi_2^{e2}, \ldots, \varphi_{10}^{e2}\}$ as in Xu et al. (2012). The designed tuning rate for value function estimator is selected as $\alpha_h = 10^{-4}$ for NCS under the unified communication protocol while initial parameters are set to zero at the beginning of the simulation.

The initial parameters of the control estimator are chosen to reflect the initial admissible control. On the other hand, the regression function for the observer is defined as Chapter 1, and the designed learning rate is defined as $\alpha_o = 10^{-3}$ for an NCS under the unified communication protocol. The simulation was run for 500 time steps, and in the first 100 time steps, the exploration noise was added to maintain the PE condition (see Remark 10.4).

In Figures 10.4 through 10.6, we evaluate the performance of the proposed value function estimator and observer-based optimal control for an NCS under Case 2 of the unified communication protocol. Even with uncertain dynamics and inaccurate reception of communication acknowledgments, the proposed value function estimator and observer-based optimal control can still force regulation errors to converge close to zero as shown in Figure 10.4.

In Figure 10.5, control inputs of the proposed scheme for an NCS under the unified communication protocol are shown. Note that there is an overshoot observed at the beginning of Figures 10.5 and 10.6 due to an initial online learning phase needed to tune the observer and value function estimator.

After a short time, the proposed scheme will generate a satisfactory performance even when the NCS system dynamics are unknown irrespective of the communication protocol. On the other hand, the performance of the proposed observer for an NCS under the unified communication protocol is evaluated in Figure 10.6. The proposed observer can force the observed system state vector to converge close to the actual state vector quickly.

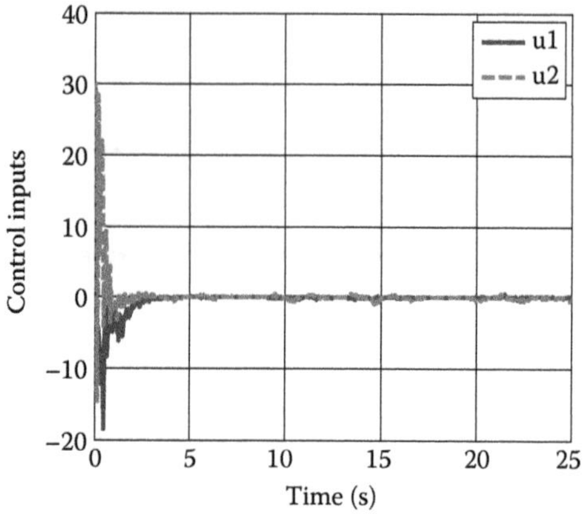

FIGURE 10.5 Control inputs $u = (u_1\ u_2)^T \in \mathbb{R}^{2\times1}$ for the proposed controller of NCS under Case 2 of the unified communication protocol.

10.3.3 NCS under TCP with Full Acknowledgment

Further, the proposed stochastic optimal controller and novel observer designs are applied to an NCS under Case 1 of the unified communication protocol, that is, TCP with full acknowledgment. As shown in Figures 10.7 and 10.8, even when the dynamics of NCS are unknown, the proposed value function estimator and observer-based optimal control can still force regulation errors and observer errors to converge close to zero. Compared with an NCS under Case 2 of the unified communication

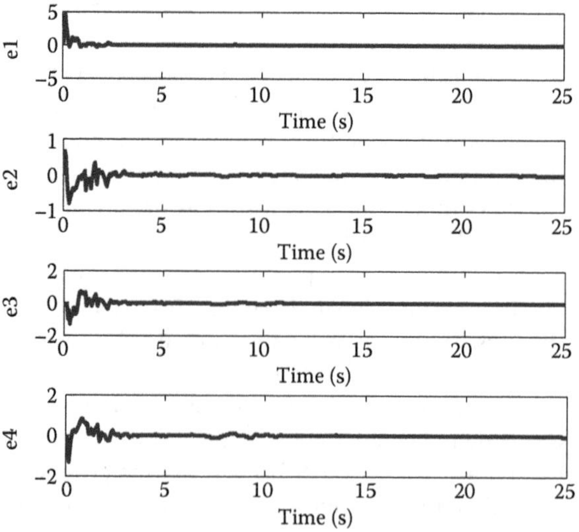

FIGURE 10.6 Performance of the proposed observer for NCS under Case 2 of the unified communication protocol.

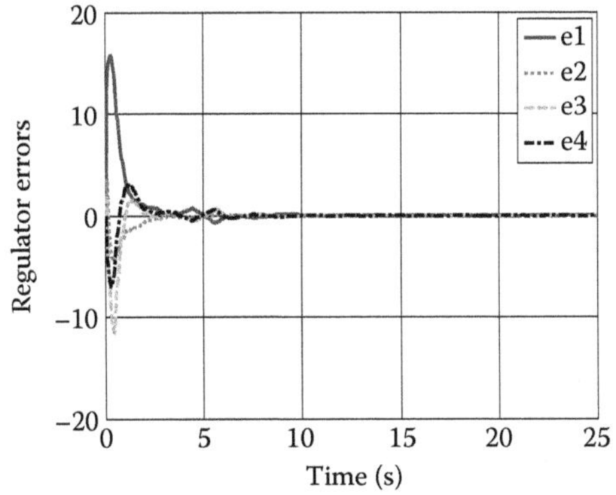

FIGURE 10.7 Performance of the proposed stochastic optimal control: state regulation errors of NCS under Case 1 of the unified communication protocol.

framework, the regulation errors and observer errors can be forced much closer to zero since full acknowledgments are received accurately.

10.3.4 NCS UNDER UDP WITH NO ACKNOWLEDGMENT

In Figures 10.9 and 10.10, we evaluate the performance of the proposed value function estimator and observer-based optimal control for an NCS under Case 3 of the unified communication protocol (i.e., UDP with no acknowledgment). Under this worst case of no acknowledgments with uncertain system dynamics of the NCS, the proposed value function estimator and observer-based optimal control can still force

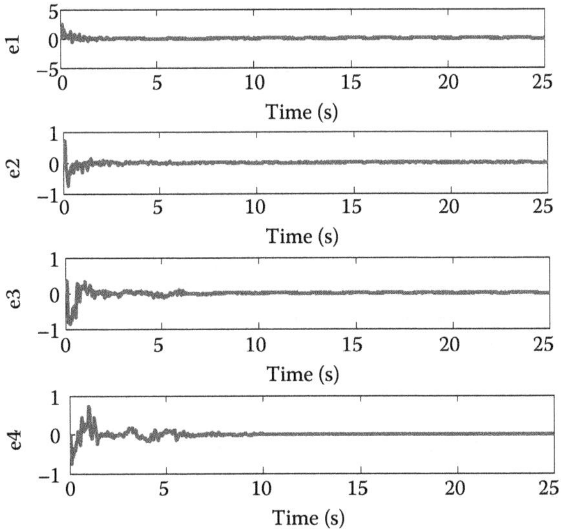

FIGURE 10.8 Performance of the proposed observer for NCS under Case 1 of the unified communication protocol.

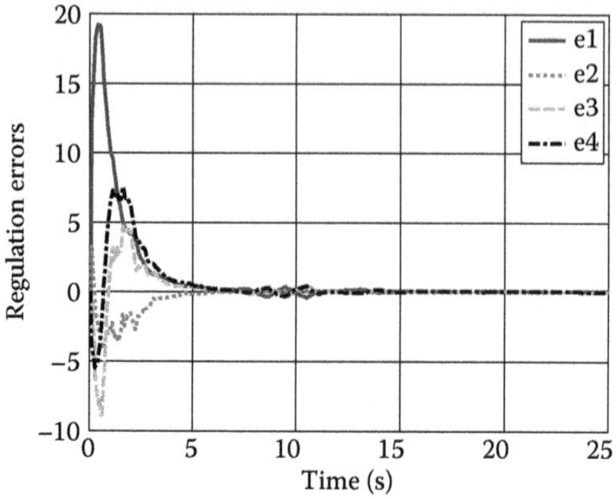

FIGURE 10.9 Performance of the proposed stochastic optimal control: State regulation errors of NCS under Case 3 of the unified communication protocol.

regulation and observation errors to converge close to zero as shown in Figures 10.9 and 10.10, respectively. It is important to note that when acknowledgments are not received, regulation and observer errors will converge slowly compared to Cases 1 and 2.

It is important to know that there is a critical domain of values for the parameters of the Bernoulli arrival processes (i.e., $\bar{\gamma}$ and $\bar{\upsilon}$) in order to ensure the stability of the observer. The three cases included in this unified communication framework have different stability regions. In Figure 10.11, these stability regions for the proposed observer under the unified communication framework are shown. For Case 1 (i.e., TCP with full acknowledgment, $\bar{\xi} = 1$), the stability region is largest since full

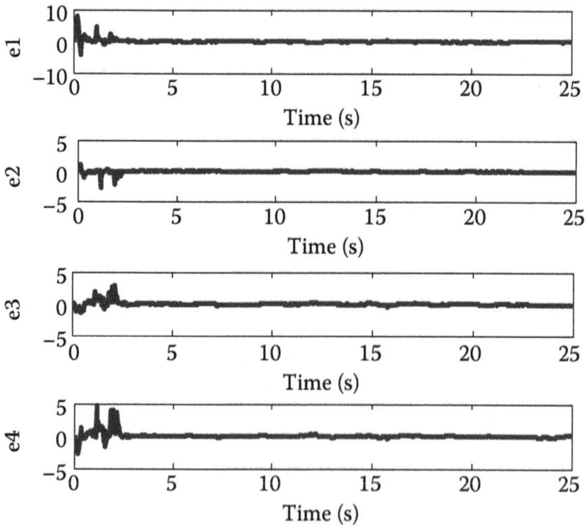

FIGURE 10.10 Performance of the proposed observer for NCS under Case 3 of the unified communication protocol.

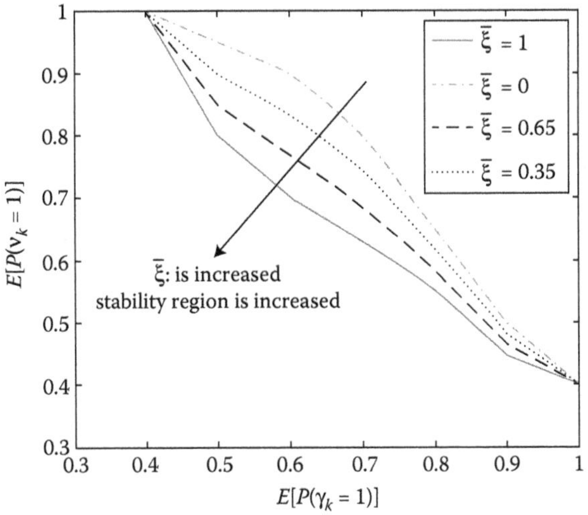

FIGURE 10.11 Comparison of the stability region of the observer for NCS under the unified communication protocol.

acknowledgments without any packet drops can provide more information for the observer design.

When intermittent acknowledgment transmission is considered (i.e., Case 2, $0 < \bar{\xi} < 1$), the stability region will decrease whereas when the delivery of acknowledgments becomes problematic (i.e., when $\bar{\xi}$ becomes small, the stability region decreases more). Further, while no acknowledgment has been transmitted (i.e., Case 3, $\bar{\xi} = 0$), the stability region becomes the smallest compared with the other two cases due to lack of information to the observer. In addition, these regions are smaller than the case when the system dynamics are known.

On the other hand, transmitting acknowledgments require more network resources such as bandwidth as shown in Figure 10.12. Since Case 1 and 2 under the unified communication framework need to transmit the acknowledgments, their requested network resources are higher than Case 3 (i.e., UDP), which does not transmit acknowledgment. Therefore, there is a trade-off between needing network resources and transmitting acknowledgments.

Based on the results presented in Figures 10.3 through 10.12, after a short initial tuning time, the proposed approach delivers performance similar to the case when the network imperfections are known. Meanwhile, since the NCS is stable even with imperfect acknowledgment, the proposed controller design is quite useful and powerful for practical industrial systems.

10.4 CONCLUSIONS

Owing to limited network resources in practical NCS, the TCP retransmission scheme is not preferred for closed-loop control environments while the acknowledgments might not have been transmitted without any packet drops. In this chapter,

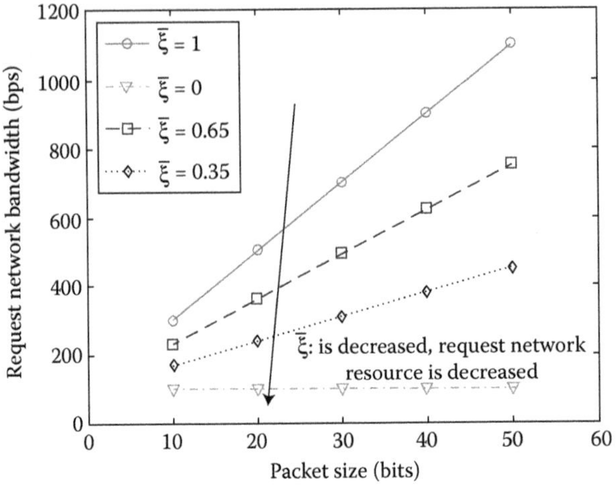

FIGURE 10.12 Comparison of the requested network bandwidth for the unified communication protocol.

a novel ADP scheme consisting of a novel observer and value function estimator is utilized to solve the Bellman equation in real time for obtaining optimal control of an NCS under the unified communication protocol, which is described in three scenarios (i.e., TCP with full acknowledgment, TCP with intermittent acknowledgment, and UDP with no acknowledgment). By using past input and estimated state vectors, the requirement on system dynamics was relaxed. By using the estimated state vector and value function, stochastic optimal control inputs were derived.

An initial admissible control ensured that the system is stable while the observer and value function estimator is tuned. Initial overshoots are observed due to the online tuning phase of the observer and the value function estimator while they disappear over time. All observer and value function estimator parameters $E[\hat{\vartheta}_k|\phi_k], \hat{\Theta}_k$ were tuned online using the proposed update law and Lyapunov theory demonstrated the *bounded* in the mean of the closed-loop system for an NCS under the unified protocol framework. Meanwhile, asymptotic stability in the mean of the closed-loop system can be achieved only when each acknowledgment is received successfully and on time (i.e., communication protocol becomes TCP with full acknowledgment).

PROBLEMS

Section 10.5

Problem 10.5.1: Repeat Example 10.1 with different initial conditions.

Problem 10.5.2: Repeat Example 10.1 with different cost function weighting matrix, that is, Q, R.

APPENDIX 10A

Proof of Theorem 10.1

Consider the Lyapunov candidate function $L_o = L_{\tilde{z}} + L_{\vartheta}$, where $L_{\tilde{z}} = tr\left\{\sum_{i=k-N_0}^{k} E\left[\tilde{z}_{i|i}^T \Xi \tilde{z}_{i|i} |\psi_i\right]\right\}$ and $L_{\vartheta} = 3tr\left\{\sum_{i=k-N_0}^{k} E\left[\tilde{\vartheta}_i^T \tilde{\vartheta}_i |\phi_i\right]\right\}$ with $\Xi = (1/((\Psi_M^2+1)(U_M-1)^2))\mathbf{I}$ being a positive definite matrix and $\mathbf{I}$ an identity matrix. Then, the first difference can be expressed as $\Delta L_o = \Delta L_{\tilde{z}} + \Delta L_{\vartheta}$. Now, take the first term $L_{\tilde{z}}$ (i.e., $\Delta L_{\tilde{z}} = tr\{\sum_{i=k-N_0}^{k} E[\tilde{z}_{i+1|i+1}^T \Xi \tilde{z}_{i+1|i+1}]|\psi_{i+1}\} - tr\{\sum_{i=k-N_0}^{k} E[\tilde{z}_{i|i}^T \Xi \tilde{z}_{i|i}]|\psi_i\}$). Using Equations 10.6 and 10.7 and applying the Cauchy–Schwartz inequality reveals

$$
\begin{aligned}
\Delta L_{\tilde{z}} &= tr\left\{\sum_{i=k-N_0}^{k} E\left[\tilde{z}_{i+1|i+1}^T \Xi \tilde{z}_{i+1|i+1} |\phi_{i+1}\right]\right\} - tr\left\{\sum_{i=k-N_0}^{k} E\left[\tilde{z}_{i|i}^T \Xi \tilde{z}_{i|i} |\phi_i\right]\right\} \\
&\le tr\left\{\sum_{i=k-N_0}^{k} \left\{[U_z - \kappa_o (D_y^o)^+] E\left[\tilde{z}_{i|i} |\phi_i\right] + [1-\kappa_o (D_y^o)^+] E\left[(\vartheta^T (M_{\xi c,i}\right.\right.\right. \\
&\quad - \bar{M}_{\xi c}) \hat{s}_i) |\phi_i\big] + [1-\kappa_o (D_y^o)^+] E\left[\tilde{\vartheta}^T M_{\xi,i} \hat{s}_i |\phi_i\right] - \frac{\chi_{\min}^2}{\chi_{\min}^2+1} \\
&\quad \times \alpha_o \gamma_{i+1} \tilde{z}_{i+1|i}\Big\}^T \Xi \Big\{[U_z - \kappa_o (D_y^o)^+] E\left[\tilde{z}_{i|i} |\phi_i\right] + [1-\kappa_o (D_y^o)^+] \\
&\quad \times E\left[\left(\vartheta^T \left(M_i - \bar{M}_{\xi,i}\right) \hat{s}_i\right) |\phi_i\right] + E\left[\tilde{\vartheta}^T \left(M_{\xi,i} + \bar{M}_{\xi c}\right) \hat{s}_i |\phi_i\right] \\
&\quad \left.\left. - \frac{\chi_{\min}^2}{\chi_{\min}^2+1} \alpha_o \gamma_{i+1} \tilde{z}_{i+1|i}\right]\right\} - tr\left\{\sum_{i=k-N_0}^{k} E\left[\tilde{z}_{i|i}^T \Xi \tilde{z}_{i|i} |\phi_i\right]\right\} \\
&\le -\Xi \sum_{i=k-N_0}^{k} \left(1 - 3\left(1 - \frac{\chi_{\min}^2}{\chi_{\min}^2+1} \alpha_o \gamma_{i+1}\right)^2 \eta^2\right) \left\|U_z - \kappa_o (D_y^o)^+\right\|^2 \left\|E\left[\tilde{z}_{i|i} |\phi_i\right]\right\|^2 \\
&\quad + \sum_{i=k-N_0}^{k} \left(3\left(1 - \frac{\chi_{\min}^2}{\chi_{\min}^2+1} \alpha_o \gamma_{i+1}\right)^2 \left\|E\left[M_{\xi,k} \hat{s}_i |\phi_i\right]\right\|^2 \left\|1 - \kappa_o (D_y^o)^+\right\|^2\right. \\
&\quad \times \Xi tr\left\{E\left[\tilde{\vartheta}_i^T \tilde{\vartheta}_i |\phi_i\right]\right\}\Big) + tr\left\{\sum_{i=k-N_0}^{k} 3\left(1 - \frac{\chi_{\min}^2}{\chi_{\min}^2+1} \alpha_o \gamma_{i+1}\right)^2 E\left[\left(M_i - \bar{M}_{\xi,i}\right)\right.\right. \\
&\quad \left.\left. \times \vartheta^T \vartheta \left(M_i - \bar{M}_{\xi,i}\right)^T |\phi_i\right]\right\}
\end{aligned} \tag{10A.1}
$$

Next, according to parameter estimation error dynamics (10.6), the term ΔL_{ϑ} can be derived as

$$\Delta L_{\vartheta} = 3tr\left\{\sum_{i=k-N_0}^{k} E\left[\tilde{\vartheta}_{i+1}^T \tilde{\vartheta}_{i+1} \middle| \phi_{i+1}\right]\right\} - 3tr\left\{\sum_{i=k-D_0}^{k} E\left[\tilde{\vartheta}_i^T \tilde{\vartheta}_i \middle| \phi_i\right]\right\}$$

$$= 3tr\left\{\sum_{i=k-N_0}^{k} E\left[\left(\tilde{\vartheta}_i - \alpha_o \gamma_{i+1} \frac{M_{\xi,k}\hat{s}_i \Gamma_{i+1} \tilde{z}_{i+1|i}^T}{(M_{\xi,k}\hat{s}_i)^T M_{\xi,k}\hat{s}_i + 1}\right)^T\right.\right. \tag{10A.2}$$

$$\left.\left.\left(\tilde{\vartheta}_i - \alpha_o \gamma_{i+1} \frac{M_{\xi,k}\hat{s}_i \Gamma_{i+1} \tilde{z}_{i+1|i}^T}{\left(M_{\xi,k}\hat{s}_i\right)^T M_{\xi,k}\hat{s}_i + 1}\right) \middle| \phi_i\right]\right\} - 3tr\left\{\sum_{i=k-N_0}^{k} E\left[\tilde{\vartheta}_i^T \tilde{\vartheta}_i \middle| \phi_i\right]\right\}$$

Substituting Equation 10.3 into Equation 10A.2, ΔL_{ϑ} can be expressed as

$$\Delta L_{\vartheta} \le -6\,tr\left\{\sum_{i=k-N_0}^{k} \alpha_o \gamma_{i+1}(1-\alpha_o \gamma_{i+1}) E\left[\frac{\tilde{\vartheta}_i^T M_{\xi,i}\hat{s}_i \Gamma_{i+1} \tilde{z}_{i|i}^T [U_z - \kappa_o (D_y^o)^+]}{(M_{\xi,i}\hat{s}_i)^T M_{\xi,i}\hat{s}_i + 1} \middle| \phi_i\right]\right\}$$

$$-6\,tr\left\{\sum_{i=k-N_0}^{k} \alpha_o \gamma_{i+1}(1-\alpha_o \gamma_{i+1}) E\left[\frac{\tilde{\vartheta}_i^T M_{\xi,i}\hat{s}_i \Gamma_{i+1} \Gamma_{i+1}^T \hat{s}_i^T M_{\xi,i}^T \vartheta}{\left(M_{\xi,i}\hat{s}_i \Gamma_{i+1}\right)^T M_{\xi,i}\hat{s}_i \Gamma_{i+1} + 1} \middle| \phi_i\right]\right\}$$

$$-3\sum_{i=k-N_0}^{k} \alpha_o \gamma_{i+1}(2-\alpha_o \gamma_{i+1}) tr\left\{E\left[\tilde{\vartheta}_i^T \tilde{\vartheta}_i \middle| \psi_i\right]\right\} + 6tr\left\{\sum_{i=k-N_0}^{k} \alpha_o^2 \gamma_{i+1}^2\right.$$

$$\left.\times E\left[\frac{[U_z - \kappa_o (D_y^o)^+]^T \Gamma_{i+1} \tilde{z}_{i|i} \hat{s}_i^T M_i \vartheta)^T}{\left(M_{\xi,i}\hat{s}_i\right)^T M_{\xi,i}\hat{s}_i + 1} \middle| \phi_i\right]\right\} + 3\alpha_o^2 \gamma_{i+1}^2 \left[\left(1-\bar{\xi}\right)^2 + \sigma_{\xi}^2\right] \sigma_{\upsilon}^2 \|\vartheta\|^2$$

$$+3tr\left\{\sum_{i=k-N_0}^{k} \alpha_o^2 \gamma_{i+1}^2 E\left[\frac{[U_z - \kappa_o (D_y^o)^+]^T \tilde{z}_{i|i} \Gamma_{i+1} \Gamma_{i+1}^T \tilde{z}_{i|i}^T [U_z - \kappa_o (D_y^o)^+]^T}{(M_{\xi,i}\hat{s}_i)^T M_{\xi,i}\hat{s}_i + 1} \middle| \phi_i\right]\right\}$$

$$\le -3\sum_{i=k-N_0}^{k} \alpha_o \gamma_{i+1}\left(\alpha_o \gamma_{i+1} - \frac{2}{\chi_{\min}^2 + 1}\right)\left\|E\left[\tilde{\vartheta}_i \middle| \phi_i\right]\right\|^2 + 3\sum_{i=k-N_0}^{k} \frac{\alpha_o \gamma_{i+1}(1+\alpha_o \gamma_{i+1})}{(\chi_{\min}^2 + 1)(U_M - 1)^2}$$

$$\times \eta^2 \left\|E[\tilde{z}_{i|i} | \phi_i]\right\|^2 + 3\alpha_o \gamma_{i+1}\left[\left(1-\bar{\xi}\right)^2 + \sigma_{\xi}^2\right] \frac{\sigma_{\upsilon}^2}{\bar{\upsilon}^2} \|\vartheta\|^2 \tag{10A.3}$$

Finally, consider the overall first difference and using Equations 10A.1 and 10A.3, ΔL_o can be expressed as

$$\Delta L_o = \Delta L_{\tilde{z}} + \Delta L_{\vartheta}$$

$$= tr\left\{ \sum_{i=k-N_0}^{k} E\left[\tilde{z}_{i+1|i+1}^T \Xi \tilde{z}_{i+1|i+1} \middle| \phi_{i+1} \right] \right\} - tr\left\{ \sum_{i=k-N_0}^{k} E\left[\tilde{z}_{i|i}^T \Xi \tilde{z}_{i|i} \middle| \phi_i \right] \right\}$$

$$+ tr\left\{ \sum_{i=k-N_0}^{k} E\left[\tilde{\vartheta}_{i+1}^T \tilde{\vartheta}_{i+1} \middle| \phi_i \right] \right\} - tr\left\{ \sum_{i=k-N_0}^{k} E\left[\tilde{\vartheta}_i^T \tilde{\vartheta}_i \middle| \phi_i \right] \right\}$$

$$\leq -\Xi \sum_{i=k-N_0}^{k} \left(1 - 3\left(1 - \frac{\chi_{\min}^2}{\chi_{\min}^2 + 1} \alpha_o \gamma_{i+1} \right)^2 \eta^2 \right) \left\| E\left[\tilde{z}_{i|i} \middle| \phi_i \right] \right\|^2$$

$$+ \sum_{i=k-N_0}^{k} \left(3\left(1 - \frac{\chi_{\min}^2}{\chi_{\min}^2 + 1} \alpha_o \gamma_{i+1} \right)^2 \left\| M_{\xi,i} s_i \right\|^2 \Xi tr\left\{ E\left[\tilde{\vartheta}_{i+1}^T \tilde{\vartheta}_{i+1} \middle| \phi_{i+1} \right] \right\} \right)$$

$$+ tr\left\{ \sum_{i=k-N_0}^{k} 3\left(1 - \frac{\chi_{\min}^2}{\chi_{\min}^2 + 1} \alpha_o \gamma_{i+1} \right)^2 \Xi E[\hat{s}_i \left(M_i - \bar{M}_{\xi,i} \right) \vartheta^T \vartheta (M_{\xi c,i} \right.$$

$$\left. - \bar{M}_{\xi c} \right)^T s_i^T \left| \phi_i \right] \Big\} - 3 \sum_{i=k-N_0}^{k} \alpha_o \gamma_{i+1} \left(\alpha_o \gamma_{i+1} - \frac{2}{\chi_{\min}^2 + 1} \right) \left\| E\left[\tilde{\vartheta}_i \middle| \phi_i \right] \right\|^2$$

$$+ 3 \sum_{i=k-N_0}^{k} \frac{\alpha_o \gamma_{i+1} (1 + \alpha_o \gamma_{i+1})}{\chi_{\min}^2 + 1} \eta^2 \left\| E\left[\tilde{z}_{i|i} \middle| \phi_i \right] \right\|^2$$

$$+ 3\alpha_o \gamma_{i+1} \left[\left(1 - \xi \right)^2 + \sigma_\xi^2 \right] \frac{\sigma_\upsilon^2}{\bar{\upsilon}^2} \left\| \vartheta \right\|^2 \tag{10A.4}$$

Since the packet loss indicator γ_{k+1} can be equal to 0 or 1, ΔL_o (10A.4) needs to be separated into two different cases for further consideration as given below:

Case 1: $\gamma_{k+1} = 1$ (no packet losses)

Substituting γ_{k+1} value into Equation 10A.4, ΔL_o can be derived as

$$\Delta L_o = \Delta L_{\tilde{z}} + \Delta L_{\vartheta}$$

$$\leq -\Xi \sum_{i=k-N_0}^{k} \left(1 - 3\left(1 - \frac{\chi_{\min}^2}{\chi_{\min}^2 + 1} \alpha_o \gamma_{i+1} \right)^2 \eta^2 \right) \left\| E\left[\tilde{z}_{i|i} \middle| \phi_i \right] \right\|^2$$

$$+\sum_{i=k-N_0}^{k}\left(3\left(1-\frac{\chi_{\min}^2}{\chi_{\min}^2+1}\alpha_o\gamma_{i+1}\right)^2\left\|M_{\xi,i}\hat{s}_i\right\|^2\Xi tr\left\{E\left[\tilde{\vartheta}_i^T\tilde{\vartheta}_i\middle|\phi_i\right]\right\}\right)$$

$$+tr\left\{\sum_{i=k-N_0}^{k}3\left(1-\frac{\chi_{\min}^2}{\chi_{\min}^2+1}\alpha_o\gamma_{i+1}\right)^2\Xi E\left[\hat{s}_i\left(M_i-\bar{M}_{\xi,i}\right)\vartheta^T\vartheta\left(M_i-\bar{M}_{\xi,i}\right)^T s_i^T\middle|\phi_i\right]\right\}$$

$$-3\sum_{i=k-N_0}^{k}\alpha_o\gamma_{i+1}\left(\alpha_o\gamma_{i+1}-\frac{2}{\chi_{\min}^2+1}\right)\left\|E\left[\tilde{\vartheta}_i\middle|\phi_i\right]\right\|^2+\sum_{i=k-N_0}^{k}\frac{3\alpha_o\gamma_{i+1}(1+\alpha_o\gamma_{i+1})}{\chi_{\min}^2+1}$$

$$\times\eta^2\left\|E\left[\tilde{z}_{i|i}\middle|\phi_i\right]\right\|^2+3\alpha_o\gamma_{i+1}\left[\left(1-\xi\right)^2+\sigma_\xi^2\right]\frac{\sigma_\upsilon^2}{\bar{\upsilon}^2}\left\|\vartheta\right\|^2$$

$$\leq-\frac{(1-3\eta^2(U_M-1)^2((1-\frac{\chi_{\min}^2}{\chi_{\min}^2+1}\alpha_o)^2-\alpha_o(1+\alpha_o)))}{(\chi_{\min}^2+1)(U_M-1)^2}\left\|E[\tilde{z}_{k|k}|\phi_k]\right\|^2$$

$$-\frac{2}{3}[(\chi_{\min}^2+1)\alpha_o-1]\left\|E\left[\tilde{\vartheta}_k\middle|\phi_k\right]\right\|^2+\varepsilon_M^o \qquad (10A.5)$$

Case 2: $\gamma_{k+1}=0$. (with packet losses)

Based on Assumption 10.1, the packet loss probability between the sensor and the controller has to satisfy the observer stability region (i.e., $P(\gamma_k=1)>(1/N_0)$). Therefore, if $\gamma_{k+1}=0$, then there exists $j\in[k-N_0,k]$ such that $\gamma_{k+1}=\gamma_k=\cdots=\gamma_{k-j+1}=0$ and $\gamma_{k-j}=1$. Therefore, using the update law for the observer estimated parameters (10.4), observer parameter estimation errors $\tilde{\vartheta}_{k+1}$ can be expressed as $\tilde{\vartheta}_{k+1}=\tilde{\vartheta}_{k-j}$. Then, substituting $\tilde{\vartheta}_{k+1}=\tilde{\vartheta}_{k-j}$ into Equation 10A.4, ΔL_o can be derived as

$$\Delta L_o=\Delta L_{\tilde{z}}+\Delta L_{\tilde{\vartheta}}$$

$$\leq-\Xi\sum_{i=k-N_0}^{k}\left(1-3\left(1-\frac{\chi_{\min}^2}{\chi_{\min}^2+1}\alpha_o\gamma_{i+1}\right)^2\eta^2\right)\left\|E\left[\tilde{z}_{i|i}\middle|\phi_i\right]\right\|^2$$

$$+\sum_{i=k-N_0}^{k}\left(3\left(1-\frac{\chi_{\min}^2}{\chi_{\min}^2+1}\alpha_o\gamma_{i+1}\right)^2\left\|M_{\xi,i}\hat{s}_i\right\|^2\Xi tr\left\{E\left[\tilde{\vartheta}_i^T\tilde{\vartheta}_i\middle|\phi_i\right]\right\}\right)$$

$$+tr\left\{\sum_{i=k-N_0}^{k}3\left(1-\frac{\chi_{\min}^2}{\chi_{\min}^2+1}\alpha_o\gamma_{i+1}\right)^2\Xi E\left[\hat{s}_i(M_i-\bar{M}_{\xi,i})\vartheta^T\vartheta\right.\right.$$

$$\left.\left.(M_i-\bar{M}_{\xi,i})^T s_i^T\middle|\phi_i\right]\right\}-3\sum_{i=k-N_0}^{k}\alpha_o\gamma_{i+1}\left(\alpha_o\gamma_{i+1}-\frac{2}{\chi_{\min}^2+1}\right)\left\|E\left[\tilde{\vartheta}_i\middle|\phi_i\right]\right\|^2$$

$$+\sum_{i=k-N_0}^{k}\frac{3\alpha_o\gamma_{i+1}(1+\alpha_o\gamma_{i+1})}{\chi_{\min}^2+1}\eta^2\left\|E\left[\tilde{z}_{i|i}\middle|\phi_i\right]\right\|^2+3\alpha_o\gamma_{i+1}\left[\left(1-\bar{\xi}\right)^2+\sigma_\xi^2\right]\frac{\sigma_\upsilon^2}{\bar{\upsilon}^2}\|\vartheta\|^2$$

$$\le-\frac{(1-3\eta^2(U_M-1)^2((1-\frac{\chi_{\min}^2}{\chi_{\min}^2+1}\alpha_o)^2-\alpha_o(1+\alpha_o)))}{(\chi_{\min}^2+1)(U_M-1)^2}\left\|E\left[\tilde{z}_{k|k}\middle|\phi_k\right]\right\|^2$$

$$-\frac{2}{3}[(\chi_{\min}^2+1)\alpha_o-1]\left\|E\left[\tilde{\vartheta}_{k-j-1}\middle|\phi_{k-j-1}\right]\right\|^2+\varepsilon_M^o \qquad (10A.6)$$

where $\varepsilon_M^o=3((1-(\chi_{\min}^2/(\chi_{\min}^2+1))\alpha_o)^2+\alpha_o)[(1-\bar{\xi})^2+\sigma_\xi^2](\sigma_\upsilon^2/\bar{\upsilon}^2)\|\vartheta\|^2$. Since Ξ is defined as $\Xi=(1/(\Psi_M^2+1))\mathbf{I}$ and α_o,η are positive constants that satisfy $(1/(\chi_{\min}^2+1))<\alpha_o<1$ and

$$0<\eta<\sqrt{\frac{1}{3(U_M-1)^2((1-\frac{\chi_{\min}^2}{\chi_{\min}^2+1}\alpha_o)^2-\alpha_o(1+\alpha_o))}}$$

and ε_M^o is a bound constant, then according to Equations 10A.5 and 10A.6, ΔL_o is less than zero provided the following inequalities hold:

$$\left\|E\left[\tilde{z}_{k|k}\middle|\phi_k\right]\right\|>\sqrt{\frac{(\chi_{\min}^2+1)\varepsilon_M^o}{1-3\eta^2(U_M-1)^2((1-\frac{\chi_{\min}^2}{\chi_{\min}^2+1}\alpha_o)^2-\alpha_o(1+\alpha_o))}}\equiv B_{eo}$$

or

$$\left\|E\left[\tilde{\vartheta}_k\middle|\phi_k\right]\right\|>\sqrt{\frac{3\varepsilon_M^o}{2[(\chi_{\min}^2+1)\alpha_o-1]}}\equiv B_\vartheta$$

Using the standard Lyapunov extension, the observer error dynamics $E[\tilde{z}_{k|k}|\phi_k]$ and its parameter estimation errors $E[\tilde{\vartheta}_k|\phi_k]$ for NCS under the unified communication protocol are *bounded* in the mean for both cases.

APPENDIX 10B

Proof of Theorem 10.2

Consider the positive definite Lyapunov candidate

$$L_J=L_\theta+L_{ao} \qquad (10B.1)$$

where $L_\theta=tr\left\{E\left[\tilde{\bar{\theta}}_k^T\Pi\tilde{\bar{\theta}}_k\middle|\phi_k\right]\right\}$ and

$$L_{ao} = tr\left\{\sum_{i=k-D_0}^{k} \vartheta_{i-1}^T\vartheta_{i-1}\vartheta_{i-1}^T\vartheta_{i-1}\left|\phi_{i-1}\right.\right\} tr\left\{\sum_{i=k-D_0}^{k} E\left[\tilde{z}_{i-1|i-1}\tilde{z}_{i-1|i-1}^T\Lambda\tilde{z}_{i-1|i-1}\tilde{z}_{i-1|i-1}^T\left|\phi_{i-1}\right.\right]\right\}$$

with

$$\Pi = \frac{(\Delta\phi_{\min}^2+1)\alpha_o^4}{2\alpha_h^2 O_e^2(\chi_{\min}^2+1)^2}\mathbf{I}$$

and

$$\Lambda = \frac{\alpha_o^4}{4(\Psi_M^2+1)^2(U_M-1)^4}\mathbf{I}$$

being positive definite matrices, **I** an identity matrix, and $\forall k = 1, 2, \ldots$. The first difference of Equation 10B.1 is given by $\Delta L_J = \Delta L_\theta + \Delta L_{ao}$. Since the Lyapunov candidate function includes observer parameters, we have to separate the proof into two cases similar to Theorem 10.1.

Case 1: $\gamma_{k+1} = 1$ (no packet losses)

Using Equations 10.6, 10.7, and 10.20, ΔL_J can be derived as

$$\begin{aligned}
\Delta L_J &= \Delta L_\theta + \Delta L_{ao} \\
&= tr\left\{E\left[\left(\tilde{\bar{\theta}}_k - \alpha_h\frac{\Delta\bar{\varphi}_k^e(e_k^a+e_k^o)^T}{\Delta\bar{\varphi}_k^{eT}\Delta\bar{\varphi}_k^e+1}\right)^T \Pi\left(\tilde{\bar{\theta}}_k - \alpha_h\frac{\Delta\bar{\varphi}_k^e(e_k^a+e_k^o)^T}{\Delta\bar{\varphi}_k^{eT}\Delta\bar{\varphi}_k^e+1}\right)\left|\phi_k\right.\right]\right\} \\
&\quad - tr\left\{E\left[\tilde{\bar{\theta}}_k^T\Pi\tilde{\bar{\theta}}_k\left|\phi_k\right.\right]\right\} + \Lambda\left\|E\left[\tilde{z}_{k|k}\left|\phi_k\right.\right]\right\|^4 \\
&\quad - \Lambda\left\|E\left[\tilde{z}_{k-1|k-1}\left|\phi_{k-1}\right.\right]\right\|^4 + \left\|E\left[\tilde{\vartheta}_k\left|\phi_k\right.\right]\right\|^4 - \left\|E\left[\tilde{\vartheta}_{k-1}\left|\phi_{k-1}\right.\right]\right\|^4
\end{aligned} \tag{10B.2}$$

According to the Cauchy–Schwartz inequality and Equation 10B.2, ΔL_J can be rewritten as

$$\begin{aligned}
\Delta L_J &= \Delta L_\theta + \Delta L_{ao} \\
&\le -\alpha_h\left(2 - 3\alpha_h - \frac{2}{\Delta\phi_{\min}^2+1}\right) tr\left\{E\left[\tilde{\bar{\theta}}_k^T\Pi\tilde{\bar{\theta}}_k\left|\phi_k\right.\right]\right\} - \frac{1}{(\chi_{\min}^2+1)^2(U_M-1)^4} \\
&\quad \times(3\alpha_o^4 - \alpha_o(1+6\alpha_o+9\alpha_o^2+\alpha_o^3)(U_M-1)^4\eta^4)\left\|E\left[\tilde{z}_{k-1|k-1}\left|\phi_{k-1}\right.\right]\right\|^4 - \alpha_o\left(\frac{1}{\chi_{\min}^2+1}\right. \\
&\quad \left.-\left(7+\frac{1}{(\chi_{\min}^2+1)^2}\right)\alpha_o - \left(\frac{6}{(\chi_{\min}^2+1)^2}+12\right)\alpha_o^2 - 5\alpha_o^3\right)\left\|E\left[\tilde{\vartheta}_{k-1}\left|\phi_{k-1}\right.\right]\right\|^4
\end{aligned}$$

$$\leq -\alpha_h\left(2-3\alpha_h-\frac{2}{\Delta\phi_{\min}^2+1}\right)\Pi\left\|E\left[\tilde{\bar{\theta}}_k\,|\phi_k\right]\right\|^2-\frac{1}{(\chi_{\min}^2+1)^2(U_M-1)^4}$$

$$(3\alpha_o^4-\alpha_o(1+6\alpha_o+9\alpha_o^2+\alpha_o^3)(U_M-1)^4\eta^4)\left\|E\left[\tilde{z}_{k-1|k-1}\,|\phi_{k-1}\right]\right\|^4-\alpha_o\left(\frac{1}{\chi_{\min}^2+1}\right.$$

$$\left.-\left(7+\frac{1}{(\chi_{\min}^2+1)^2}\right)\alpha_o-\left(\frac{6}{(\chi_{\min}^2+1)^2}+12\right)\alpha_o^2-5\alpha_o^3\right)\left\|E\left[\tilde{\vartheta}_{k-1}\,|\phi_{k-1}\right]\right\|^4+\varepsilon_M^{AE} \qquad (10B.3)$$

Case 2: $\gamma_{k+1}=0$ (with packet lost)

Based on Assumption 10.1, the packet loss probability between sensor and controller has to satisfy the observer stability region (i.e., $P(\gamma_k=1)>(1/N_0)$). Therefore, if $\gamma_{k+1}=0$, then there exists $j\in[k-N_0,k]$ such that $\gamma_{k+1}=\gamma_k=\cdots=\gamma_{k-j+1}=0$ and $\gamma_{k-j}=1$. Therefore, using the update law for the estimated parameters of the observer (10.4), the observer parameter estimation errors $\tilde{\vartheta}_{k+1}$ can be expressed as $\tilde{\vartheta}_{k+1}=\tilde{\vartheta}_{k-j}$. Then, substituting $\tilde{\vartheta}_{k+1}=\tilde{\vartheta}_{k-j}$ into ΔL_J, we have

$$\Delta L_J=\Delta L_\theta+\Delta L_{ao}$$

$$\leq -\alpha_h\left(2-3\alpha_h-\frac{2}{\Delta\phi_{\min}^2+1}\right)\Pi\left\|E\left[\tilde{\bar{\theta}}_k\,|\phi_k\right]\right\|^2-\frac{1}{(\chi_{\min}^2+1)^2(U_M-1)^4}(3\alpha_o^4$$

$$-\alpha_o(1+6\alpha_o+9\alpha_o^2+\alpha_o^3)(U_M-1)^4\eta^4)\sum_{i=k-N_0}^{k}\left\|E\left[\tilde{z}_{i-1|i-1}\,|\phi_{i-1}\right]\right\|^4-\alpha_o\left(\frac{1}{\chi_{\min}^2+1}\right.$$

$$\left.-(7+\frac{1}{(\chi_{\min}^2+1)^2}\right)\alpha_o-\left(\frac{6}{(\chi_{\min}^2+1)^2}+12\right)\alpha_o^2-5\alpha_o^3)\left\|E[\tilde{\vartheta}_{k-j-2}\,|\phi_{k-j-2}]\right\|^4+\varepsilon_M^{AE}$$

(10B.4)

where

$$\varepsilon_M^{AE}=\left(\frac{3\alpha_o^2(2+3\alpha_o)}{(\chi_{\min}^2+1)^2}+1+\alpha_o^4\right)\left[\left(1-\xi\right)^2+\sigma_\xi^2\right]^2\|\vartheta\|^4\frac{\sigma_\upsilon^4}{\bar{\upsilon}^4}$$

Since

$$0<\alpha_h<\frac{2\Delta\phi_{\min}^2}{3(\Delta\phi_{\min}^2+1)},\quad 0<\alpha_o<\frac{(\chi_{\min}^2+1)}{21(\chi_{\min}^2+1)^2+3}\text{ and }0<\eta$$

$$<\sqrt[4]{\frac{3\alpha_o^3}{(1+6\alpha_o+9\alpha_o^2+\alpha_o^3)(U_M-1)^4}}$$

Then, according to Equations 10B.3 and 10B.4, the first difference ΔL_J is less than zero provided the following inequalities hold:

$$\left\| E\left[\tilde{\tilde{\theta}}_k \middle| \phi_k\right]\right\| > \sqrt{\frac{\varepsilon_M^{AE}}{\alpha_h\left(2-3\alpha_h-(2/(\Delta\phi_{\min}^2+1))\right)\Pi}} \equiv B_\theta$$

or

$$\left\| E\left[\tilde{z}_{k-1|k-1} \middle| \phi_{k-1}\right]\right\| > \sqrt[4]{\frac{(\chi_{\min}^2+1)^2\varepsilon_M^{AE}}{(3\alpha_o^4-\alpha_o(1+6\alpha_o+9\alpha_o^2+\alpha_o^3)(U_M-1)^4\eta^4)}}$$

or

$$\left\| E\left[\tilde{\tilde{\vartheta}}_{k-1} \middle| \phi_{k-1}\right]\right\| > \sqrt[4]{\frac{\varepsilon_M^{AE}}{\alpha_o(\frac{1}{\chi_{\min}^2+1}-(7+\frac{1}{(\chi_{\min}^2+1)^2})\alpha_o-(\frac{6}{(\chi_{\min}^2+1)^2})+12)\alpha_o^2-5\alpha_o^3)}}$$

Therefore, based on the Lyapunov theory, value function estimator parameter estimation errors are *bounded* in the mean.

APPENDIX 10C

Proof of Theorem 10.3

Consider the following positive definite Lyapunov function candidate:

$$L = L_D + L_J + L_o \tag{10C.1}$$

where L_D is defined as $L_D = tr\left\{E\left[z_k^T\Omega z_k \middle| \phi_k\right]\right\}$ with $\Omega = (1/(8B_M^2K_M^2(\chi_{\min}^2+1)))\mathbf{I}$ is a positive definite matrix, L_J (10A.3) with positive matrix Ξ is defined in Theorem 10.1, L_o (10B.1) with positive matrices Λ, Π are given by Theorem 10.2 and $\mathbf{I}$ is an identity matrix. The first difference of Equation 10C.1 can be expressed as $\Delta L = \Delta L_D + \Delta L_J + \Delta L_o$. Consider the first part $\Delta L_D = tr\left\{E\left[z_{k+1}^T\Omega z_{k+1} \middle| \phi_{k+1}\right]\right\} - tr\left\{E\left[z_k^T\Omega z_k \middle| \phi_k\right]\right\}$ and by applying NCS under the unified communication protocol and the Cauchy–Schwartz inequality, we have

$$\begin{aligned}\Delta L_D &= tr\left\{E\left[z_{k+1}^T\Omega z_{k+1} \middle| \phi_{k+1}\right]\right\} - tr\left\{E\left[z_k^T\Omega z_k \middle| \phi_k\right]\right\}\\ &\le \Omega\left\|E\left[A_{zk}z_k + B_{zk}u_k^* + B_{zk}K_k\tilde{z}_{k|k} - B_{zk}\tilde{u}_k \middle| \phi_k\right]\right\|^2 - \Omega\left\|E\left[z_k \middle| \phi_k\right]\right\|^2\\ &\le 2\Omega\left\|E\left[A_{zk}z_k + B_{zk}u_k^* \middle| \phi_k\right]\right\|^2 + 4\Omega\left\|E\left[B_{zk}K_k\tilde{z}_{k|k} \middle| \phi_k\right]\right\|^2\\ &\quad + 4\Omega\left\|E\left[B_{zk}\tilde{u}_k \middle| \phi_k\right]\right\|^2 - \Omega\left\|E\left[z_k \middle| \phi_k\right]\right\|^2\end{aligned} \tag{10C.2}$$

Using Lemma 10.1, Equation 10C.2 can be expressed as

$$\Delta L_D \le -(1-2k_a)\Omega\left\|E\left[z_k|\phi_k\right]\right\|^2 + 4\Omega\left\|E\left[B_{zk}\tilde{u}_k|\phi_k\right]\right\|^2 + 4\Omega\left\|E\left[B_{zk}K_k\tilde{z}_{k|k}|\phi_k\right]\right\|^2 \tag{10C.3}$$

Similar to Theorems 10.1 and 10.2, we separate the proof into two cases: $\gamma_k+1 = 1$ and $\gamma_{k+1} = 0$ as follows:

Case 1: $\gamma_{k+1} = 1$ (no packet losses)

Combining Equations 10A.5, (10B.3, and 10C.3, ΔL can be derived as

$$\begin{aligned}
\Delta L &= \Delta L_D + \Delta L_J + \Delta L_o \\
&\le -(1-2k_a)\Omega\left\|E\left[z_k|\phi_k\right]\right\|^2 - \alpha_h\left(1-3\alpha_h-\frac{2}{\Delta\phi_{\min}^2+1}\right)\left\|E\left[\tilde{\bar{\theta}}_k|\phi_k\right]\right\|^2 \\
&\quad -\frac{(3\alpha_o^4-\alpha_o(1+6\alpha_o+9\alpha_o^2+\alpha_o^3)(U_M-1)^4\eta^4)}{(\chi_{\min}^2+1)^2(U_M-1)^4}\left\|E\left[\tilde{z}_{k-1|k-1}|\phi_{k-1}\right]\right\|^4 \\
&\quad -\alpha_o\left(\frac{1}{\chi_{\min}^2+1}-\left(7+\frac{1}{(\chi_{\min}^2+1)^2}\right)\alpha_o-\left(\frac{6}{(\chi_{\min}^2+1)^2}+12\right)\alpha_o^2-5\alpha_o^3\right) \\
&\quad \times\left\|E\left[\tilde{\vartheta}_{k-1}|\phi_{k-1}\right]\right\|^4 + \left(\frac{3\alpha_o^2(2+3\alpha_o)}{(\chi_{\min}^2+1)^2}+1+\alpha_o^4\right)\left[\left(1-\bar{\xi}\right)^2+\sigma_\xi^2\right]^2\|\vartheta\|^4\frac{\sigma_\upsilon^4}{\bar{\upsilon}^4} \\
&\quad -\frac{\left(1-6\eta^2(U_M-1)^2\left(\left(1-\frac{\chi_{\min}^2}{\chi_{\min}^2+1}\alpha_o\right)^2-2\alpha_o(1+\alpha_o)\right)\right)}{2(\chi_{\min}^2+1)(U_M-1)^2}\left\|E\left[\tilde{z}_{k|k}|\phi_k\right]\right\|^2 \\
&\quad -\frac{2}{3}[(\chi_{\min}^2+1)\alpha_o-1]\left\|E\left[\tilde{\vartheta}_k|\phi_k\right]\right\|^2 + 3\left(\left(1-\frac{\chi_{\min}^2}{\chi_{\min}^2+1}\alpha_o\right)^2+\alpha_o\right) \\
&\quad \times\left[\left(1-\bar{\xi}\right)^2+\sigma_\xi^2\right]\frac{\sigma_\upsilon^2}{\bar{\upsilon}^2}\|\vartheta\|^2 \\
&\le -(1-2k_a)\Omega\left\|E\left[z_k|\phi_k\right]\right\|^2 - \frac{\alpha_h}{2}\left(2-3\alpha_h-\frac{2}{\Delta\phi_{\min}^2+1}\right)\left\|E\left[\tilde{\bar{\theta}}_k|\phi_k\right]\right\|^2 \\
&\quad -\frac{(3\alpha_o^4-\alpha_o(1+6\alpha_o+9\alpha_o^2+\alpha_o^3)(U_M-1)^4\eta^4)}{(\chi_{\min}^2+1)^2(U_M-1)^4}\left\|E\left[\tilde{z}_{k-1|k-1}|\phi_{k-1}\right]\right\|^4 \\
&\quad -\alpha_o\left(\frac{1}{\chi_{\min}^2+1}-\left(7+\frac{1}{(\chi_{\min}^2+1)^2}\right)\alpha_o-\left(\frac{6}{(\chi_{\min}^2+1)^2}+12\right)\alpha_o^2\right.
\end{aligned}$$

$$-5\alpha_o^3)\left\|E\left[\tilde{\vartheta}_{k-1}|\psi_{k-1}\right]\right\|^4-\frac{2}{3}\left(\frac{1}{\chi_{\min}^2+1}\alpha_o-\frac{1}{2}\right)\left\|E\left[\tilde{\vartheta}_k|\phi_k\right]\right\|^2$$

$$-\frac{(1-6\eta^2(U_M-1)^2((1-\frac{\chi_{\min}^2}{\chi_{\min}^2+1}\alpha_o)^2-2\alpha_o(1+\alpha_o)))}{2(\chi_{\min}^2+1)(U_M-1)^2}\left\|E\left[\tilde{z}_{k|k}|\phi_k\right]\right\|^2+\varepsilon_{TM} \quad (10C.4)$$

Case 2: $\gamma_{k+1}=0$ (with packet losses)

After applying Assumption 10.1, Equations 10A.6, 10B.4, and 10C.3, ΔL can be expressed as

$$\Delta L=\Delta L_D+\Delta L_J+\Delta L_o$$

$$\le-(1-2k_a)\Omega\left\|E\left[z_k|\phi_k\right]\right\|^2-\frac{\alpha_h}{2}\left(2-3\alpha_h-\frac{2}{\Delta\phi_{\min}^2+1}\right)\left\|E\left[\tilde{\bar{\theta}}_k|\phi_k\right]\right\|^2$$

$$-\frac{(3\alpha_o^4-\alpha_o(1+6\alpha_o+9\alpha_o^2+\alpha_o^3)(U_M-1)^4\eta^4)}{(\chi_{\min}^2+1)^2(U_M-1)^4}\sum_{i=k-N_0}^{k}\left\|E\left[\tilde{z}_{i-1|i-1}|\phi_{i-1}\right]\right\|^4$$

$$-\alpha_o\left(\frac{1}{\chi_{\min}^2+1}-\left(7+\frac{1}{(\chi_{\min}^2+1)^2}\right)\alpha_o-\left(\frac{6}{(\chi_{\min}^2+1)^2}+12\right)\alpha_o^2-5\alpha_o^3\right)$$

$$\times\left\|E\left[\tilde{\vartheta}_{k-j-2}|\phi_{k-j-2}\right]\right\|^4-\frac{2}{3}[(\chi_{\min}^2+1)\alpha_o-1]\left\|E\left[\tilde{\vartheta}_{k-j-1}|\phi_{k-j-1}\right]\right\|^2$$

$$-\frac{(1-6\eta^2(U_M-1)^2((1-\frac{\chi_{\min}^2}{\chi_{\min}^2+1}\alpha_o)^2-2\alpha_o(1+\alpha_o)))\sum_{i=k-N_0}^{k}\left\|E\left[\tilde{z}_{i|i}|\varphi_i\right]\right\|^2}{2(\chi_{\min}^2+1)(U_M-1)^2}+\varepsilon_{TM} \quad (10C.5)$$

where

$$\varepsilon_{TM}=3\left(\left(1-\frac{\chi_{\min}^2}{\chi_{\min}^2+1}\alpha_o\right)^2+\alpha_o\right)[(1-\bar{\xi})^2+\sigma_\xi^2]\frac{\sigma_\upsilon^2}{\bar{\upsilon}^2}\|\vartheta\|^2$$

$$+\left(\frac{3\alpha_o^2(2+3\alpha_o)}{(\chi_{\min}^2+1)^2}+1+\alpha_o^4\right)\left[\left(1-\bar{\xi}\right)^2+\sigma_\xi^2\right]^2\|\vartheta\|^4\frac{\sigma_\upsilon^4}{\bar{\upsilon}^4}$$

is a positive constant. Therefore, the first difference ΔL is less than zero while the inequalities

$$\left\|E\left[z_k|\phi_k\right]\right\|>\sqrt{\frac{\varepsilon_{TM}}{(1-2k_a)}}\equiv b_z$$

or

$$\left\| E\left[\tilde{z}_{k|k} \middle| \phi_k \right] \right\| > \sqrt{\frac{2(\chi_{\min}^2+1)\varepsilon_{TM}}{(1-6\eta^2(U_M-1)^2((1-\frac{\chi_{\min}^2}{\chi_{\min}^2+1}\alpha_o)^2-2\alpha_o(1+\alpha_o)))}} \equiv b_e$$

or

$$\left\| E\left[\tilde{z}_{k-1|k-1} \middle| \phi_{k-1} \right] \right\| > \sqrt[4]{\frac{(\chi_{\min}^2+1)^2\varepsilon_{TM}}{(3\alpha_o^4-\alpha_o(1+6\alpha_o+9\alpha_o^2+\alpha_o^3)(U_M-1)^4\eta^4)}}$$

or

$$\left\| E\left[\tilde{\bar{\vartheta}}_k \middle| \phi_k \right] \right\| > \sqrt{\frac{3\varepsilon_{TM}}{2[(\chi_{\min}^2+1)\alpha_o-1]}} \equiv b_\vartheta \tag{10C.6}$$

or

$$\left\| E\left[\tilde{\bar{\vartheta}}_{k-1} \middle| \phi_{k-1} \right] \right\| > \sqrt[4]{\frac{\varepsilon_{TM}}{\alpha_o(\frac{1}{\chi_{\min}^2+1}-(7+\frac{1}{(\chi_{\min}^2+1)^2})\alpha_o-(\frac{6}{(\chi_{\min}^2+1)^2}+12)\alpha_o^2-5\alpha_o^3)}}$$

or

$$\left\| E\left[\tilde{\bar{\theta}}_k \middle| \psi_k \right] \right\| > \sqrt{\frac{2\varepsilon_{TM}}{\alpha_h(2-3\alpha_h-\frac{2}{\Delta\phi_{\min}^2+1})}} \equiv b_\theta$$

hold. According to the Lyapunov theory, the system states, observed states, and its parameter estimation errors and value function estimator parameter estimation errors are *bounded* in the mean.

REFERENCES

Åstrom, K.J. 1970. *Introduction to Stochastic Control Theory*. Academic Press, New York.

Carnevale, D., Teel, A.R., and Nesic, D. 2007. A Lyapunov proof of improved maximum allowable transfer interval for networked control systems. *IEEE Transactions on Automatic Control*, 52, 892–897.

Dacic, D.B. and Nesic, D. 2007. Quadratic stabilization of linear networked control system via simultaneous protocol and controller design. *Automatica*, 43, 1145–1155.

Dierks, T. and Jagannathan, S. 2012. Online optimal control of affine nonlinear discrete-time systems with unknown internal dynamics by using time-based policy update. *IEEE Transactions on Neural Networks and Learning Systems*, 23, 1118–1129.

Halevi, Y., and Ray, A. 1988. Integrated communication and control systems: Part I—Analysis. *Journal of Dynamic Systems, Measurement, and Control*, 110, 367–373.

Lian, F., Moyne, J., and Tilbury, D. 2002. Optimal controller design and evaluation for a class of networked control systems with distributed constant delays. *Proceeding of American Control Conference*, pp. 3009–3014.

Nilsson, J., Bernhardsson, B., and Wittenmark, B. 1998. Stochastic analysis and control of real-time systems with random time delays. *Automatica*, 1, 57–64.

Schenato, L., Sinopoli, B., Franceschetti, M., Poolla, K., and Sastry, S. 2007. Foundations of control and estimation over lossy networks. *Proceeding of IEEE*, 95, 163–187.

Stevens, R. 1995. *TCP/IP Illustrated: The Protocol*. 2nd edition. Wiley, New York.

Werbos, P.J. 1983. A menu of designs for reinforcement learning over time. *Journal of Neural Networks Control*, 3, 835–846.

Werbos, P.J. 1992. Approximate dynamic programming for real-time and neural modeling. In *Handbook of Intelligent Control: Neural, Fuzzy, and Adaptive Approaches*, D.A. White, D.A. Sofge, (eds.), Van Nostrand Reinhold, New York, pp. 493–525.

Xu, H., Jagannathan, S., and Lewis, F.L. 2012. Stochastic optimal control of unknown linear networked control system in the presence of random delays and packet losses. *Automatica*, 48, 1017–1030.

Zhang, W., Branicky, M.S., and Phillips, S. 2001. Stability of networked control systems. *IEEE Control Systems Magazine*, 21, 84–99.

Index

A

B

C

D

K

L

M

N

O